AF522318

Corrosion Engineering

Corrosion Engineering

Yogesh Kumar

Corrosion Engineering

ISBN 978-93-5111-409-3

Published in 2014 in India by

RANDOM PUBLICATIONS

4376-A/4B, Gali Murari Lal, Ansari Road
New Delhi-110 002
Phone : +91-11-43580356, +91-11-23289044
e-mail: randomexports@gmail.com, sales@randompublications.com, info@randompublications.com
Reprinted 2021

Type Setting by : Keystoneprintads, Delhi-110051
Digitally Printed at : Replika Press Pvt. Ltd.

Preface

Corrosion engineers research and develop methods to lower the environmental impact on roadways, bridges, power plants, windmills, manufacturing plants and pipelines. Corrosion is a natural and costly process of destruction like earthquakes, tornados, ?oods and volcanic eruptions, with one major difference. Whereas we can be only a silent spectator to the above processes of destruction, corrosion can be prevented or at least controlled.

Corrosion is the gradual destruction of materials, (usually metals), by chemical reaction with its environment. In the most common use of the word, this means electrochemical oxidation of metals in reaction with an oxidant such as oxygen. Rusting, the formation of iron oxides, is a well-known example of electrochemical corrosion. This type of damage typically produces oxide(s) or salt(s) of the original metal. Corrosion can also occur in materials other than metals, such as ceramics or polymers, although in this context, the term degradation is more common. Corrosion degrades the useful properties of materials and structures including strength, appearance and permeability to liquids and gases.

Corrosion is the degradation of any material, (ie any naturally occurring or man-made materials), through its contact with the environment. This occurs naturally as the material will always return to its lowest energy state. So, in terms of a piece of steel, which is man-made and has been processed, it now would like to return to its natural state (iron oxide). After being processed, it is now in an unstable form, and thus with the right combination of environment, it will return to its natural state of iron ore.

Corrosion in the oil and gas sector is one of the major flow assurance issues. From an economic point of view, the efficient management of corrosion, ensures that oil and gas can be recovered from wells for longer. Corrosion is also a major cause of hydrocarbon leaks and so safety drives the technological advances in corrosion control. There is currently an increasingly high demand for qualified corrosion engineers with specific expertise in oilfield operations.

This course helps satisfy this demand by providing engineers and physical scientists with skills in corrosion measurement, asset integrity assessment, corrosion prediction and corrosion management.

Thus, this book will take a more practical approach and make it especially useful as a basic text and reference for professional engineers.

I thank all members of my team who have helped in the preparation of the book. My special thanks go to "Random Publications" who have published the book.

– ***Yogesh Kumar***

Contents

1

Introduction

Corrosion Engineering is the specialist discipline of applying scientific knowledge, natural laws and physical resources in order to design and implement materials, structures, devices, systems and procedures to manage the natural phenomenon known as corrosion. Generally related to Metallurgy, Corrosion Engineering also relates to non-metallics including ceramics. Corrosion Engineers often manage other not-strictly-corrosion processes including (but not restricted to) cracking, brittle fracture, crazing, fretting, erosion and more. Corrosion engineering groups have formed around the world in order to prevent, slow and manage the effects of corrosion. Examples of such groups are the National Association of Corrosion Engineers (NACE) and the European Federation of Corrosion (EFC), see Corrosion societies. The corrosion engineers main task is to economically and safely manage the effects of corrosion on materials. Corrosion Engineering Masters degree courses are available worldwide and are concerned with the control and understanding of corrosion.

CORROSION

Corrosion is the gradual destruction of materials, (usually metals), by chemical reaction with its environment. In the most common use of the word, this means electrochemical oxidation of metals in reaction with an oxidant such as oxygen. Rusting, the formation of iron oxides, is a well-known example of electrochemical corrosion. This type of damage typically produces oxide(s) or salt(s) of the original metal. Corrosion can also occur in materials other than metals, such as ceramics or polymers, although in this context, the term degradation is more common. Corrosion degrades the useful properties of materials and structures including strength, appearance and permeability to liquids and gases. Many structural alloys corrode merely from exposure to moisture in air, but the process can be strongly affected by exposure to certain substances. Corrosion can be concentrated locally to form a pit or crack, or it can extend across a wide area more or less uniformly corroding the surface. Because corrosion is a diffusion-controlled process, it occurs on exposed surfaces. As a result, methods to reduce the activity of the exposed surface, such as passivation and chromate conversion, can increase a material's corrosion resistance. However, some corrosion mechanisms are less visible and less predictable.

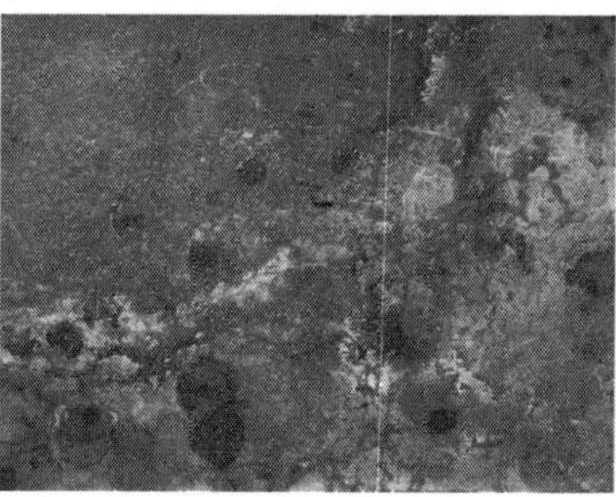

Fig. Rust, the most familiar example of corrosion.

Fig. Volcanic gases have accelerated the corrosion of this abandoned mining machinery.

Fig. Corrosion on exposed metal.

GALVANIC CORROSION

Galvanic corrosion occurs when two different metals have physical or electrical contact with each other and are immersed in a common electrolyte, or when the same metal is exposed to electrolyte with different concentrations. In a galvanic couple, the more active metal (the anode) corrodes at an accelerated rate and the more noble metal (the cathode) corrodes at a retarded rate. When immersed separately, each metal corrodes at its own rate. What type of metal(s) to use is readily determined by following thegalvanic series.

For example, zinc is often used as a sacrificial anode for steel structures. Galvanic corrosion is of major interest to the marine industry and also anywhere water (containing salts) contacts pipes or metal structures.

Fig. Galvanic corrosion of aluminium. A 5-mm-thick Al alloy plate is physically (and hence, electrically) connected to a 10-mm-thick mild steel structural support. Galvanic corrosion occurred on the Al plate along the joint with the steel. Perforation of Al plate occurred within 2 years.

Factors such as relative size of anode, types of metal, and operating conditions (temperature, humidity, salinity, etc.) affect galvanic corrosion. The surface area ratio of the anode and cathode directly affects the corrosion rates of the materials. Galvanic corrosion is often utilized in sacrificial anodes.

Galvanic Series

In a given environment (one standard medium is aerated, room-temperature seawater), one metal will be either more noble or more active than others, based on how strongly its ions are bound to the surface. Two metals in electrical contact share the same electrons, so that the "tug-of-war" at each surface is analogous to competition for free electrons between the two materials. Using the electrolyte as a host for the flow of ions in the same direction, the noble metal will take electrons from the active one. The resulting mass flow or electrical current can be measured to establish a hierarchy of materials in the medium of interest. This hierarchy is called a galvanic series and is useful in predicting and understanding corrosion. This method is expensive but offers maximum protection against corrosion.

CORROSION REMOVAL

Often it is possible to chemically remove the products of corrosion. For example [phosphoric acid] in the form of [naval jelly] is often applied to ferrous tools or surfaces to remove rust. Corrosion removal should not be confused with electropolishing, which removes some layers of the underlying metal to make a smooth surface. For example, phosphoric acid may also be used to electropolish copper but it does this by removing copper, not the products of copper corrosion.

RESISTANCE TO CORROSION

Some metals are more intrinsically resistant to corrosion than others (for some examples, see galvanic series). There are various ways of protecting metals from corrosion including painting, hot dip galvanizing, and combinations of these.

Intrinsic Chemistry

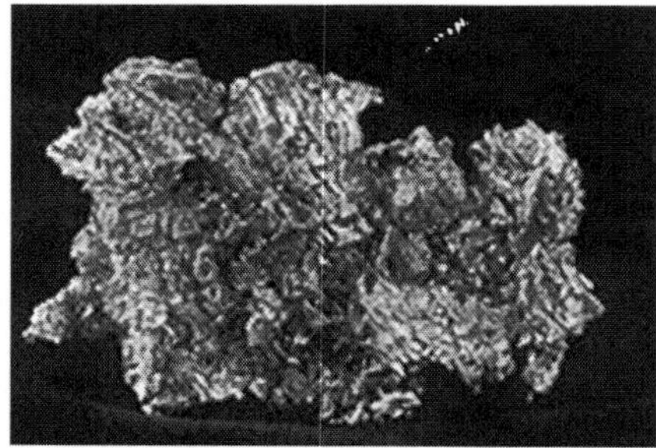

Fig. Gold nuggets do not naturally corrode, even on a geological time scale.

The materials most resistant to corrosion are those for which corrosion is thermodynamically unfavorable. Any corrosion products of gold or platinum tend to decompose spontaneously into pure metal, which is why these elements can be found in metallic form on Earth and have long been valued. More common "base" metals can only be protected by more temporary means.

Some metals have naturally slow reaction kinetics, even though their corrosion is thermodynamically favorable. These include such metals as zinc, magnesium, and cadmium. While corrosion of these metals is continuous and ongoing, it happens at an acceptably slow rate. An extreme example is graphite, which releases large amounts of energy upon oxidation, but has such slow kinetics that it is effectively immune to electrochemical corrosion under normal conditions.

Passivation

Passivation refers to the spontaneous formation of an ultrathin film of corrosion products known as passive film, on the metal's surface that act as a barrier to further oxidation. The chemical composition and microstructure of a passive film are different from the underlying metal. Typical passive film thickness on aluminium, stainless steels and alloys is within 10 nanometers. The passive film is different from oxide layers that are formed upon heating and are in the micrometer thickness range – the passive film recovers if removed or damaged whereas the oxide layer does not. Passivation in natural environments such as air, water and soil at moderate pH is seen in such materials as aluminium, stainless steel, titanium, and silicon.

Passivation is primarily determined by metallurgical and environmental factors. The effect of pH is summarized using Pourbaix diagrams, but many other factors are influential. Some conditions that inhibit passivation include high pH for aluminium and zinc, low pH or the presence of chloride ions for stainless steel, high temperature for titanium (in which case the oxide dissolves into the metal, rather than the electrolyte) and fluoride ions for silicon. On the other hand, unusual conditions may result in passivation of materials that are normally unprotected, as the alkaline environment of concrete does for steel rebar. Exposure to a liquid metal such as mercury or hot solder can often circumvent passivation mechanisms. Passivation is primarily determined by metallurgical and environmental factors.

CORROSION IN PASSIVATED MATERIALS

Passivation is extremely useful in mitigating corrosion damage, however even a high-quality alloy will corrode if its ability to form a passivating film is hindered. Proper selection of the right grade of material for the specific environment is important for the long-lasting performance of this group of materials. If breakdown occurs in the passive film due to chemical or mechanical factors, the resulting major modes of corrosion may include pitting corrosion, crevice corrosion and stress corrosion cracking.

Pitting Corrosion

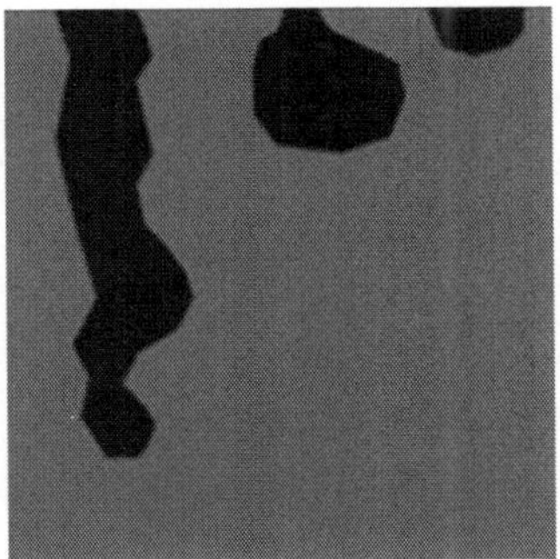

Fig. The scheme of pitting corrosion

Certain conditions, such as low concentrations of oxygen or high concentrations of species such as chloride which complete as anions, can interfere with a given alloy's ability to re-form a passivating film. In the worst case, almost all of the surface will remain protected, but tiny local fluctuations will degrade the oxide film in a few critical points. Corrosion at these points will be greatly amplified, and can causecorrosion pits of several types, depending upon conditions. While the corrosion pits only nucleate under fairly extreme circumstances, they can continue to grow even when conditions return to normal, since the interior of a pit is naturally deprived of oxygen and locally the pH decreases to very low values and the corrosion rate increases due to an autocatalytic process. In extreme cases, the sharp tips of extremely long and narrow corrosion pits can cause stress concentration to the point that otherwise tough alloys can shatter; a thin film pierced by an invisibly small hole can hide a thumb sized pit from view. These problems are especially dangerous because they are difficult to detect before a part or structure fails. Pitting remains among the most common and damaging forms of corrosion in passivated alloys[citation needed], but it can be prevented by control of the alloy's environment.

Weld decay and knifeline attack

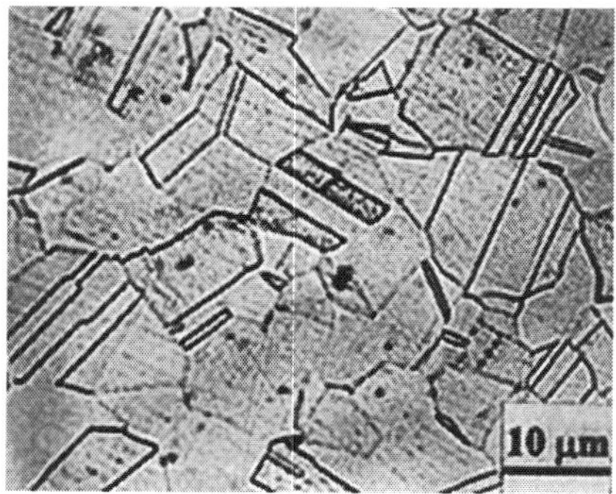

Fig. Normal microstructure

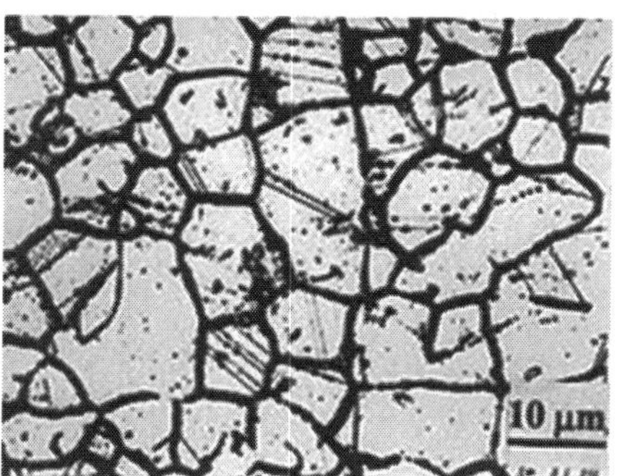

Fig. Sensitized microstructure

Stainless steel can pose special corrosion challenges, since its passivating behavior relies on the presence of a major alloying component (chromium, at least 11.5 per cent). Because of the elevated temperatures of welding and heat treatment, chromium carbidescan form in the grain boundaries of stainless alloys. This chemical reaction robs the material of chromium in the zone near the grain boundary, making those areas much less resistant to corrosion. This creates a galvanic couple with the well-protected alloy nearby, which leads to weld decay (corrosion of the grain boundaries in the heat affected zones) in highly corrosive environments. A stainless steel is said to be sensitized if chromium carbides are formed in the microstructure. A typical microstructure of a normalized type-304 stainless steel shows no signs of sensitization while a heavily sensitized steel shows the presence of grain boundary precipitates.

The dark lines in the sensitized microstructure are networks of chromium carbides formed along the grain boundaries.

Special alloys, either with low carbon content or with added carbon "getters" such as titanium and niobium (in types 321 and 347, respectively), can prevent this effect, but the latter require special heat treatment after welding to prevent the similar phenomenon of knifeline attack. As its name implies, corrosion is limited to a very narrow zone adjacent to the weld, often only a few micrometers across, making it even less noticeable.

Crevice corrosion

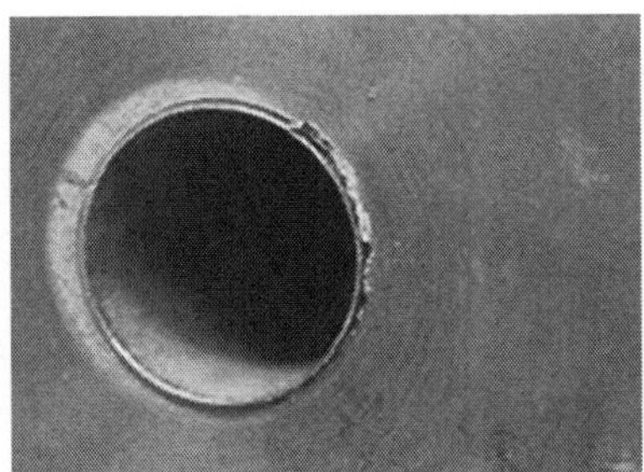

Fig. Corrosion in the crevice between the tube and tube sheet (both made of type-316 stainless steel) of a heat exchanger in a seawater desalination plant.

Crevice corrosion is a localized form of corrosion occurring in confined spaces (crevices), to which the access of the working fluid from the environment is limited. Formation of a differential aeration cell leads to corrosion inside the crevices. Examples of crevices are gaps and contact areas between parts, under gaskets or seals, inside cracks and seams, spaces filled with deposits and under sludge piles. Crevice corrosion is influenced by the crevice type (metal-metal, metal-nonmetal), crevice geometry (size, surface finish), and metallurgical and environmental factors. The susceptibility to crevice corrosion can be evaluated with ASTM standard procedures. A critical crevice corrosion temperature is commonly used to rank a material's resistance to crevice corrosion.

MICROBIAL CORROSION

Microbial corrosion, or commonly known as microbiologically influenced corrosion (MIC), is a corrosion caused or promoted bymicroorganisms, usually chemoautotrophs. It can apply to both metallic and non-metallic materials, in the presence or absence of oxygen. Sulfate-reducing bacteria are active in the absence of oxygen (anaerobic); they produce hydrogen sulfide, causing sulfide stress cracking. In the presence of oxygen (aerobic), some bacteria may directly oxidize iron to iron oxides and hydroxides, other bacteria oxidize sulfur and produce sulfuric acid causing biogenic sulfide corrosion. Concentration cells can form in the deposits of corrosion products, leading to localized corrosion.

Accelerated low-water corrosion (ALWC) is a particularly aggressive form of MIC that affects steel piles in seawater near the low water tide mark. It is characterized by an orange sludge, which smells of hydrogen sulfide when treated with acid. Corrosion rates can be very high and design corrosion allowances can soon be exceeded leading to premature failure of the steel pile.[5] Piles that have been coating and have cathodic protection installed at the time of construction are not susceptible to ALWC.

For unprotected piles, sacrificial anodes can be installed local to the affected areas to inhibit the corrosion or a complete retrofitted sacrificial anode system can be installed. Affected areas can also be treated electrochemically by using an electrode to first produce chlorine to kill the bacteria, and then to produced a calcareous deposit, which will help shield the metal from further attack.

HIGH-TEMPERATURE CORROSION

High-temperature corrosion is chemical deterioration of a material (typically a metal) as a result of heating. This non-galvanic form of corrosion can occur when a metal is subjected to a hot atmosphere containing oxygen, sulfur or other compounds capable of oxidizing (or assisting the oxidation of) the material concerned. For example, materials used in aerospace, power generation and even in car engines have to resist sustained periods at high temperature in which they may be exposed to an atmosphere containing potentially highly corrosive products of combustion.

The products of high-temperature corrosion can potentially be turned to the advantage of the engineer. The formation of oxides on stainless steels, for example, can provide a protective layer preventing further atmospheric attack, allowing for a material to be used for sustained periods at both room and high temperatures in hostile conditions. Such high-temperature corrosion products, in the form of compacted oxide layer glazes, prevent or reduce wear during high-temperature sliding contact of metallic (or metallic and ceramic) surfaces.

Metal Dusting

Metal dusting is a catastrophic form of corrosion that occurs when susceptible materials are exposed to environments with high carbon activities, such as synthesis gas and other high-CO environments. The corrosion manifests itself as a break-up of bulk metal to metal powder. The suspected mechanism is firstly the deposition of a graphite layer on the surface of the metal, usually from carbon monoxide (CO) in the vapour phase. This graphite layer is then thought to form metastable M3C species (where M is the metal), which migrate away from the metal surface. However, in some regimes no M3C species is observed indicating a direct transfer of metal atoms into the graphite layer.

PROTECTION FROM CORROSION

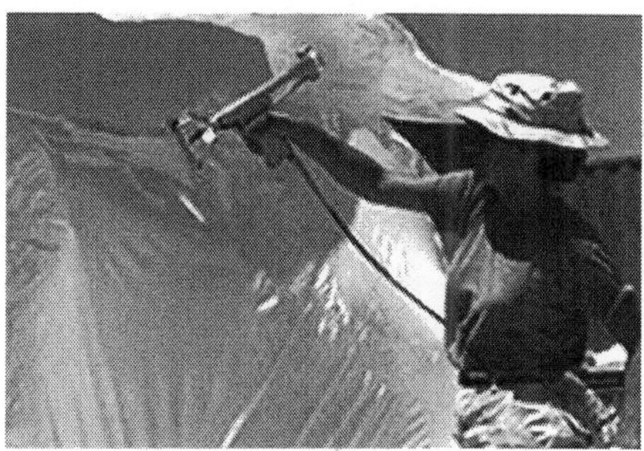

Fig. US Army shrink wraps equipment such as helicopters to protect it from corrosion and thus save millions of dollars.

Surface Treatments

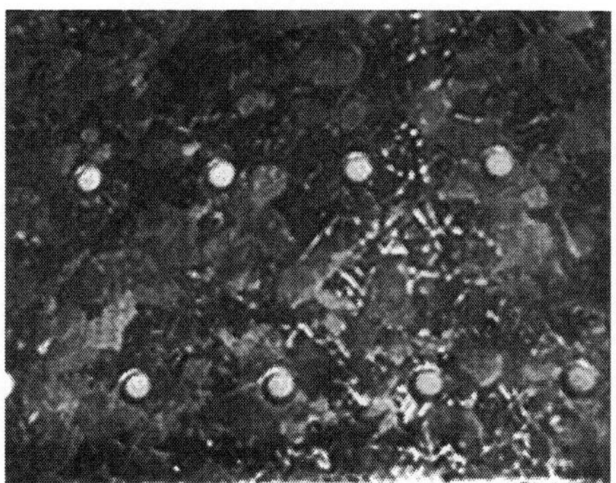

Fig. Galvanized surface

Plating, painting, and the application of enamel are the most common anti-corrosion treatments. They work by providing a barrier of corrosion-resistant material between the damaging environment and the structural material. Aside from cosmetic and manufacturing issues, there may be tradeoffs in mechanical flexibility versus resistance to abrasion and high temperature. Platings usually fail only in small sections, and if the plating is more noble than the substrate (for example, chromium on steel), a galvanic couple will cause any exposed area to corrode much more rapidly than an unplated surface would. For this reason, it is often wise to plate with active metal such as zinc or cadmium. Painting either by roller or brush is more desirable for tight spaces; spray would be better for larger coating areas such as steel decks and waterfront applications. Flexible polyurethane coatings, like Durabak-M26 for example, can provide an anti-corrosive seal with a highly durable slip resistant membrane. Painted coatings are relatively easy to apply and have fast drying times although temperature and humidity may cause dry times to vary.

Reactive coatings

If the environment is controlled (especially in recirculating systems), corrosion inhibitors can often be added to it. These form an electrically insulating or chemically impermeable coating on exposed metal surfaces, to suppress electrochemical reactions. Such methods obviously make the system less sensitive to scratches or defects in the coating, since extra inhibitors can be made available wherever metal becomes exposed. Chemicals that inhibit corrosion include some of the salts in hard water (Roman water systems are famous for their mineral deposits), chromates, phosphates, polyaniline, other conducting polymers and a wide range of specially-designed chemicals that resemble surfactants (i.e. long-chain organic molecules with ionic end groups).

Anodization

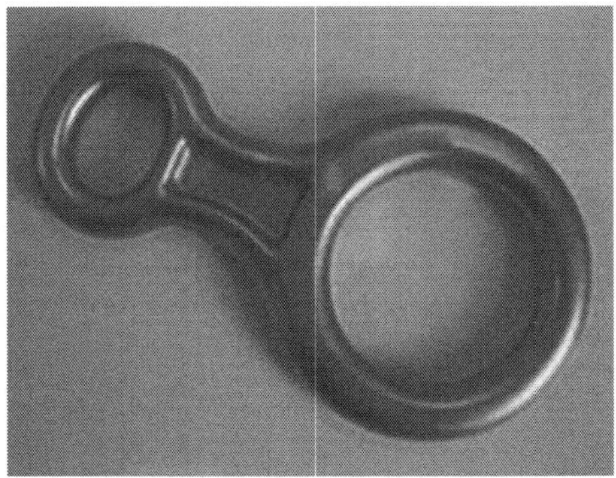

Fig. This climbing descender is anodized with a yellow finish.

Aluminium alloys often undergo a surface treatment. Electrochemical conditions in the bath are carefully adjusted so that uniform pores, several nanometers wide, appear in the metal's oxide film. These pores allow the oxide to grow much thicker than passivating conditions would allow. At the end of the treatment, the pores are allowed to seal, forming a harder-than-usual surface layer. If this coating is scratched, normal passivation processes take over to protect the damaged area. Anodizing is very resilient to weathering and corrosion, so it is commonly used for building facades and other areas that the surface will come into regular contact with the elements. Whilst being resilient, it must be cleaned frequently. If left without cleaning,panel edge staining will naturally occur.

Biofilm Coatings

A new form of protection has been developed by applying certain species of bacterial films to the surface of metals in highly corrosive environments. This process increases the corrosion resistance substantially. Alternatively, antimicrobial-producingbiofilms can be used to inhibit mild steel corrosion from sulfate-reducing bacteria.

Controlled Permeability Formwork

Controlled permeability formwork (CPF) is a method of preventing the corrosion of reinforcement by naturally enhancing the durability of the cover during concrete placement. CPF has been used in environments to combat the effects of carbonation, chlorides, frost and abrasion.

Cathodic Protection

Cathodic protection (CP) is a technique to control the corrosion of a metal surface by making that surface the cathode of an electrochemical cell. Cathodic protection systems are most commonly used to protect steel, water, and fuel pipelines and tanks; steel pier piles, ships, and offshore oil platforms.

Sacrificial Anode Protection

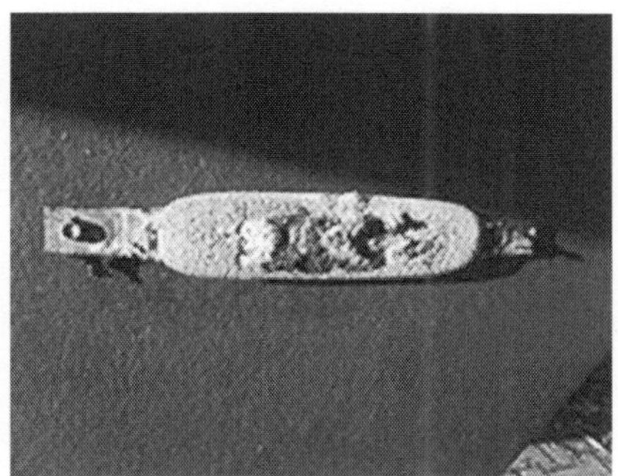

Fig. Sacrificial anode in the hull of a ship.

For effective CP, the potential of the steel surface is polarized (pushed) more negative until the metal surface has a uniform potential. With a uniform potential, the driving force for the corrosion reaction is halted. For galvanic CP systems, the anode material corrodes under the influence of the steel, and eventually it must be replaced. The polarization is caused by the current flow from the anode to the cathode, driven by the difference in electrochemical potential between the anode and the cathode.

Impressed Current Cathodic Protection

For larger structures, galvanic anodes cannot economically deliver enough current to provide complete protection. Impressed current cathodic protection (ICCP) systems use anodes connected to a DC power source (such as a cathodic protection rectifier). Anodes for ICCP systems are tubular and solid rod shapes of various specialized materials. These include high silicon cast iron, graphite, mixed metal oxide or platinum coated titanium or niobium coated rod and wires.

Anodic Protection

Anodic protection impresses anodic current on the structure to be protected (opposite to the cathodic protection). It is appropriate for metals that exhibit passivity (e.g., stainless steel) and suitably small passive current over a wide range of potentials. It is used in aggressive environments, e.g., solutions of sulfuric acid.

Rate of Corrosion

A simple test for measuring corrosion is the weight loss method.[citation needed] The method involves exposing a clean weighed piece of the metal or alloy to the corrosive environment for a specified time followed by cleaning to remove corrosion products and weighing the piece to determine the loss of weight. The rate of corrosion (R) is calculated as

$$R = KW/(\rho At)$$

where k is a constant, W is the weight loss of the metal in time t, A is the surface area of the metal exposed, and ρ is the density of the metal (in g/cm^3).

Economic Impact

In 2002, the US Federal Highway Administration released a study titled Corrosion Costs and Preventive Strategies in the United States on the direct costs associated with metallic corrosion in the U.S. industry. In 1998, the total annual direct cost of corrosion in the U.S. was ca. $276 billion (ca. 3.2 per cent of the US gross domestic product).

Rust is one of the most common causes of bridge accidents. As rust has a much higher volume than the originating mass of iron, its build-up can also cause failure by forcing apart adjacent parts. It was the cause of the collapse of the Mianus river bridge in 1983, when the bearings rusted internally and pushed one corner of the road slab off its support. Three drivers on the roadway at the time died as the slab fell into the river below. The following NTSB investigation showed that a drain in the road had been blocked for road re-surfacing, and had not been unblocked; as a result, runoff water penetrated the support hangers.

Rust was also an important factor in the Silver Bridge disaster of 1967 in West Virginia, when a steel suspension bridge collapsed within a minute, killing 46 drivers and passengers on the bridge at the time. Similarly, corrosion of concrete-covered steel and iron can cause the concrete to spall, creating severe structural problems. It is one of the most common failure modes of reinforced concrete bridges. Measuring instruments based on the half-cell potential can detect the potential corrosion spots before total failure of the concrete structure is reached.

Until 20–30 years ago; galvanized steel pipe was used extensively in the potable water systems for single and multi-family residents as well as commercial and public construction. Today, these systems have long consumed the protective zinc and are corroding internally resulting in poor water quality and pipe failures.[8] The economic impact on homeowners, condo dwellers, and the public infrastructure is estimated at 22 billion dollars as insurance industry braces for a wave of claims due to pipe failures.

CORROSION IN NONMETALS

Most ceramic materials are almost entirely immune to corrosion. The strong chemical bonds that hold them together leave very little free chemical energy in the structure; they can be thought of as already corroded. When corrosion does occur, it is almost always a simple dissolution of the material or chemical reaction, rather than an electrochemical process. A common example of corrosion protection in ceramics is the lime added to soda-lime glass to reduce its solubility in water; though it is not nearly as soluble as pure sodium silicate, normal glass does form sub-microscopic flaws when exposed to moisture. Due to its brittleness, such flaws cause a dramatic reduction in the strength of a glass object during its first few hours at room temperature.

Corrosion of Polymers

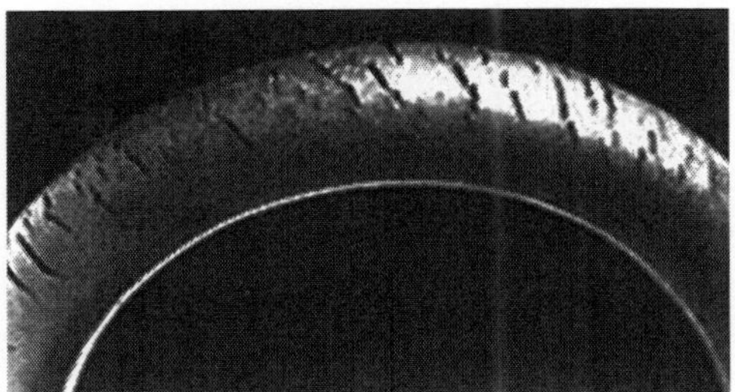

Fig. Ozone cracking in natural rubber tubing

Polymer degradation involves several complex and often poorly-understood physiochemical processes. These are strikingly different from the other processes discussed here, and so the term "corrosion" is only applied to them in a loose sense of the word. Because of their large molecular weight, very little entropy can be gained by mixing a given mass of polymer with another substance, making them generally quite difficult to dissolve. While dissolution is a problem in some polymer applications, it is relatively simple to design against. A more common and related problem is swelling, where small molecules infiltrate the structure, reducing strength and stiffness and causing a volume change. Conversely, many polymers (notably flexible vinyl) are intentionally swelled with plasticizers, which can be leached out of the structure, causing brittleness or other undesirable changes. The most common form of degradation, however, is a decrease in polymer chain length. Mechanisms which break polymer chains are familiar to biologists because of their effect on DNA: ionizing radiation (most commonly ultraviolet light), free radicals, and oxidizers such as oxygen, ozone, and chlorine. Ozone cracking is a well-known problem affecting natural rubber for example.

Additives can slow these process very effectively, and can be as simple as a UV-absorbing pigment (i.e., titanium dioxide or carbon black). Plastic shopping bags often do not include these additives so that they break down more easily aslitter.

Corrosion of Glasses

Fig. Glass corrosion

Glass disease is the corrosion of silicate glasses in aqueous solutions. It is governed by two mechanisms: diffusion-controlled leaching (ion exchange) and hydrolytic dissolution of the glass network. Both mechanisms strongly depend on the pH of contacting solution: the rate of ion exchange decreases with pH as $10^{-0.5pH}$ whereas the rate of hydrolytic dissolution increases with pH as $10^{0.5pH}$. Mathematically, corrosion rates of glasses are characterized by normalized corrosion rates of elements NRi ($g/cm^2 \cdot d$) which are determined as the ratio of total amount of released species into the water Mi (g) to the water-contacting surface area S (cm^2), time of contact t (days) and weight fraction content of the element in the glass fi:

$$NR_i = \frac{M_i}{Sf_it}$$

The overall corrosion rate is a sum of contributions from both mechanisms (leaching + dissolution) $NR_i = Nrx_i$+NRh. Diffusion-controlled leaching (ion exchange) is characteristic of the initial phase of corrosion and involves replacement of alkali ions in the glass by a hydronium (H_3O^+) ion from the solution. It causes an ion-selective depletion of near surface layers of glasses and gives an inverse square root dependence of corrosion rate with exposure time. The diffusion-controlled normalized leaching rate of cations from glasses (g/cm^2.d) is given by:

$$NRx_i = 2\rho\sqrt{\frac{D_i}{\pi t}},$$

where t is time, D_i is the i-th cation effective diffusion coefficient (cm²/d), which depends on pH of contacting water as $D_i = D_{i0} \cdot 10^{-pH}$, and ñ is the density of the glass (g/cm³). Glass network dissolution is characteristic of the later phases of corrosion and causes a congruent release of ions into the water solution at a time-independent rate in dilute solutions (g/cm²·d):

$$NRh = \rho r_h,$$

where rh is the stationary hydrolysis (dissolution) rate of the glass (cm/d). In closed systems the consumption of protons from the aqueous phase increases the pH and causes a fast transition to hydrolysis. However, a further saturation of solution with silica impedes hydrolysis and causes the glass to return to an ion-exchange, e.g. diffusion-controlled regime of corrosion. In typical natural conditions normalized corrosion rates of silicate glasses are very low and are of the order of 10^{-7}–10^{-5} g/(cm²·d). The very high durability of silicate glasses in water makes them suitable for hazardous and nuclear waste immobilisation.

Glass corrosion tests

There exist numerous standardized procedures for measuring the corrosion (also called chemical durability) of glasses in neutral, basic, and acidic environments, under simulated environmental conditions, in simulated body fluid, at high temperature and pressure, and under other conditions. The standard procedure ISO 719 describes a test of the extraction of water-soluble basic compounds under neutral conditions: 2 g of glass, particle size 300–500 μm, is kept for 60 min in 50 ml de-ionized water of grade 2 at 98 °C; 25 ml of the obtained solution is titrated against 0.01 mol/l HCl solution. The volume of HCl required for neutralization is classified according to the table below.

Amount of 0.01M HCl needed to neutralize extracted Basic Oxides, ml	Extracted Na2O Equivalent, μg	Hydrolytic Class
< 0.1	< 31	1
0.1-0.2	31-62	2
0.2-0.85	62-264	3
0.85-2.0	264-620	4
2.0-3.5	620-1085	5
> 3.5	> 1085	> 5

TYPES OF CORROSION

INTERNAL VS. EXTERNAL CORROSION

Corrosion can occur on the outside of a pipe (due to corrosive soil) or on the inside of a pipe (due to corrosive water.) We will be most concerned with internal corrosion, although external corrosion is a similar process and can also cause problems in the distribution system. Either outside or inside a pipe, corrosion can have one of several causes. Each cause somehow sets up an anode and a cathode so that corrosion can occur. The creation of the corrosion cell can be through electrolysis, oxygen concentration cells, or through galvanic action.

Electrolysis

In electrolysis, a D.C. electric current enters a metal pipe and causes flow of electrons through the pipe and to the ground. The pipe, fueled by the electric current, becomes the anode while the soil becomes the cathode. The outside of the pipe corrodes, with the metal from the pipe plating out in the surrounding soil.

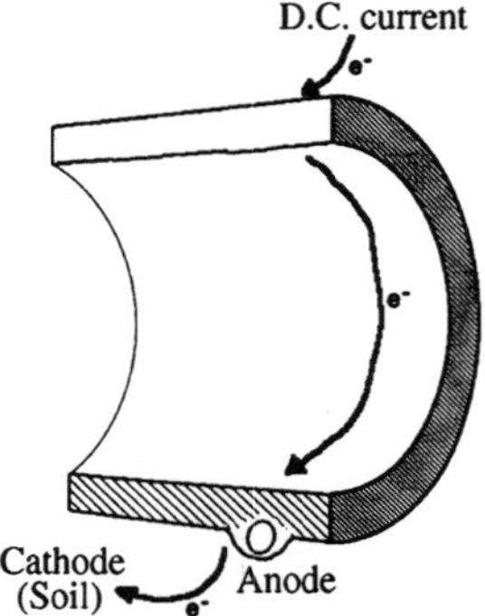

Electrolysis can occur when D.C. electric currents are grounded onto pipes. Nearby electric transit systems can also cause electrolysis.

OXYGEN CONCENTRATION CELL

More commonly, the water and its constituents may set up a corrosion cell within the pipe. These corrosion cells, known as oxygen concentration cells, result from varying oxygen concentration in the water. The portion of the pipe touching water with a low oxygen concentration becomes the anode while the part of the pipe in contact with a high oxygen concentration becomes the cathode. Oxygen concentration cells are probably the primary cause of corrosion in the distribution system. They may occur at dead ends in the distribution system where water is stagnant and loses its dissolved oxygen. Alternatively, oxygen concentration cells may begin in annular spaces, which are ring-shaped spaces between two pipes or between a pipe and a pipe lining. In every case, oxygen becomes depleted in these regions since they are cut off from the normal flow of water, so a difference in oxygen concentration is set up between the dead end or annular space and the main flow of water. Oxygen concentration cells can also be caused by bits of dirt or bacteria. Both of these can become attached to the pipe walls, shielding the metal from dissolved oxygen in the water and setting up an anode.

GALVANIC CORROSION

Metals themselves can also set up corrosion cells. When a pipe consists of only one type of metal, impurities in the pipe wall can develop into anodes and cathodes. Alternatively, when two dissimilar metals come into contact, galvanic corrosion will occur. Galvanic corrosion is often set up in the distribution system in meter installations and at service connections and couplings. The galvanic series, shown below, arranges metals according to their tendency to corrode. This series can be used to determine whether galvanic corrosion is likely to occur and how strong the corrosion reaction will be.

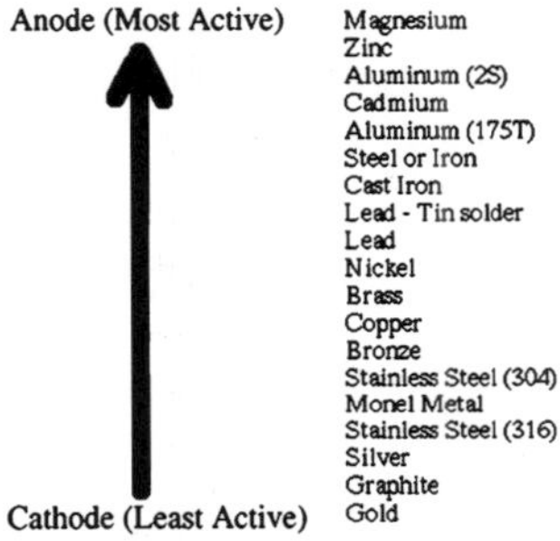

As you can see on the series, some metals (such as gold and silver) are very inactive and unlikely to corrode. Many of these metals have been traditionally used as jewelry because of their low tendency to corrode even when in the presence of salts (in sweat) and oils found on the human body. Although these inactive metals would make non-corrosive pipes, they are usually too expensive to use in the distribution system. At the other end of the galvanic series are metals which are very active and have a high tendency to corrode. These metals can be used as sacrificial anodes, which we will discuss later. They should not be used for distribution system pipes. Most of the metals used in piping - iron, steel, and copper - are found in the middle of the galvanic series. These metals have some tendency to corrode, with those higher on the galvanic series (such as iron and steel) tending more toward corrosion. The distance on the galvanic series between two metals will also influence the likelihood of galvanic corrosion when the two metals are placed in conjunction with each other. For example, if aluminum is brought in contact with a steel pipe, the likelihood of corrosion is low since aluminum and steel are close together on the galvanic series. However, if a stainless steel fitting is used on an iron pipe, the likelihood of corrosion is much higher. When galvanic corrosion occurs, the more active metals always become the anodes. This means that they are corroded, and in extreme cases can begin to leak. The less active metal becomes the cathode and is not damaged.

CHARACTERISTICS INFLUENCING CORROSION

Corrosion in the distribution system is a very complex situation which is influenced by many water characteristics, by the metals used, and by any stray electrical current. We will briefly describe the influence of each characteristic in the following sections. You may want to refer back to the explanation of the chemistry behind corrosion in order to understand some of these factors better.

Primary Water Characteristics

The chemical characteristics of the water flowing through a pipe will influence whether the water is stable and will also affect the extent of any corrosive reaction. Primary factors include alkalinity, hardness, and pH, but oxidizing agents, carbon dioxide, and dissolved solids can also influence corrosion and will be discussed in the next section. Alkalinity, hardness, and pH interact to determine whether the water will produce scale or corrosion or will be stable. The table below summarizes characteristics of corrosive water and of scale-forming water.

Corrosive Water	Scale-forming Water
• low pH	• high pH
• soft or with primarily noncarbonate hardness	• hard with primarily carbonate hardness
• low alkalinity	• high alkalinity

In general, corrosion is the result of water with a low pH. Acidic waters have lots of H^+ ions in the water to react with the electrons at the cathode, so corrosion is enhanced. In contrast, water with a higher pH (basic water) lowers the solubility of calcium carbonate so that the calcium carbonate is more likely to precipitate out as scale. Scaling, as mentioned in the last lesson, tends to be the result of water with a high hardness. Hard water typically contains a lot of calcium compounds which can precipitate out as calcium carbonate. However, if the hardness in the water is primarily noncarbonate, the chlorate and sulfate ions will tend to keep the calcium in solution and will prevent scale formation. Alkalinity is a measure of how easily the pH of the water can be changed, so it can be considered to be a mitigating influence with regards to pH. Water with a high alkalinity is more likely to be scale-forming even at a relatively low pH. In contrast, low alkalinity waters lack the buffering capacity to deal with acids, so they can easily become acidic and corrosive.

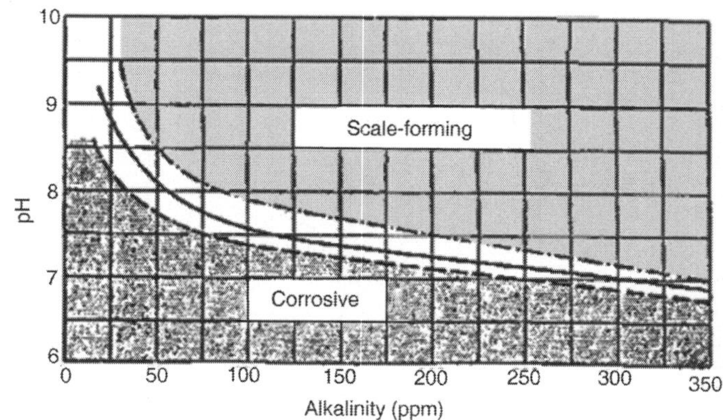

The graph above is known as the Baylis Curve. It shows the relationship between pH, alkalinity, and water stability. Water above the lines is scale-forming while water below the lines is corrosive. Stable water is found in the white area between the lines.

Secondary Water Characteristics

Other chemicals and compounds found in water also influence the corrosion process. The most common of these are oxygen, carbon dioxide, and dissolved solids. Oxygen, as you will remember, reacts with hydrogen gas at the cathode, causing depolarization and speeding up the corrosion. As a result, water with a high D.O. (dissolved oxygen) will tend to be corrosive. Other oxidizing agents can perform the same function, although they are less common. Nitrates and chlorine are two other oxidizing agents found in water. Carbon dioxide in water also tends to cause corrosion. The carbon dioxide gas can combine with water to form carbonic acid, which lowers the pH of the water. As mentioned in the last section, a low pH promotes corrosion. Dissolved solids are typically present in water as ions. These ions increase the electrical conductivity of the water, making the electrolyte more effective. Thus, they will increase the rate of corrosion.

Physical Water Characteristics

In addition to the chemical properties of water, physical characteristics will influence corrosion. The most important of these physical characteristics are temperature and velocity of flow. Temperature speeds up the rate of corrosion just as it does most other reactions. However, the effect of temperature on corrosion can be more complex. A high water temperature reduces the solubility of calcium carbonate in water, which promotes scale formation and slows corrosion. Temperature also alters the form of corrosion. Pits and tubercles tend to form in cold water while hot water promotes uniform corrosion. Uniform corrosion spreading across the entire surface of a pipe is far less problematic than tuberculation, so high temperatures can actually seem to slow the corrosive process.The influence of flow velocity on corrosion is also rather complex. Moderate flow rates are the most beneficial since they promote the formation of scale without breaking loose tubercles. At low flow velocities, corrosion is increased and tends to be in the form of tuberculation due to the prevalence of oxygen concentration cell corrosion. At very high flow velocities, abrasion of the water against the pipe tends to wear the pipe away in a very different form of corrosion. High flow velocities also remove protective scale and tubercles and increase the contact of the pipe with oxygen, all of which will increase the rate of corrosion.

Bacteria

Bacteria can both cause and accelerate the rate of corrosion. In general, bacterial colonies on pipe walls accelerate corrosion below them due to oxygen cell concentration, causing increased pitting and tuberculation. Like humans, some bacteria produce carbon dioxide, which can combine with water to become carbonic acid and accelerate corrosion. The bacterial colonies also block the deposition of calcium carbonate scale on the pipe walls.

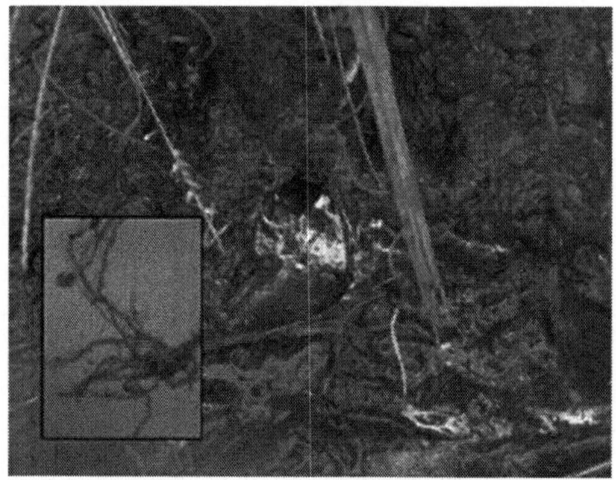

Fig. A colony of iron bacteria.

There are two main types of corrosion-related bacteria, each of which causes its own set of additional corrosion problems. Iron bacteria use the ferrous iron created at the anode, converting it into rust which they deposit in the slime around their cells. Since they use up the ferrous iron, this increases the rate of corrosion. Their slime can also come loose during high flow velocities, causing red water complaints and a bad smell. Sulfate-reducing bacteria use up sulfate in the water to produce hydrogen sulfide. Hydrogen sulfide is an acid which can react with metals, causing corrosion. In addition, the sulfides produce a distinctive rotten egg smell.

Other Factors

Factors other than water characteristics and bacteria can also influence corrosion. Characteristics of the metal pipe and electrical currents are common causes of corrosion. We have already discussed many corrosion-related characteristics of metal in the section on galvanic corrosion. To summarize, metals higher on the galvanic series tend to be more corrosive while metals further apart on the series are more likely to cause galvanic corrosion. In galvanic corrosion, the size of the cathode in relation to the anode has a large influence on corrosion as well. Larger cathodes tend to promote corrosion by speeding the electrical current's flow. When a system has very small anodes and large cathodes, corrosion is so rapid at the anodes that pinholes tend to form all the way through the metal. Stray electrical current can cause electrolytic corrosion. Electrolysis usually causes problems on the outsides of pipes.

CORROSION THEORY

Corrosion is a natural phenomenon which attacks metal by chemical or electrochemical action and converts it into a metallic compound, such as an oxide, hydroxide, or sulfate. Corrosion is to be distinguished from erosion, which is primarily destruction by mechanical action. The corrosion occurs because of the tendency for metals to return to their natural state. Noble metals, such as gold and platinum, do not corrode since they are chemically uncombined in their natural state. Four conditions must exist before corrosion can occur:

- Presence of a metal that will corrode (anode);
- Presence of a dissimilar conductive material (cathode) which has less tendency to corrode;
- Presence of a conductive liquid (electrolyte); and
- Electrical contact between the anode and cathode (usually metal to metal contact, or a fastener.

Elimination of any one of these conditions will stop corrosion. An example would be a paint film on the metal surface. Some metals (such as stainless steel and titanium), under the right conditions, produce corrosion products that are so tightly bound to the corroding metal that they form an invisible oxide film (called a passive film), which prevents further corrosion. When the film of corrosion products is loose and porous (such as those of aluminum and magnesium), an electrolyte can easily penetrate and continue the corrosion process, producing more extensive damage than surface appearance would show.

DEVELOPMENT OF CORROSION

All corrosive attack begins on the surface of the metal. The corrosion process involves two chemical changes. The metal that is attacked or oxidized undergoes an anodic change, with the corrosive agent being reduced and undergoing a cathodic change. The tendency of most metals to corrode creates one of the major problems in the maintenance of the aircraft, particularly in areas where adverse environmental or weather conditions exist. Paint coatings can mask the initial stages of corrosion. Since corrosion products occupy more volume than the original metal, paint surfaces should be inspected often for irregularities such as blisters, flakes, chips, and lumps.

FACTORS INFLUENCING CORROSION.

Some factors which influence metal corrosion and the rate of corrosion are the:

- Type of metal;
- Heat treatment and grain direction;
- Presence of a dissimilar, less corrodible metal (galvanic corrosion);
- Anode and cathode surface areas (in galvanic corrosion);
- Temperature;
- Presence of electrolytes (hard water, salt water, battery fluids, etc.);
- Availability of oxygen;

- Presence of different concentrations of the same electrolyte;
- Presence of biological organisms;
- Mechanical stress on the corroding metal; and
- Time of exposure to a corrosive environment.

Most pure metals are not suitable for aircraft construction and are used only in combination with other metals to form alloys. Most alloys are made up entirely of small crystalline regions, called grains. Corrosion can occur on surfaces of those regions which are less resistant and also at boundaries between regions, resulting in the formation of pits and intergranular corrosion. Metals have a wide range of corrosion resistance. The most active metals (those which tend to lose electrons easily), such as magnesium and aluminum, corrode easily. The most noble metals (those which do not lose electrons easily), such as gold and silver, do not corrode easily.

Corrosion is accelerated by higher temperature environments which accelerate chemical reactions and allow greater moisture content at saturation in air. Electrolytes (electrically conducting solutions) form on surfaces when condensation, salt spray, rain, or rinse water accumulate. Dirt, salt, acidic gases, and engine exhaust gases can dissolve on wet surfaces, increasing the electrical conductivity of the electrolyte, thereby increasing the rate of corrosion. When some of the electrolyte on a metal surface is partially confined (such as between faying surfaces or in a deep crevice), metal in this confined area corrodes more rapidly than other metal surfaces of the same part outside this area.

This type of corrosion is called an oxygen concentration cell. Corrosion occurs more rapidly than would be expected, because the reduced oxygen content of the confined electrolyte causes the adjacent metal to become anodic to other metal surfaces on the same part immersed in electrolyte exposed to the air. Slimes, molds, fungi, and other living organisms (some microscopic) can grow on damp surfaces. Once they are established, the area tends to remain damp, increasing the possibility of corrosion.

Manufacturing processes such as machining, forming, welding, or heat treatment can leave stresses in aircraft parts. This residual stress can cause cracking in a corrosive environment when the threshold for stress corrosion is exceeded. Corrosion, in some cases, progresses at the same rate no matter how long the metal has been exposed to the environment. In other cases, corrosion can decrease with time, due to the barrier formed by corrosion products, or increase with time if a barrier to corrosion is being broken down.

FORMS OF CORROSION

There are many different types of corrosive attack and these will vary with the metal concerned, corrosive media location, and time exposure. For descriptive purposes, the types are discussed under what is considered the most commonly accepted titles.

Uniform Etch Corrosion

Uniform etch corrosion results from a direct chemical attack on a metal surface and involves only the metal surface. On a polished surface, this type of corrosion is first seen as a general dulling of the surface, and if the attack is allowed to continue, the surface becomes rough and possibly frosted in appearance. The discoloration or general dulling of metal created by exposure to elevated temperatures is not to be considered as uniform etch corrosion.

Pitting Corrosion

The most common effect of corrosion on aluminum and magnesium alloys is called pitting. It is first noticeable as a white or gray powdery deposit, similar to dust, which blotches the surface. When the deposit is cleaned away, tiny pits or holes can be seen in the surface. Pitting corrosion may also occur in other types of metal alloys. The combination of small active anodes to large passive cathodes causes severe pitting. The principle also applies to metals which have been passivated by chemical treatments, as well as for metals which develop passivation due to environmental condition.

Galvanic Corrosion

Galvanic corrosion occurs when two dissimilar metals make electrical contact in the presence of an electrolyte. The rate which corrosion occurs depends on the difference in the activities. The greater the difference in activity, the faster corrosion occurs. For example, magnesium would corrode very quickly when coupled with gold in a humid atmosphere, but aluminum would corrode very slowly in contact with cadmium. The rate of galvanic corrosion also depends on the size of the parts in contact. If the surface area of the corroding metal (the anode) is smaller than the surface area of the less active metal (the cathode), corrosion will be rapid and severe. When the corroding metal is larger than the less active metal, corrosion will be slow and superficial. For example, an aluminum fastener in contact with a relatively inert Monel structure may corrode severely, while a Monel bracket secured to a large aluminum member would result in a relatively superficial attack on the aluminum sheet.

Concentration Cell Corrosion

Concentration cell corrosion is corrosion of metals in a metal to metal joint, corrosion at the edge of a joint even though joined metals are identical, or corrosion of a spot on the metal surface covered by a foreign material. Another term for this type of corrosion is crevice corrosion. Metal ion concentration cells, oxygen concentration cells, and active passive cells are the three general types of concentration cell corrosion.

Metal ion concentration cells

The solution may consist of water and ions of the metal which is in contact with water. A high concentration of the metal ions will normally exist under faying surfaces where the solution is stagnant, and a low concentration of metal ions will exist adjacent to the crevice which is created by the faying surface. An electrical potential will exist between the two points; the area of the metal in contact with the low concentration of metal ions will be anodic and corrode, and the area in contact with the high metal ions concentration will be cathodic and not show signs of corrosion.

Oxygen concentration cells

The solution in contact with the metal surface will normally contain dissolved oxygen. An oxygen cell can develop at any point where the oxygen in the air is not allowed to diffuse into the solution, thereby creating a difference in oxygen concentration between two points. Typical locations of oxygen concentration cells are under either metallic or nonmetallic deposits on the metal surface and under faying surfaces such as riveted lap joints. Oxygen cells can also develop under gaskets, wood, rubber, and other materials in contact with the metal surface. Corrosion will occur at the area of low oxygen concentration (anode). Alloys, such as stainless steel, which owe their corrosion resistance to surface passivity, are particularly susceptible to this type of crevice corrosion.

Active passive cells.

Metals which depend on a tightly adhering passive film, usually an oxide for corrosion protection, such as corrosion resistant steel, are prone to rapid corrosive attack by active passive cells. The corrosive action usually starts as an oxygen concentration cell. As an example, salt deposits on the metal surface in the presence of water containing oxygen can create the oxygen cell. The passive film will be broken beneath the dirt particle. Once the passive film is broken, the active metal beneath the film will be exposed to corrosive attack. An electrical potential will develop between the large area of the cathode (passive film) and the small area of the anode (active metal).

Intergranular Corrosion

Intergranular corrosion is an attack along the grain boundaries of a material. Each grain has a clearly defined boundary which, from a chemical point of view, differs from the metal within the grain center. The grain boundary and grain center can react with each other as anode and cathode when in contact with an electrolyte. Rapid selective corrosion at the grain boundary can occur with subsequent delamination. High strength aluminum alloys such as 2014 and 7075 are more susceptible to intergranular corrosion if they have been improperly heat treated and are then exposed to a corrosive environment.

Exfoliation Corrosion

Exfoliation corrosion is an advanced form of intergranular corrosion where the surface grains of a metal are lifted up by the force of expanding corrosion products occurring at the grain boundaries just below the surface. The lifting up or swelling is visible evidence of exfoliation corrosion. Exfoliation is most prone to occur in wrought products such as extrusions, thick sheet, thin plate and certain die forged shapes which have a thin, highly elongated platelet type grain structure. This is in contrast with other wrought products and cast products that tend to have an equiaxed grain structure.

Filiform Corrosion

Filiform corrosion is a special form of oxygen concentration cell corrosion or crevice corrosion which occurs on metal surfaces having an organic coating system. It is recognized by its characteristic wormlike trace of corrosion products beneath the paint film. Filiform occurs when the relative humidity of the air is between 78 and 90 percent and the surface is slightly acidic. Corrosion starts at breaks in the coating system and proceeds underneath the coating due to the diffusion of water vapor and oxygen from the air through the coating. Filiform corrosion can attack steel and aluminum surfaces. The traces never cross on steel, but they will cross under one another on aluminum which makes the damage deeper and more severe for aluminum. If filiform corrosion is not removed and the area treated and a protective finish applied, the corrosion can lead to intergranular corrosion, especially around fasteners and at seams. Filiform corrosion can be removed using glass bead blasting material with portable abrasive blasting equipment and/or mechanical means such as buffing or sanding. Filiform corrosion can be prevented by storing aircraft in an environment with a relative humidity below 70 percent, using coating systems having a low rate of diffusion for oxygen and water vapors, and by washing aircraft to remove acidic contaminants from the surface, such as those created by pollutants in the air.

CORROSION AND MECHANICAL FACTORS

Corrosive attack is often aggravated by mechanical factors that are either within the part (residual) or applied to the part (cyclic service loads). Erosion by sand and/or rain and mechanical wear will remove surface protective films and contribute to corrosive attack of underlying metal surfaces. Corrosive attack that is aided by some mechanical factor usually causes the part to degenerate at an accelerated rate compared to the rate at which the same part would deteriorate if it were subjected solely to corrosive attack. Environmental conditions and the composition of the alloy also influence the extent of attack. Examples of this kind of alliance are stress corrosion cracking, corrosion fatigue, and fretting corrosion.

Stress Corrosion Cracking

Stress corrosion cracking is an intergranular cracking of the metal which is caused by a combination of stress and corrosion. Stress may be caused by internal or external loading. Internal stresses are produced by nonuniform deformation during cold working, by unequal cooling from high temperatures, and by internal structural rearrangement involving volume changes. Internal stresses are induced when a piece of structure is deformed during an assembly operation, (i.e., during pressing in bushings, shrinking a part for press fit, installing interference bolts, installing rivets, etc.). Concealed stress is more important than design stress, because stress corrosion is difficult to recognize before it has overcome the design safety factor. The level of stress varies from point to point within the metal. Stresses near the yield strength are generally necessary to promote stress corrosion cracking, but failures may occur at lower stresses. Specific environments have been identified which cause stress corrosion cracking of certain alloys. Salt solutions and seawater may cause stress corrosion cracking of high strength heat treated steel and aluminum alloys. Methyl alcohol hydrochloric acid solutions will cause stress corrosion cracking of some titanium alloys. Magnesium alloys may stress corrode in moist air. Stress corrosion may be reduced by applying protective coatings, stress relief heat treatment, using corrosion inhibitors, or controlling the environment. Shot peening a metal surface increases resistance to stress corrosion cracking by creating compressive stresses on the surface which should be overcome by applied tensile stress before the surface sees any tension load. Therefore, the threshold stress level is increased.

Corrosion Fatigue

Corrosion fatigue is caused by the combined effects of cyclic stress and corrosion. No metal is immune to some reduction in its resistance to cyclic stressing if the metal is in a corrosive environment. Damage from corrosion fatigue is greater than the sum of the damage from both cyclic stresses and corrosion. Corrosion fatigue failure occurs in two stages. During the first stage, the combined action of corrosion and cyclic stress damages the metal by pitting and crack formation to such a degree that fracture by cyclic stressing will ultimately occur, even if the corrosive environment is completely removed. The second stage is essentially a fatigue stage in which failure proceeds by propagation of the crack (often from a corrosion pit or pits) and is controlled primarily by stress concentration effects and the physical properties of the metal. Fracture of a metal part, due to fatigue corrosion, generally occurs at a stress level far below the fatigue limit in laboratory air, even though the amount of corrosion is relatively small. For this reason, protection of all parts subject to alternating stress is particularly important, even in environments that are only mildly corrosive.

Fretting Corrosion

Damage can occur at the interface of two highly loaded surfaces which are not supposed to move against each other. However, vibration may cause the surfaces to rub together resulting in an abrasive wear known as fretting. The protective film on the metallic surfaces is removed by the rubbing action. The continued rubbing of the two surfaces prevents formation of protective oxide film and exposes fresh active metal to the atmosphere. Fretting can cause severe pitting. Dampening of vibration, tightening of joints, application of a lubricant, or installation of a fretting resistant material between the two surfaces can reduce fretting corrosion.

Heat Treatment

Heat treatment of airframe materials should be rigidly controlled to maintain their corrosion resistance as well as to improve their essential mechanical properties. For example, improper heat treatment of clad aluminum alloy may cause the cladding to incur excessive diffusion because the solution heat treatment is too long or at too high a temperature. This degrades the inherent resistance of the cladding itself, and reduces its ability to provide protection to the core aluminum alloy. Aluminum alloys which contain appreciable amounts of copper and zinc are highly vulnerable to intergranular corrosion attack if not quenched rapidly during heat treatment or other special treatment. Stainless steel alloys are susceptible to carbide sensitization when slowly cooled after welding or high temperature heat treatment. Post weld heat treatments are normally advisable for reduction of residual stress.

Hydrogen Embrittlement

Environmentally induced failure processes may often be the result of hydrogen damage rather than oxidation. Atomic hydrogen is a cathodic product of many electrochemical reactions, forming during naturally occurring corrosion reactions as well as during many plating or pickling processes. Whether hydrogen is liberated as a gas, or atomic hydrogen is absorbed by the metal, depends on the surface chemistry of the metal. Atomic hydrogen, due to its small size and mass, has very high diffusivity in most metals. It will therefore penetrate most clean metal surfaces easily and migrate rapidly to favorable sites where it may remain in solution, precipitate as molecular hydrogen to form small pressurized cavities, cracks or large blisters, or it may react with the base metal or with alloying elements to form hydrides.

The accumulation of hydrogen in high strength alloys often leads to cracking, and this often occurs in statically loaded components several hours or even days after the initial application of the load or exposure to the source of hydrogen. Cracking of this type is often referred to as hydrogen stress cracking, hydrogen delayed cracking, or hydrogen induced cracking. Similar fracture processes can occur in new and unused parts when heat treatments or machining have left residual stresses in the parts, and have then been exposed to a source of hydrogen. For this reason, all processes such as pickling or electroplating must be carried out under well controlled conditions to minimize the amount of hydrogen generated.

COMMON CORROSIVE AGENTS

Substances that cause corrosion of metals are called corrosive agents. The most common corrosive agents are acids, alkalies, and salts. The atmosphere and water, the two most common media for these agents, may act as corrosive agents too.

Acids

In general, moderately strong acids will severely corrode most of the alloys used in airframes. The most destructive are sulfuric acid (battery acid), halogen acids (i.e., hydrochloric, hydrofluoric, and hydrobromic), nitrous oxide compounds, and organic acids found in the wastes of humans and animals.

Alkalies

Although alkalies, as a group, are generally not as corrosive as acids, aluminum and magnesium alloys are exceedingly prone to corrosive attack by many alkaline solutions unless the solutions contain a corrosion inhibitor. Particularly corrosive to aluminum are washing soda, potash (wood ashes), and lime (cement dust). Ammonia, an alkali, is an exception because aluminum alloys are highly resistant to it.

Salts

Most salt solutions are good electrolytes and can promote corrosive attack. Some stainless steel alloys are resistant to attack by salt solutions but aluminum alloys, magnesium alloys, and other steels are extremely vulnerable. Exposure of airframe materials to salts or their solutions is extremely undesirable.

The Atmosphere

The major atmospheric corrosive agents are oxygen and airborne moisture, both of which are in abundant supply. Corrosion often results from the direct action of atmospheric oxygen and moisture on metal, and the presence of additional moisture often accelerates corrosive attack, particularly on ferrous alloys. However, the atmosphere may also contain other corrosive gases and contaminants, particularly industrial and marine environments, which are unusually corrosive. Industrial atmospheres contain many contaminants, the most common of which are partially oxidized sulfur compounds. When these sulfur compounds combine with moisture, they form sulfur based acids that are highly corrosive to most metals. In areas where there are chemical industrial plants, other corrosive atmospheric contaminants may be present in large quantities, but such conditions are usually confined to a specific locality. Marine atmospheres contain chlorides in the form of salt particles or droplets of salt saturated water. Since salt solutions are electrolytes, they corrosively attack aluminum and magnesium alloys which are vulnerable to this type of environment.

Water

The corrosivity of water will depend on the type and quantity of dissolved mineral and organic impurities and dissolved gasses (particularly oxygen) in the water. One characteristic of water which determines its corrosivity is the conductivity or its ability to act as an electrolyte and conduct a current. Physical factors, such as water temperature and velocity, also have a direct bearing on the corrosivity.

- The most corrosive of natural waters (sea and fresh waters) are those that contain salts. Water in the open sea is extremely corrosive due to the presence of chloride ions, but waters in harbors are often even more so because they are contaminated by industrial waste.
- The corrosive effects of fresh water varies from locality to locality due to the wide variety of dissolved impurities that may be present in any particular area. Some municipal waters (potable water) to which chlorine and fluorides have been added can be quite corrosive. Commercially softened water and industrially polluted rain water are usually considered to be very corrosive.

MICRO-ORGANISMS

Microbial attack includes actions of bacteria, fungi, or molds. Micro-organisms occur nearly everywhere. Those organisms causing the greatest corrosion problems are bacteria and fungi. Bacteria may be either aerobic or anaerobic. Aerobic bacteria require oxygen to live. They accelerate corrosion by oxidizing sulfur to produce sulfuric acid. Bacteria living adjacent to metals may promote corrosion by depleting the oxygen supply or by releasing metabolic products. Anaerobic bacteria, on the other hand, can survive only when free oxygen is not present. The metabolism of these bacteria requires them to obtain part of their sustenance by oxidizing inorganic compounds, such as iron, sulfur, hydrogen, and carbon monoxide. The resultant chemical reactions cause corrosion.

Fungi are the growths of micro-organisms that feed on organic materials. While low humidity does not kill microbes, it slows their growth and may prevent corrosion damage. Ideal growth conditions for most micro-organisms are temperatures between 68 and 104 °F (20 and 40 °C) and relative humidity between 85 and 100 percent. It was formerly thought that fungal attack could be prevented by applying moisture proofing coatings to nutrient materials or by drying the interiors of compartments with desiccants. However, some moisture proofing coatings are attacked by mold, bacteria, or other microbes, especially if the surfaces on which they are used are contaminated. Microbial growth occurs at the interface of water and fuel, where the fungus feeds on fuel. Organic acids, alcohols, and esters are produced by growth of the fungus. These byproducts provide even better growing conditions for the fungus. The fungus typically attaches itself to the bottom of the tank and looks like a brown deposit on the tank coating when the tank is dry. The fungus growth may start again when water and fuel are present. The spore form of some micro-organisms can remain dormant for long periods while dry, and can become active when moisture is available. When desiccants become saturated and unable to absorb moisture passing into the affected area, micro-organisms can begin to grow. Dirt, dust, and other airborne contaminants are the least recognized contributors to microbial attack. Unnoticed, small amounts of airborne debris may be sufficient to promote fungal growth.

Fungi nutrients have been considered to be only those materials that have been derived from plants or animals. Thus, wool, cotton, rope, feathers, and leather were known to provide sustenance for microbes, while metals and minerals were not considered fungi nutrients. To a large extent this rule of thumb is still valid, but the increasing complexity of synthetic materials makes it difficult or impossible to determine from the name alone whether a material will support fungus. Many otherwise resistant synthetics are rendered susceptible to fungal attack by the addition of chemicals to change the material's properties. Damage resulting from microbial growth can occur when any of three basic mechanisms, or a combination of these, is brought into play. First, fungi are damp and have a tendency to hold moisture, which contributes to other forms of corrosion. Second, because fungi are living organisms, they need food to survive. This food is obtained from the material on which the fungi are growing. Third, these micro-organisms secrete corrosive fluids that attack many materials, including some that are not fungi nutrient. Microbial growth must be removed completely to avoid corrosion. Microbial growth should be removed by hand with a firm nonmetallic bristle brush and water. Removal of microbial growth is easier if the growth is kept wet with water. Microbial growth may also be removed with steam at 100 psi and steam temperatures not exceeding 150 °F (66 °C). Protective clothing must be used when using steam for removing microbial growth.

METALLIC MERCURY CORROSION ON ALUMINUM ALLOYS

Spilled mercury on aluminum should be cleaned immediately because mercury causes corrosion attack which is rapid in both pitting and intergranular attack and is very difficult to control. The most devastating effect of mercury spillage on aluminum alloys is the formation of an amalgam which proceeds rapidly along grain boundaries, causing liquid metal embrittlement. If the aluminum alloy part is under tension stress, this embrittlement will result in splitting with an appearance similar to severe exfoliation. X-ray inspection may be an effective method of locating the small particles of spilled mercury because the dense mercury will show up readily on the X-ray film.

CORROSION : DEGRADATION OF A MATERIAL DUE TO ENVIRONMENTAL EXPOSURE

The aim of this introduction is to provide some of the vocabulary of corrosion along with a few examples of different corrosion forms that are commonly encountered. The scale of corrosion and how it affects everyday processes such as industrial production, maintenance, transportation, safety, and other issues with costs associated will be introduced. Corrosion can be defined as "degradation of a material due to environmental exposure". This is a very wide definition, which includes wood rotting and many other areas beyond the scope of these notes. A much narrower definition will be employed here. For the purposes of this book the definition of corrosion is "degradation of engineering materials by exposure to a wet environment". This definition includes corrosion of metallic materials by a wide range of processes as well as degradation of plastic materials, for example carbon fiber based composites, and steel reinforced concrete. In many cases corrosion is obvious. An example would be rust on steel. The picture below shows oil storage tanks where the paint is not fully protecting the underlying steel from corrosion. Rust can be seen on both tanks starting to show through. Unless some maintenance is conducted on these tanks they will continue to corrode.

Another world wide examples of corrosion is degradation of bridges. In this case the concrete surrounding steel reinforcement has "spalled", leaving the steel reinforcement or "rebars" exposed. The structural strength of the pilings is being compromised. These examples are very clear and simply observed and corrected. However, corrosion does not necessarily involve such clear visual changes or material loss. A bright, shiny high strength steel can break at a stress level much below that predicted and the process responsible for failure can be a corrosion process. In this case it could be stress corrosion cracking or more likely hydrogen embrittlement. Therefore one important aspect of corrosion is that is does not always produce a visual change such as rust on steels. Indeed the most dangerous forms of corrosion are those which cannot be easily detected. These would be "pitting corrosion" from the inside of a tank to the outside or "crevice corrosion" under bolt heads or in threads. An example of "pitting corrosion" is shown below and on the cover. In this type of corrosion, which is termed "localized" as it does not spread laterally over surface, a pit or hole forms in a metal locally while the remainder of the surface shows no evidence of corrosion. This form of corrosion of often very difficult to detect, for example if it starts on the inside of a tank due to the combination of material and environment.

"Crevice corrosion" is another example of a localized form of corrosion. The example in the figure below is a stainless steel bolt that was used to hold the keel of a boat in place. The area of crevice corrosion is in the threaded region where the nut was in contact down in the bilge area.

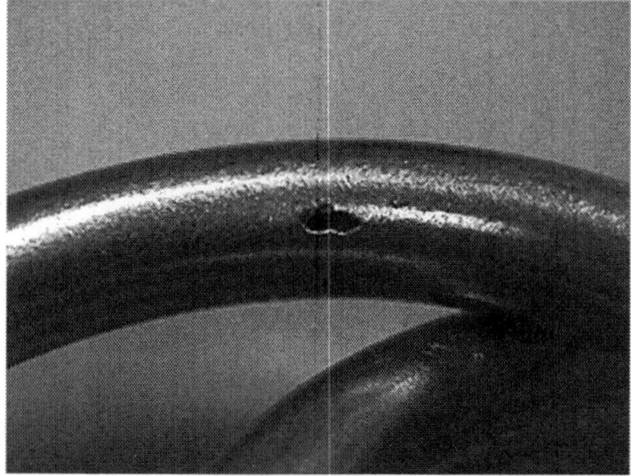

Fig. A pit in a stainless steel ring.

Other forms of corrosion reduce the stress bearing capability of the material, such as stress corrosion cracking, corrosion fatigue, fretting fatigue and hydrogen embrittlement. In these cases the material will fail at stress levels below those expected. In addition, if these forms of corrosion are not detected, "catastrophic failure" can occur, in that there is no warning before a devastating failure takes place. For non-metals, degradation can take the form of blisters on the surface and removal of polymer in carbon fiber based composite materials. In this case the reactions responsible are within the composite and do not necessarily directly lead to damage. It is a secondary process that produces damage such as a reaction between corrosion reaction products to form an osmotic pressure buildup in the polymer at the carbon fiber to polymer matrix interface.

Fig. Crevice corrosion shown above is usually hidden.

For steel reinforced concrete, it is often corrosion of the steel reinforcing bars in the structure which causes problems. The volume of the iron oxides resulting from corrosion are several times that of the iron which was dissolved to form it. This large volume expansion imposes a tensile stress in the concrete at the concrete to rebar interface. Concrete will crack and rust "bleed" form this crack. Eventually, that damage will propagate and become sever enough to "spall" concrete from the rebar. Many examples of this process exist on the bridges and highways throughout the world.

Prevention and reduction measures for corrosion are available. An article from the Times newspaper on June 22, 1815. The main news in the paper was the dispatches from the Duke of Wellington to George III on the battle of Waterloo several days earlier on June 17. A vivid description of the battle against Napoleon and his forces was provided along with a casualty list of officers. Tucked in a corner was an article on anti-corrosive paints. It describes the advantages of the paint system, the cost, colors and amounts. Today industry is still advancing with the task of producing improved anti-corrosive paints. Other measures are widely used today, for example galvanized steel sheet, or sacrificial anodes where a metal corrodes instead of the all important component or structure, and impressed current cathodic protection where an electric current is applied for corrosion protection. These processes will be described later in the course.

The aim of this introduction is to generally describe the process of corrosion. Many estimates of the cost to corrosion were made. These include 5 per cent of the US gross national product and other numbers too large to understand. Other methods involve the cost of a system can be associated with 33 per cent going to maintenance. A more rational feel for corrosion is provided by the time a student has to spend repairing the rust on cars or watching rust take over the car. Corrosion is expensive and stopping or retarding also costs money. However stopping corrosion is a worthwhile exercise if it keeps industry operating. The remainder of this book outlines in more detail the types of corrosion shown above, along with testing methods and methods to protect and reduce the effects of corrosion.

2

Corrosion Control and Monitoring Engineering

CORROSION CONTROL METHODS

Painting

The plate is suitably prepared, primed and finished off with a suitable top coat. This method of protection excludes water and oxygen from the plate surface and as long as the paint remains intact, this method of corrosion control is very effective and comparatively cheap. The problem however lies at the plate preparation stage. Invariably, traces of rust, mill scale, and dirt and water moisture get trapped beneath the paint coat and consequently corrosion commences below the film of paint.

Cathodic Protection

Cathodic Protection technique can only be used where the concerned metals are immersed in an electrolyte. The principle involved in this process is that the anodic corrosion reaction is either suppressed by the application of an opposing current or the anode is of a type which allows it to be consumed. Two means of cathodic protection are in general use on modern ships, mainly for the shipâ€™s sides or in tanks.

Sacrificial Anode Type

Metals such as aluminum or Zinc are fitted to the ship side or in the tank at strategic positions which form the anode of the corrosion cell making the steel of the shipâ€™s plating the cathode. The sacrificial anodes are gradually corroded away and have a life of about 2-4 years when they have to be replaced, usually at dry dock. This system is suitable for shipâ€™s protection as well as in cargo and ballast tanks. Anodes can be typically seen in the vicinity of the propeller, in the vicinity of Engine Room inlets and discharges and in the forward portion near the bow thruster. In tanks, they are usually fitted in the bottom portion of the tank for two reasons.

- The concentration of dissimilar metals i.e. valves, pipe lines etc is only in the bottom portion of the tanks.
- When the anode has corroded away, often it drops down to the bottom of the tank. It is possible that when it falls, it would cause an intrinsic spark and possible explosion.

Impressed current type

These systems are fitted for protection of the immersed portions of the extent hull only. A voltage of difference is maintained between hull and fitted anodes, which will protect the hull against corrosion. The system consists of a source of direct current, anodes and apparatus foe measuring and controlling the current. The amount of the impressed current will vary with the area to be protected, ship speed. Salinity and condition of the paint work.

For normal operating conditions, the potential difference is maintained by means of an extremely mounted silver Chloride cell detecting the voltage difference between itself and the hull. The potential difference is fed back to a controller which impresses the same amount of current onto several anodes positioned on the hull. The anodes used are usually of permanent noble metals such as lead or silver. In this way any corrosion cells which might exist are exactly cancelled and corrosion is controlled. The propeller shaft and rudder stock have to be specially grounded in order to ensure continuity with the hull. Care is to be exercised that the controller only impresses the exact amount of current required. If the impressed current is too little,corrosion protection will be inadequate. If the current impressed is too high, the paint work on the hull will be damaged and blistering occurs.

DESCRIPTION OF APPLICABLE CORROSION CONTROL AND CORROSION PROTECTION METHODS

In most cases, effective corrosion control is obtained by combining two or more of these methods. Corrosion control should be considered at the design stage of a given facility or system. The methods selected must be appropriate for the materials used, for the configurations, and for the types and forms of corrosion which must be controlled.

MATERIALS RESISTANT TO CORROSION

There are no materials that are immune to corrosion in all environments. Materials must be matched to the environment that they will encounter in service. Corrosion Resistance Data are used to assess the suitability of a material in an environment.

PROTECTIVE COATINGS

Protective coatings are the most widely used corrosion control technique. Essentially, protective coatings are a means for separating the surfaces that are susceptible to corrosion from the factors in the environment which cause corrosion to occur. Remember, however, that protective coatings can never provide 100 percent protection of 100 percent of the surface. If localized corrosion at a coating defect is likely to cause rapid catastrophic failure, additional corrosion control measures must be taken. Coatings are particularly useful when used in combination with other methods of corrosion control such as cathodic protection or galvanic corrosion

CATHODIC PROTECTION

Cathodic protection interferes with the natural action of the electrochemical cells that are responsible for corrosion. Cathodic protection can be effectively applied to control corrosion of surfaces that are immersed in water or exposed to soil. Cathodic protection in its classical form cannot be used to protect surfaces exposed to the atmosphere. The use of anodic metallic coatings such as zinc on steel (galvanizing) is, however, a form of cathodic protection, which is effective in the atmosphere. There are two basic methods of supplying the electrical currents required to interfere with the electrochemical cell action. The first method, cathodic protection with galvanic anodes, uses the corrosion of an active metal, such as magnesium or zinc, to provide the required electrical current. In this method, called sacrificial or galvanic anode cathodic protection, the active metal is consumed in the process of protecting the surfaces where corrosion is controlled and the anodes must be periodically replaced. In the second method, impressed current cathodic protection, an alternative source of direct electrical current, usually a rectifier that converts alternating current to direct current, is used to provide the required electrical current. In this system, the electrical circuit is completed through an inert anode material that is not consumed in the process.

CORROSION INHIBITORS - MODIFY THE OPERATING ENVIRONMENT

Another method of corrosion control often neglected is modifying the operating environment. Using a selective backfill around a buried structure, using corrosion inhibitors in power plant or in engine cooling systems, and modifying structures to provide adequate drainage are all examples of the use of this method of corrosion control. Although best employed during the design stage, in some cases, actions taken to correct corrosion problems through modifying the environment can be taken after a system is built. Careful identification and characterization of corrosion problems will often reveal opportunities for changing the environment to control corrosion.

CORROSION CONTROL

TESTING

Corrosion Indicators

Every treatment plant should have a corrosion control plan for its distribution system. This system may be as simple as long-term monitoring of the water to determine if water is corrosive, or it can include a complex array of chemicals or equipment. Here, we will consider methods used to monitor the stability of water. The most common indicators of corrosion in the distribution system are red water complaints and leaks. If the incidence of these problems increases in a certain area of the distribution system, then some sort of corrosion control may need to be undertaken. Red water is usually caused by tuberculation and iron bacteria while leaks are caused by the pitting below tubercles. However, the operator should be aware of other possible causes of these problems. High iron concentrations in the source water can cause red water problems while leaks can be caused by corrosive soil acting on the outside of the pipes as well as by corrosive water acting on the inside of the pipes. During routine maintenance of the distribution system, the operator should watch out for signs of corrosion and scale. When pipes are removed and replaced, the old pipes should be visually examined for signs of tubercles, pitting, or uniform corrosion, and for excessive scaling.

Long-Term Testing

More active forms of corrosion monitoring include coupons and tests for flow, dissolved oxygen, and heavy metals. These tests will determine whether the treated water is corrosive over a span of a few months (in the case of coupons), weeks (for flow tests), or immediately.

Coupons, like the one shown above, are small pieces of the same type of metal used in the distribution system piping. These coupons are inserted into pipes at various locations in the distribution system and are left in place for about three months to give adequate time for corrosion to occur. By weighing the coupon before and after the test period, the amount of metal lost from the coupon due to corrosion can be determined. This is a simple method of corrosion monitoring which is widely used in many distribution systems. Flow monitoring can also be used to detect corrosion. A new piece of pipe is placed in service and the flow of water through the pipe is measured over time. If the flow becomes lower after a few weeks, then either tubercles or scale have formed on the inside of the pipe, decreasing the area available to carry water.

Short-Term Testing

Dissolved oxygen and toxic heavy metals in the distribution system can be used as indicators of corrosion over a much shorter time frame. There are also a range of tests done at the water treatment plant to determine whether water is stable. Dissolved oxygen is tested at various points in the distribution system at the same time. If the dissolved oxygen concentration becomes lower further from the treatment plant, then the oxygen is probably being used up by corrosion. However, the operator should be aware of the possibility that D.O. is being used to oxidize organic matter.

Toxic heavy metals, such as copper and lead, are tested at the consumer's tap. High concentrations of these metals in the water indicate corrosion in the distribution system, although in a few cases the metals may have originated in the source water. Finally, water can be tested directly to determine whether it is stable. Both the Langelier Index and the Marble Test are laboratory tests which can determine the degree of calcium carbonate saturation in the water at the treatment plant. Water which is just saturated with calcium carbonate or which is slightly supersaturated with calcium carbonate is considered stable and safe to release into the distribution system.

TREATMENT

Chemical Treatment

Treatment of corrosive water can be either chemical or physical. In this section, we will discuss chemical methods of corrosion control. These chemical are either meant to stabilize the water, to form a protective film on the pipe surface, or to kill problematic bacteria. Stabilizing the water is often the simplest form of corrosion control. When stabilizing corrosive water, the operator usually adds alkalinity in the form of lime, soda ash, or caustic soda. The goal is to saturate or slightly supersaturate the water with calcium carbonate so that it is stable or slightly scale-forming. When these chemicals are used to stabilize water, they should be fed after filtration to prevent cementing of the filter sand and may be fed before, during, or after chlorination.

Corrosion inhibitors are used to form thin protective films on pipe walls, which will prevent corrosion. The chemicals used for this purpose are more expensive than lime, but also prevent scale which can be a problem when feeding stabilizing chemicals into the water. Sodium silicate is sometimes used by individual customers as an inhibitor but is not widely used by utilities. Glassy phosphates such as sodium hexametaphosphate or tetrasodium pyrophosphate are more widely used, but can increase corrosion rates. Both types of inhibitors require continual application into the water, so dead ends in the distribution system must be flushed at intervals to ensure that fresh water containing the inhibitors reaches these areas as well. A large amount of the inhibitor chemicals ends up forming the film on the pipe walls, but some ends up in the drinking water, though this is not a problem since all inhibitor chemicals are considered safe. If bacteria are a major component of the corrosion problem, then proper disinfection may be part or all of the answer.

Maintaining an adequate chlorine residual in the distribution system will kill the bacteria and prevent corrosion.

Physical Protection

Physical protection against corrosion may be very simple or very complex. On the simple end of the spectrum, corrosion can be prevented by breaking the corrosion cell circuit in some manner. Metal pipes can be replaced with nonmetals which are non-conductive and will not corrode. Alternatively, pipes may be lined with portland cement or bituminous or asphaltic compounds to prevent the water from reaching the metal, serving the same purpose. If galvanic corrosion is a problem, then the two metals can be separated by dielectric couplings. Dielectric couplings are plastic, ceramic, or other non-conductive sections used between the two different types of metal. Since electrons cannot flow through the dielectric coupling, it breaks the circuit and prevents corrosion.

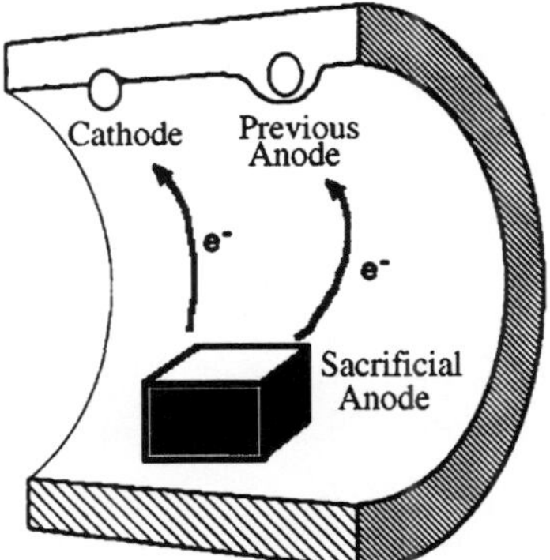

Cathodic protection using a sacrificial anode.

At the more expensive and complicated end of the protection spectrum is cathodic protection, which is the introduction of a different electrical circuit into the pipe. Some cathodic protection systems operate as shown in the picture above, by introducing a sacrificial anode into the pipe. A sacrificial anode is a piece of very active metal (usually zinc or magnesium) which is more galvanically active than any other metal in the system. The sacrificial anode will be the only metal corroded, and even previously active anodes on the pipe wall will become cathodes and will thus be protected. Since the sacrificial anodes slowly corrode away, they must be replaced at intervals, which is the only form of maintenance required on the protection system.

Alternatively, some cathodic protection systems involve the introduction of an external direct current source, known as a rectifier. The rectifier creates a very strong anode since it is constantly producing electrons (an electric current.) This turns the rest of the pipe into a cathode, which prevents any corrosion in the pipe. To complete the circuit, the pipe must be connected back to the rectifier. Direct current cathodic protection systems have been developed which are fully automatic and will compensate for any changes without operator control. However, they also tend to be very expensive to install.

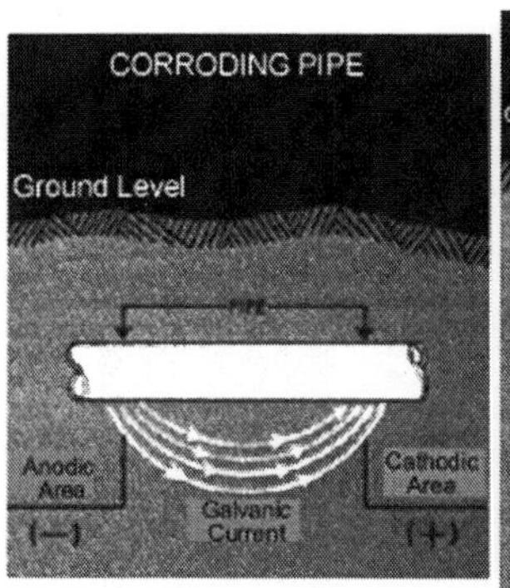

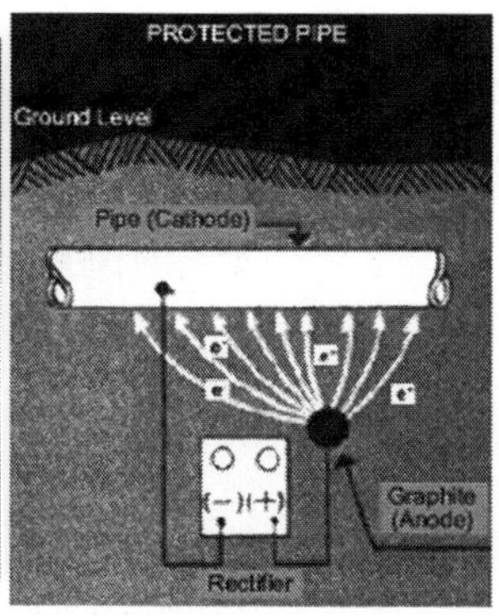

Review

Stable water is neither scale-forming nor corrosive, both of which characteristics create problems in the distribution system. Scale forms when calcium carbonate precipitates out of hard water. Corrosion occurs when an anode, cathode, conductive connection, and electrolyte create a corrosive cell. In the corrosive cell, the metal of the pipe is oxidized in a series of reactions, producing rust. Corrosion inside a pipe can be caused by electrolysis, oxygen concentration cells, or galvanic corrosion. Many factors can influence the corrosion, including pH, hardness, alkalinity, oxidizing agents, carbon dioxide, dissolved solids, temperature, velocity of flow, bacteria, metal characteristics, and stray electric currents. Corrosion testing includes monitoring red water complaints and leaks; inspecting old pipes; using coupons; testing flow, dissolved oxygen, and heavy metals; and using the Langelier Index and Marble Test. Chemical treatment involves addition of chemicals to stabilize the water, use of inhibitors to form a protective film on pipes, or addition of disinfectants. Physical protection either breaks the corrosive cell or consists of cathodic protection.

CORROSION MONITORING

A corrosion monitoring program provides comprehensive monitoring of all critical components of industrial objects, assets, facilities and plants for signs of corrosion. For reliable operation it is important to identify the location, rate, and underlying causes of corrosion. A corrosion monitoring program identifies any non-conforming alloycomponents, as these are generally susceptible to accelerated corrosion and can give relatively frequent cause for catastrophic failure. Corrosion Monitoring can provide significant advantages when integrated into both preventative maintenance and the processes inherent to safety management programs. Based on the results of the Corrosion Monitoring program, informed decisions can be made, not only regarding the remaining life of the object affected by corrosion but also regarding life extension strategies, prospective material selection, and cost-effective methods for remedy of corrosion issues and problems.

ACTIVITIES

An effective corrosion monitoring program includes a wide range of activities:

- Identification of all critical components
- Identification of component alloy composition
- Measurement of the location and extent of corrosion
- Prediction of remaining life
- Identification of failure mechanisms
- Determination of fitness for service condition
- Inspection scheduling
- Development of recommendations for remedy and correction of problems
- Development of corrosion prevention strategies

Why Monitor Corrosion

Corrosion is a major problem in many industries, particularly in the petrochemical industry. Corrosion is one of the most serious ageing mechanisms impacting the equipment and assets of refineries and plants. Uncontrolled corrosion can cause leaks and component failures, bringing about a reduction in both the performance and reliability of important equipment. In extreme cases, corrosion can lead to unexpected failures that can be costly in terms of repair costs, environmental damage and potential harm to humans.

Methods

Corrosion Monitoring uses a wide range of measurement techniques. Non-Destructive Testing (NDT) methods are the most effective and broadly applied testing methods. Suitable NDT methods for the monitoring of corrosion include:

- Ultrasonic Testing
- Radiographic Testing
- Guided Wave Testing
- Electromagnetic Testing.

The selection of the appropriate method as well as the detection and monitoring of corrosion requires knowledgeable and experienced personnel.

WHAT IS CORROSION MONITORING?

The field of corrosion measurement, control, and prevention covers a very broad spectrum of technical activities. Within the sphere of corrosion control and prevention, there are technical options such as cathodic and anodic protection, materials selection, chemical dosing and the application of internal and external coatings. Corrosion measurement employs a variety of techniques to determine how corrosive the environment is and at what rate metal loss is being experienced. Corrosion measurement is the quantitative method by which the effectiveness of corrosion control and prevention techniques can be evaluated and provides the feedback to enable corrosion control and prevention methods to be optimized.

A wide variety of corrosion measurement techniques exists, including:

Non Destructive Testing

Analytical Chemistry

- Ultrasonic testing
- Radiography
- Thermography
- Eddy current/magnetic flux
- Intelligent pigs

Operational Data

- pH
- Flow rate (velocity)
- Pressure
- Temperature

Corrosion Monitoring

- Weight loss coupons
- Electrical resistance
- Linear polarization
- Hydrogen penetration
- Galvanic current

Analytical Chemistry

- pH measurement
- Dissolved gas (O_2, CO_2, H_2S)
- Metal ion count (Fe^{2+}, Fe^{3+})
- Microbiological analysis

Fluid Electrochemistry

- Potential measurement
- Potentiostatic measurements
- Potentiodynamic measurements
- A.C. impedance

Some corrosion measurement techniques can be used on-line, constantly exposed to the process stream, while others provide off-line measurement, such as that determined in a laboratory analysis. Some techniques give a direct measure of metal loss or corrosion rate, while others are used to infer that a corrosive environment may exist. Corrosion monitoring is the practice of measuring the corrosivity of process stream conditions by the use of "probes" which are inserted into the process stream and which are continuously exposed to the process stream condition. Corrosion monitoring "probes" can be mechanical, electrical, or electrochemical devices. Corrosion monitoring techniques alone provide direct and online measurement of metal loss/ corrosion rate in industrial process systems. Typically, a corrosion measurement, inspection and maintenance program used in any industrial facility will incorporate the measurement elements provided by the four combinations of on-line/offline, direct/indirect measurements.

- Corrosion Monitoring Direct, On-line
- Non Destructive Testing Direct, Off-line
- Analytical Chemistry Indirect, Off-line
- Operational Data Indirect, On-line

In a well controlled and coordinated program, data from each source will be used to draw meaningful conclusions about the operational corrosion rates with the process system and how these are most effectively minimized.

The Need for Corrosion Monitoring

The rate of corrosion dictates how long any process plant can be usefully and safely operated. The measurement of corrosion and the action to remedy high corrosion rates permits the most cost effective plant operation to be achieved while reducing the life-cycle costs associated with the operation.

Corrosion monitoring techniques can help in several ways:

- by providing an early warning that damaging process conditions exist which may result in a corrosion-induced failure.
- by studying the correlation of changes in process parameters and their effect on system corrosivity.
- by diagnosing a particular corrosion problem, identifying its cause and the rate controlling parameters, such as pressure, temperature, pH, flow rate, etc.
- by evaluating the effectiveness of a corrosion control/prevention technique such as chemical inhibition and the determination of optimal applications.
- by providing management information relating to the maintenance requirements and ongoing condition of plant.

Corrosion Monitoring Techniques

A large number of corrosion monitoring techniques exist. The following list details the most common techniques which are used in industrial applications:

- Corrosion Coupons (weight loss measurements)
- Electrical Resistance (ER)
- Linear Polarization Resistance (LPR)
- Galvanic (ZRA) / Potential
- Hydrogen Penetration
- Microbial
- Sand/Erosion

Other techniques do exist, but almost all require some expert operation, or otherwise are not sufficiently rugged or adaptable to plant applications. Of the techniques listed above, corrosion coupons, ER, and LPR form the core of industrial corrosion monitoring systems. The four other techniques are normally found in specialized applications which are discussed later. These corrosion monitoring techniques have been successfully applied and are used in an increasing range of applications because:

- The techniques are easy to understand and implement.
- Equipment reliability has been demonstrated in the field environment over many years of operational application.

- Results are easy to interpret.
- Measuring equipment can be made intrinsically safe for hazardous area operation.
- Users have experienced significant economic benefit through reduced plant down time and plant life extension.

Corrosion Coupons (Weight Loss)

The Weight Loss technique is the best known and simplest of all corrosion monitoring techniques. The method involves exposing a specimen of material (the coupon) to a process environment for a given duration, then removing the specimen for analysis. The basic measurement which is determined from corrosion coupons is weight loss; the weight loss taking place over the period of exposure being expressed as corrosion rate.

The simplicity of the measurement offered by the corrosion coupon is such that the coupon technique forms the baseline method of measurement in many corrosion monitoring programs. The technique is extremely versatile, since weight loss coupons can be fabricated from any commercially available alloy. Also, using appropriate geometric designs, a wide variety of corrosion phenomena may be studied which includes, but is not limited to:

- Stress-assisted corrosion
- Bimetallic (galvanic) attack
- Differential aeration
- Heat-affected zones

Advantages of weight loss coupons are that:

- The technique is applicable to all environments - gases, liquids, solids/particulate flow.
- Visual inspection can be undertaken.
- Corrosion deposits can be observed and analyzed.
- Weight loss can be readily determined and corrosion rate easily calculated.
- Localized corrosion can be identified and measured.
- Inhibitor performance can be easily assessed.

In a typical monitoring program, coupons are exposed for a 90-day duration before being removed for a laboratory analysis. This gives basic corrosion rate measurements at a frequency of four times per year. The weight loss resulting from any single coupon exposure yields the "average" value of corrosion occurring during that exposure. The disadvantage of the coupon technique is that, if a corrosion upset occurs during the period of exposure, the coupon alone will not be able to identify the time of occurrence of the upset, and depending upon the peak value of the upset and its duration, may not even register a statistically significant increased weight loss. Therefore, coupon monitoring is most useful in environments where corrosion rates do not significantly change over long time periods. However, they can provide a useful correlation with other techniques such as ER and LPR measurements.

Electrical Resistance (ER) Monitoring

ER probes can be thought of as "electronic" corrosion coupons. Like coupons, ER probes provide a basic measurement of metal loss, but unlike coupons, the value of metal loss can be measured at any time, as frequently as required, while the probe is in-situ and permanently exposed to the process stream. The wire loop represented here is new with its total diameter of 40 mils (total useful probe life of 10 mils). Here the element has experienced about 5 mils penetration or about half of its useful life. The increased electrical resistance of the element will register as 5 mils metal penetration into the piping or process system.

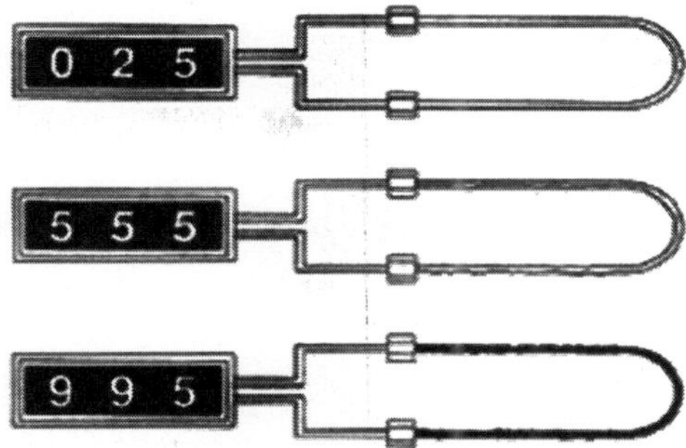

Here the element has measured 10 mils of penetration and requires replacement. The ER technique measures the change in Ohmic resistance of a corroding metal element exposed to the process stream.

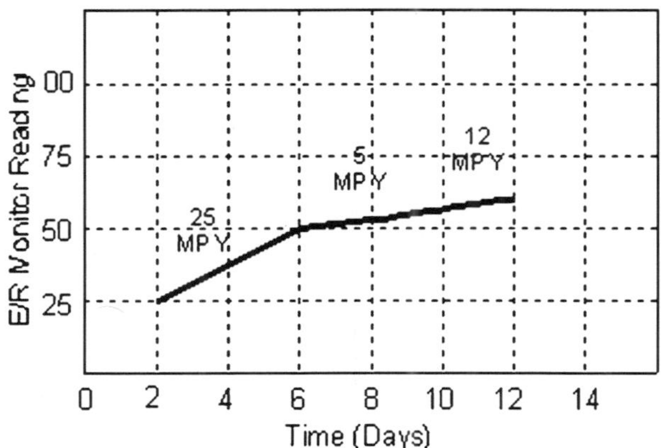

The action of corrosion on the surface of the element produces a decrease in its cross-sectional area with a corresponding increase in its electrical resistance. The increase in resistance can be related directly to metal loss and the metal loss as a function of time is by definition the corrosion rate. Although still a time averaged technique, the response time for ER monitoring is far shorter than that for weight loss coupons. The graph below shows typical response times.

ER probes have all the advantages of coupons, plus:

- They are applicable to all working environments gases, liquids, solids, particulate flows.
- Direct corrosion rates can be obtained.
- Probe remains installed in-line until operational life has been exhausted.
- They respond quickly to corrosion upsets and can be used to trigger an alarm.

ER probes are available in a variety of element geometries, metallurgies and sensitivities and can be configured for flush mounting such that pigging operations can take place without the necessity to remove probes. The range of sensitivities allows the operator to select the most dynamic response consistent with process requirements.

Linear Polarization Resistance (LPR) Monitoring

The LPR technique is based on complex electro-chemical theory. For purposes of industrial measurement applications it is simplified to a very basic concept. In fundamental terms, a small voltage (or polarization potential) is applied to an electrode in solution. The current needed to maintain a specific voltage shift (typically 10 mV) is directly related to the corrosion on the surface of the electrode in the solution. By measuring the current, a corrosion rate can be derived.

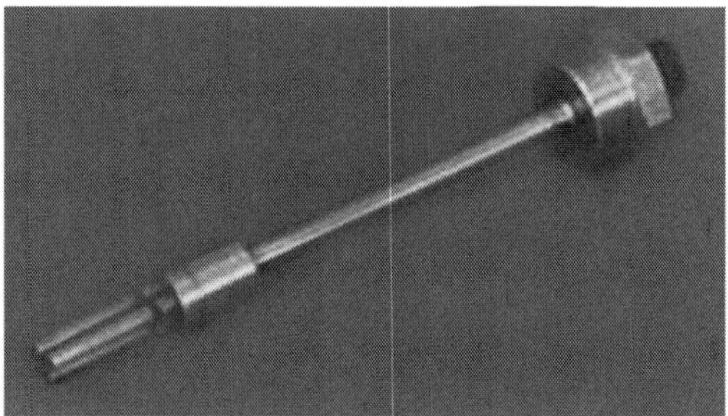

The advantage of the LPR technique is that the measurement of corrosion rate is made instantaneously. This is a more powerful tool than either coupons or ER where the fundamental measurement is metal loss and where some period of exposure is required to determine corrosion rate. The disadvantage to the LPR technique is that it can only be successfully performed in relatively clean aqueous electrolytic environments. LPR will not work in gases or water/ oil emulsions where fouling of the electrodes will prevent measurements being made.

Galvanic/Potential Monitoring

The galvanic monitoring technique, also known as Zero Resistance Ammetry (ZRA) is another electrochemical measuring technique. With ZRA probes, two electrodes of dissimilar metals are exposed to the process fluid. When immersed in solution, a natural voltage (potential) difference exits between the electrodes. The current generated due to this potential difference relates to the rate of corrosion which is occurring on the more active of the electrode couple. Galvanic/Potential monitoring is applicable to the following electrode couples:

- Bimetallic corrosion
- Crevice and pitting attack
- Corrosion assisted cracking
- Corrosion by highly oxidizing species
- Weld decay

Galvanic current measurement has found its widest applications in water injection systems where dissolved oxygen concentrations are a primary concern. Oxygen leaking into such systems greatly increases galvanic currents and thus the corrosion rate of steel process components. Galvanic monitoring systems are used to provide an indication that oxygen may be invading injection waters through leaking gaskets or deaeration systems.

SPECIALIZED MONITORING

Biological Monitoring

Biological monitoring and analysis generally seeks to identify the presence of Sulphate Reducing Bacteria - SRB's. This is a class of anaerobic bacteria which consume sulphate from the process stream and generate sulphuric acid, a corrosive which attacks production plant materials.

Sand/Erosion Monitoring

These are devices which are designed to measure erosion in a flowing system. They find wide application in oil/gas production systems where particulate matter is present.

Hydrogen Penetration Monitoring

In acidic process environments, hydrogen is a by-product of the corrosion reaction. Hydrogen generated in such a reaction can be absorbed by steel particularly when traces of sulphide or cyanide are present. This may lead to hydrogen induced failure by one or more of several mechanisms. The concept of hydrogen probes is to detect the amount of hydrogen permeating through the steel by mechanical or electrochemical measurement and to use this as a qualitative indication of corrosion rate.

Instrumentation

There exists a variety of instrument options associated with the various corrosion monitoring techniques. Three classifications are:

- Portable
- Continuous Single Channel
- Continuous Multi-Channel

In some applications such as those experienced in oil/gas production and refining, instrumentation is required to be certified for use in "hazardous areas". For portable instruments this is most often achieved by having the equipment certified as "intrinsically safe" by a recognized authority such as BASEEFA (U.K.), U.L. (U.S.A.), or CENELEC (Europe). For hardwired continuous monitoring electronics, isolation barriers can be used to ensure that, in the event of a fault condition, insufficient energy is transmitted to the hazardous field area so that an explosive spark cannot be produced.

Probe Fitting Styles

There are two fundamental fitting styles for corrosion probes: fixed and removable under pressure. Fixed styles of probes/sensors have typically a threaded or flanged attachment to the process plant. For fixed styles of sensors, removal can only be accomplished during system shut down or by isolation and depressurization of the sensor location. From time to time, corrosion coupons and probes require removal and replacement. It is sometimes more convenient to be able to remove and install sensors while the process system is operational. To facilitate this, there are two distinct systems which permit removal/installation under pressure. In refinery and process plant environments where pressures are normally less than 1500 psi, a Retractable System is used. This consists of a packing gland (stuffing box) and valve arrangement. For environments such as those experienced in oil/gas production where pressures of several thousand psi are experienced, a special High Pressure Access System is used. This permits the safe and easy installation/removal of corrosion monitoring devices at working pressures up to 3600 psi.

APPLICATIONS OF CORROSION MONITORING TECHNIQUES

Corrosion monitoring is typically used in the following situations:

- Where risks are high - high pressure, high temperature, flammable, explosive, toxic processes.

- Where process upsets can cause high corrosivity.
- Where changes in operating conditions can cause significant changes in corrosion rate.
- Where corrosion inhibitors are in use.
- In batch processes, where corrosive constituents are concentrated due to repeated cycling.
- Where process feedstock is changed.
- Where plant output or operating parameters are changed from design specifications.
- In the evaluation of corrosion behavior of various alloys.
- Where induced potential shifts are used to protect systems and/or structures.
- Where product contamination due to corrosion is a vital concern.

Corrosion monitoring may be used in virtually any industry where corrosion prevention is a primary requirement. Some examples of industries and specific areas of interest include, but are not limited to:

Oil/Gas Production

- Flowlines
- Gathering Systems
- Transport Pipelines
- Water Injection Facilities
- Vessels
- Processing
- Water Systems
- Chemical Injection Systems
- Drilling Mud Systems
- Water Wash Systems
- Desalters

Refining

- Crude Overheads
- Visbreakers
- Vacuum Towers
- Sour Water Strippers
- Amine Systems
- Cooling Systems

Pulp and Paper

- Digesters
- White Liquor
- Boiler Systems

Utilities

- Cooling Systems
- Effluent Systems
- Make-Up Water Systems
- Boiler Water Systems

Petrochemicals/Chemicals/Processing

- Process Systems
- Cooling Systems

In any corrosion monitoring system, it is common to find two or more of the techniques combined to provide a wide base for data gathering. The exact techniques which can be used depend on the actual process fluid, alloy system, and operating parameters. Corrosion monitoring offers an answer to the question of whether more corrosion is occurring today as compared to yesterday. Using this information it is possible to qualify the cause of corrosion and quantify its effect. Corrosion monitoring remains a valuable weapon in the fight against corrosion, thereby providing substantial economic benefit to the user.

CORROSION MONITORING BASICS

Corrosion can lead to failures in plant infrastructure and machines which are usually costly to repair, costly in terms of lost or contaminated product, in terms of environmental damage, and ultimately it may be costly in terms of human safety. Decisions regarding the future integrity of a structure or its components depend entirely upon an accurate assessment of the conditions affecting its corrosion and rate of deterioration. With this information, an owner can make an informed decision as to the type, cost and urgency of remedial measures. Monitoring corrosion characteristics of a proposed or existing structure can lead to the proper selection of longer life materials, durable and protective coatings and corrosion control measures. Modern corrosion monitoring technologies can emphasize the highly time-dependent nature of corrosion damage. The integration of corrosion monitoring technology in existing systems can also provide early warning of costly corrosion damage and provide information on where the damage is taking place.

CORROSION INSPECTION AND MONITORING

In a modern business environment, successful enterprises cannot tolerate major corrosion failures, especially when these involve personal injuries, fatalities, unscheduled shutdowns and environmental contamination. For this reason considerable effort must be expended in corrosion control at all stages of a system's life, from the design table to the last day of operation. Typically, once a system, a plant or any piece of equipment is put into service, maintenance is required to keep it operating safely and efficiently. This is particularly true for aging systems and structures that will be required to operate beyond their original design life.

Corrosion Inspection and Monitoring

Correct and effective corrosion monitoring strategies should be used as a proactive tool to assist with operating a plant more effectively, thereby prolonging its life and gaining optimum throughput. It also enables continuous monitoring of actual corrosion rates, allowing for timely preventative action if a variance is observed. Current corrosion inspection and monitoring typically requires planned periodic shutdowns to inspect equipment. Scheduled shutdowns are costly in terms of productivity losses, restart energy and material costs. Unscheduled shutdowns are disruptive and often quite expensive. Internal corrosion failures result in costly cross contaminations of product and process streams. External corrosion leaks put process fluids into the plant environment and can create significant safety hazards. Corrosion monitoring systems vary significantly in complexity, from simple coupon exposures or hand held data loggers to fully integrated plant process surveillance units with remote data access and data management capabilities. CLIMAT devices, for example, have been used for more than three decades to monitor atmospheric corrosivity. These units have been utilized successfully around the world in marine and industrial type atmospheres.

Other sensors or devices will produce a signal that needs to be recorded and analyzed further for an estimation of the severity of a situation. By combining such signal with the knowledge of a process or environment it allows an operator to be proactive. Experience has shown that the potential cost savings resulting from the implementation of corrosion monitoring programs generally increase with the sophistication level (and cost) of the monitoring system. It is therefore helpful to have previous history or even a rough estimate of the types of corrosion problems to be investigated. It is also advisable to use several complementary techniques rather than rely on a single monitoring method. Real time monitoring of pipelines, vessels and other static equipment enables a near instantaneous appraisal of the corrosivity of produced and transported fluids. On-line nature monitoring means that corrosion information is immediately available to the operator. If corrosion activity increases as a result of process non-conformities, the corrosion information can be viewed alongside process variables (including chemical injection data) such that cause-and-effect can be determined and rapid action can be taken to overcome the problem. Effectiveness of remedial action or treatment can be similarly proven.

MATERIAL SELECTION PROCESS

In Material Science Engineering understanding the material selection process is the key to engineering any application and/or part design. Material selection is the foundation of all engineering applications and design. This selection process can be defined by application requirements, possible materials, physical principles, and selection. The design or function of the part/application is the application requirements. The application requirements are specific given the application. The possible materials are simply the only materials that can be used in the application. Possible Materials are defined by the application requirements. For an example you cannot use clothe to build a bicycle. Physical principles are methods of changing a material that are learned through material science techniques. Using material science physical principles we can change material properties. Three common physical principles we can use for functional material strengthening are densification, composites, and alloying. There many manufacturing techniques used to strengthen and form materials as well. Densification is the most common and necessary way to strengthen any material. In general, this increases the tensile strength by reducing the porosity of the material.

The standard composite rule of mixtures is when the standard matrix is soft/pliable and the reinforcing material is tensile strong. One the major reasons for the prevalent use of composite materials in construction is the adaptability of the composite to many kinds of applications. The selection of mixture proportions can be aimed to achieve optimum mechanical behavior of the harden product. Selection can result in the change of the strength, consistency, density, appearance, and durability. The alloying of metals is one of the oldest and most fundamental material processing techniques. An Alloy is a solid solution that is composed of two or more elements. There is a solvent (majority composition) and a solute. The Solute element can strengthen the overall solid solution by different element size, density, and other material properties.

Given the application requirements, possible materials, and physical principles we can select the best material. Thus in the selection of a material: First, we decide on the requirements of the application. Second, we decide on the possible materials we can use in the application. Third, we decide what changes in the material properties are needed. Lastly, we decide which material out of the possible materials best fulfills the requirements of the application given possible changes in the material properties.

PRINCIPLES USED IN SELECTING AND EDITING MATERIALS

GENERAL PRINCIPLES OF SELECTION

We now have media-videodisc, compact disc, hard disc - which will hold hitherto inconceivably large sets of visual, aural and textual information. Nevertheless, in practice there is always too little space, time, energy or interest, so that only part of the materials that have survived can be transferred. We have already described in a preliminary way how the minds and interests of the observers absorbed only a small part of the Naga world into their camera lenses and writings. A second distortion then occurs by the action of history and accidents of storage; a great deal has been lost, or is buried in private or public collections in such a way that we were unable to use it. Nevertheless, even with these two filters, a very great deal has survived and this has meant that we have had to make some selections from what we have found. Since such a selection is inevitable in any project of this nature and affects the meaning and value of what is available, we may briefly describe the principles of selection in each major type of material.

PHOTOGRAPHS

Since there are a finite number of these, and they are of great value, we have tried to include practically all the photographs we have located. Only a few hundred out of the roughly seven thousand black and white photographs we discovered have been omitted. These were left out on the following grounds: they were duplicates of, or very similar to, other images; their quality was poor; they were outside the delimited geographical area; they were outside our time span; or they fell on the side of "private experience" as opposed to "public experience", in the sense meant by John Berger (About Looking, 1980, p.51). We did not "censor" any photographs because their content was embarrassing or shocking in any way, or might do damage to the reputation of individuals, the British, anthropology as a discipline, or for other such reasons. This latitude obviously raises important issues, given that the videodisc system is intended for use in a variety of educational contexts.

Many of the images from the colonial era do, after all, portray the people concerned in a way which is objectifying, decontextualizing, or exoticizing. It is undeniably the case that the camera did contribute to the "normalising gaze", in Foucault's terms, by which a subtle form of power was exercised over others by classifying them and making them visible. The Naga videodisc does not attempt to avoid these issues, but, rather, hopes through the associated tutorials and courseware to encourage a critical attitude on the part of users to the historical interaction of anthropology and administration. It must also be said that the antithesis to decontextualized images is also contained on the disc itself, in the shape of some of the earliest, and best, examples of a recognisably "modern" era of empathetic and contextualized anthropological photography.

MOVING FILM

There were two problems here. First, we found very little film before 1947; but since film is so important, we decided to break the temporal boundaries and include movie film taken on two visits to Nagaland in 1963 and 197O by an anthropologist who had worked in the area in 1936-7. When we included these, we now had some six hours of film. One side of a videodisc will hold only 36 minutes of moving film (in interactive mode), and since we needed to allocate at least six minutes to the photographs, we only had about thirty minutes free. This meant that we had to reduce the film at a ratio of 1:12.

We spent many weeks going through again and again, whittling away material, trying to minimise the loss of valuable archival footage. The principles we evolved and acted on were as follows. The first consideration was the content of the moving images. We tried to include that material which was most intellectually and academically interesting. This is a subjective matter, but includes that material which portrayed events and processes which were most representative, most revealing, most unusual, and illuminated the other still images and texts in the most significant way. The visual images which were unusual included those which were only preserved in this medium, for instance the eating of dried rats, or a shared joke between anthropologist and Naga. We concentrated on subjects where movement and action were important, for instance dance, games, postures and gestures, rituals, agricultural labour. If long and repeated sequences of film on the same subject were present, we selected only one or two sequences. The second set of principles was concerned with form. Taking into account the interest of the content, we rejected badly filmed sequences, that is to say the few sequences which were out of focus, badly composed, unsteady, from too great a distance and so on. We rejected film that was damaged, the colour fading, or otherwise unsatisfactory, unless it was particularly interesting.

We preferred close-up shots of detail in most cases to the wider shots of a more static kind, which could be preserved with one or two selected stills. Close-ups are more effective on the intimate television screen. In considering the selection for the videodisc, there was one set of factors which gave us a considerable advantage over the person who is undertaking normal editing down from raw film.

These are connected to the absolute precise controllability of this new medium. With normal film, one is constrained to save and knit together reasonably long sequences, five seconds at least, and often three times that length, just to capture the run up to something important happening, for instance an immobile man before he starts to run up to do a jump. Again, the action itself, if it is to be appreciated by the viewer and happens rapidly, may need to be shown at some length, or from several different angles. In other words, a great deal of redundancy has to be built in to normal editing, since the viewer will only see the images flashing past once.

With videodisc the user becomes the editor, master of the medium. One effect of this is that one is not constrained in selecting shots for inclusion on the disc by the usual film conventions, such as keeping the stage line or matching shots. There is a sense in which film that has been shrunk or contracted in the editing process can be expanded again at the viewing stage. This is because it is possible to treat each frame in a precise way. One can take just one shot of a view, or non-moving group of people, and hold it as an establishing shot for a number of seconds; if necessary one can easily go back to the start of the sequence and play through in slow motion to explain in detail what is happening, and so on. In other words, instead of merely cutting a long set of moving sequences into shorter pieces and sticking them in a precise order, one is creating a set of images, some of them still frames, some of them sequences of still frames taken every few seconds, some of them moving sequences. The boundaries between still and moving films blur, and one is able to abstract precisely a great deal of the visual information without apparently losing much. It should be stressed, however, that there is a different 'feel' as between a frozen frame and a 'still' filmed sequence.

In this way we created out of six hours of film approximately one hundred and fifty moving sequences, each lasting between three and about twenty-five seconds, the average being about eight seconds. We also abstracted about one thousand "stills" taken out of moving sequences where there was little movement, or randomly every few seconds to capture a series of events, for instance transactions in the market.

ARTEFACTS

Here there is again a problem of selection. There are known to be well over fifteen thousand Naga artefacts in European museums and private collections, at least twelve thousand of them in Britain. To have located and photographed them all would not only have absorbed much of the effort of the team, but the final photographs would have used up over a quarter of the total space for visual images on the videodisc. It would also have given, in some artefacts, hundreds of almost exact duplicates. One general criterion was accessibility. That is to say we confined our work to certain major private and public collections in England, though we knew of others in other parts of the British Isles and Europe, not to mention America and, of course, India and Nagaland itself. From the brief descriptions of other collections, it did appear that we were able to see and select a fairly representative sample of artefacts.

Within the ten thousand or so artefacts which we either examined in themselves, or through catalogue descriptions, we used a number of criteria. We sought a representative selection, in terms of the types and functions of artefacts and their origins in different Naga groups. We attempted to use Naga criteria of significance rather than our own-that is, to ensure the chosen artefacts did reveal the key features of Naga social structure and belief (status, head-taking, kinship organisation and so on). Where there was duplication, we tended to choose artefacts with superior documentation, and left out those where the condition of the artefact was very poor. We sought to photograph as many nineteenth century artefacts as possible, on the grounds of their rarity.

Such artefacts also were crucial in throwing light on a key research interest - the colonial encounter. What can we learn from the types of artefacts collected, about how the Nagas were perceived and classified as the colonial era developed? We also bore in mind the need to photograph artefacts which would be interesting from a comparative perspective - for example, artefacts indicative of trade amongst Naga groups or between Nagas and neighbouring peoples; artefacts bearing strong resemblances to other South East Asian hill tribes; artefacts which, when compared with others, revealed continuity or change over time. We emphasised artefacts which we knew would tie in well with other material on the disc, such as artefacts collected by administrators or ethnographers whose writings feature prominently in the textual database. We sought to ensure that mundane, everyday items were as well represented as the more exotic and aesthetically exciting artefacts. Where there was obvious interest, we photographed from two angles, but constraints of time and difficulty in placing artefacts left this task incomplete.

PAINTINGS AND SKETCHES

Here there was relatively little difficulty with selection. Since the absolute number was not great, in the hundreds, we included all those pictures which we thought could possibly be of some interest. Many very simple line drawings were included when they tied up with textual descriptions, and only very occasionally did we leave out an illustration because it duplicated something else or was so minor and badly documented that it seemed confusing to include it.

SOUND

The 72 minutes of sound data included on the two tracks of the disc, are an attempt to provide very different kinds of data. We sought to combine recordings of a time breadth to match the photographs. Thus we included early wax cylinder recordings (1919) and present-day (1987) recordings of songs, with their considerable Christian influence. Examples of several kinds of instruments have been included, such as drums, jew's-harps, stringed instruments, as well as singing. Field recordings of conversation (1970) have also been included. This is referred to as 'sound' in the indexes.

MAPS

We had hoped that we would merely have to photograph the various maps of Nagaland from the earliest times up to 1947, including the very detailed Survey Of India maps of c.1910-1945. However, when we experimented with the maps by looking at photographs of the ordnance survey maps on the screen, we discovered that this was impossible. However much we magnified the maps by photographing them in tiny sections, the mountainous nature of Nagaland meant that all we could see were blurry pictures of contour lines, with the odd village name, almost unreadable, dotted among them.

We found therefore that we had to re-draw the maps, ending up with 165 sketch maps on the videodisc. All contours were left off, but otherwise they included rivers, major mountains, borders and the location and names of some 1400 villages and towns which were mentioned in one or more of our photographic or textual sources. The maps were mainly based on the 1:2 Survey of India maps of the area. All maps and mapping were subject to considerable errors, compounded in this case by the difficult terrain, the shifting character of many Naga villages and the immense complexity of village names, which can vary radically from author to author, or even within the same author. The maps are therefore, very much to be thought of as sketch maps. They form the basis of "map-walking" software, developed for the project, which allows the user to move north or west to the next map.

GENERAL PRINCIPLES IN SELECTING TEXTS

The selection of texts was done with two overall considerations in mind, which relate to the innovative nature of Interactive Video as a technology. The system can be thought of as essentially an archive, in which case one would probably wish to exercise fairly limited editorial control; on the other hand, it is also a teaching resource, and in this sense does require some attention to including material which is likely to be of use in diverse educational contexts. The main problem here is not so much one of total available space, but of the effort of data entry. It will not be long before standard microcomputers have hundreds of megabytes of storage available, so there is unlikely to be a long-term difficulty in holding materials. Yet if we consider the labour not only of inputting material, but then of checking it and indexing it very precisely to make it useful, it becomes clear that unless one has very large funds and a team of workers over a long period, one is bound to reach the limit of what can be fed into the computer.

In effect, if we add up all the different parts of people's lives that have gone into this project, it is probable that we had the equivalent of five or six person-years of human labour available. Thus one has to select from the surviving materials. A few of the broader principles of such selection need to be mentioned. Firstly, we have concentrated, with the major exception of one very long diary in German, entirely on texts in English. Perhaps a tenth of the writings on the Nagas before 1947 has thus been excluded for consideration. Secondly, we have tended to concentrate our efforts mainly on the period between about 191O and 1947, leaving the nineteenth century more selectively treated. Thirdly, we have directed much of our efforts to the more intimate and detailed accounts by specific individuals whose visual materials are available on the videodisc.

MAJOR TEXTUAL COLLECTIONS IN THE DATABASE

Thus we have attempted to collect and put into the computer all the relevant surviving materials of R.G.Woodthorpe, J.H.Hutton, J.P.Mills, C. von Furer-Haimendorf, Ursula Graham Bower and W.G.Archer. Added to these are smaller collections from other twentieth century authors. For the nineteenth century we have only selected a few texts, for instance part of the ethnographic survey by Dalton, the diary of Godwin-Austen, Woodthorpe's papers and some manuscripts of Colonel J.Butler. At present,the gaps are as follows. Of the several hundred articles written about the Nagas between 1832 and 1947, we have only included a few. Secondly, as far as the official records are concerned, we have only sampled a few of the very extensive files. We have included a file on a military expedition, a gazetteer, some official reports. Since the technique we have used is deliberately open-ended, it will always be possible to incorporate more of the materials which undoubtedly exist in the India Office Library and other archives both in Britain and India. Little use was made of missionary archives, or of materials on Burma.

OMISSIONS AND EDITING

This summary describes the major editorial decisions, to include or exclude a whole class of documents, or whole lengthy text. There are more subtle ways in which one is forced to edit documents, however. As far as internal editing of texts is concerned, we have only applied two rules. If the material is likely to cause personal offence or political embarrassment to living persons, we have omitted it. In all the files we have included, this has led us to omit probably less than a couple of pages of material. Secondly, if the material is repetitive or apparently trivial and of only personal interest, we have omitted this. With regard to the actual materials, we have occasionally corrected spelling or minor grammatical errors which obscure the meaning, but otherwise we have not changed the documents. The fieldwork diaries of Professor Furer-Haimendorf were especially translated for us by Dr Ruth Barnes, an anthropologist and native German speaker. They were checked for accuracy by Professor Furer-Haimendorf.

CORROSION TESTING AND CORROSION MONITORING

WebCorr has NACE certified Corrosion Specialist providing corrosion advisory services, corrosion diagnosis, corrosion testing and corrosion monitoring, in-house training, online and distance learning corrosion courses, corrosion consulting, corrosion failure analysis and corrosion expert witness in litigation and arbitration cases related to corrosion, materials, metallurgy, paints and metallic coatings including thermal spray metallizing, galvanizing, anodizing, chromating, phosphating, electroplating, electroless plating, mechanical plating, and sheradizing or diffusion coating.

In addition to Corrosion Literature Search and Supply (CLSS), our NACE certified corrosion specialist can also provide corrosion testing and monitoring services in accordance with international standards such as ASTM, ISO, BSI, DIN, JIS, NACE and SIS. Our capability covers both DC polarization techniques and AC impedance measurements and EIS equivalent circuit modeling with popular softwares such as Boukamp's Equivcirt, Scribnere's ZView and CorrView and Mansfield's EIS Modules. Our NACE certified corrosion specialist will analyze the raw (or processed) corrosion data, write and sign the technical report. Expert witness and litigation support are also available if required. Verification or Certification Test: Verification of whether materials or products have been produced in accordance with the relevant international standards or to contract specifications may be established by corrosion test. WebCorr has access to a wide range of equipment and expert personnel for testing physical, mechanical and corrosion resistance properties of materials.

3

Corrosion of Electronic Process Control Systems

INTRODUCTION

WHAT GASES CAUSE ODORS?

Wastewater plants are home to many different gases depending on the location within the plant and the processes taking place. Many times the gases considered "odorous" are those that contain sulfur, i.e. reduced sulfur compounds such as those shown in Table I (ASTM, 1978; Braker and Mossman, 1980). Probably the most common and most familiar one of these is hydrogen sulfide (H2S) which is associated with rotten egg odor. H2S is used as a target compound when investigating potential odor problems in wastewater systems.

Table. Reduced Sulfur Compounds and Respective Odor Thresholds

Reduced Sulfur Compound	Odor Thresholds		Odor Description
	Low	High	
hydrogen sulfide	0.47 ppb	789 ppb	rotten eggs
methyl mercaptan	0.02 ppb	560 ppb	rotten cabbage
carbon disulfide	210 ppb	835 ppb	ether odor
carbonyl sulfide	*n/a	n/a	similar to rotten eggs
dimethyl disulfide	n/a	n/a	unpleasant odor
dimethyl sulfide	3 ppb	n/a	characteristic
disagreeable odor			

Other types of gases that can be the cause of odor problems are volatile organic compounds (VOCs) and ammonia. VOCs may have the same source(s) as the reduced sulfur compounds and total VOCs are sometimes monitored to gauge the strength of an odor source. Ammonia is less of a problem contaminant, but may be significant in some facilities.

What Gases Cause Corrosion?

Corrosion can be thought of in many different ways. Corrosion appears as rust on iron due to reactions with water and oxygen. Corrosion appears on motor vehicle tail pipes as a result of reactions with the air and sulfur-containing gases emitted in the engine exhaust. What some refer to as "tarnish" is also a form of corrosion. Wastewater facility engineers are familiar with corrosion of the plant infrastructure, but not as much with corrosive attacks on process control computers and other electronic equipment.

Figure 1 displays two common types of corrosion. One is an example of damage to plant infrastructure showing corrosion of a handrail. The other shows corrosion of a component on a printed circuit board (PCB) which is often undetected with visual inspection. However, both are results of chemical attack with one being quite obvious and the other type that may go undetected for months or even years. Gases can cause corrosion at very low concentrations. Standard 71.04 indicates that the most aggressive gases are classified as reactive sulfides and inorganic chlorides (ISA, 1985). Active sulfides would include many of the reduced sulfur compounds described above (hydrogen sulfide, mercaptans, etc) and inorganic chlorides would include chlorine, chlorine dioxide, hydrogen chloride, chloride ions from sea salt, etc. Inorganic chlorides are more aggressive, but occur less often than the active sulfides in wastewater treatment applications. Both of these types are known to cause corrosion of electronic circuits at low parts per billion (ppb) levels.

EXAMPLES OF CORROSIVE ATTACK

Purafil has performed investigations on multiple process control circuit boards that have been exposed to aggressive gases such as those discussed above. Real evidence of corrosive attack has been shown on many boards. The corrosion is not easily seen to the naked eye, so it can be easily overlooked. Figure 2 below displays some graphics of circuit boards from the naked eye view and then at a magnified view. These pictures speak for themselves. Aggressive environments will attack the circuits of control equipment.

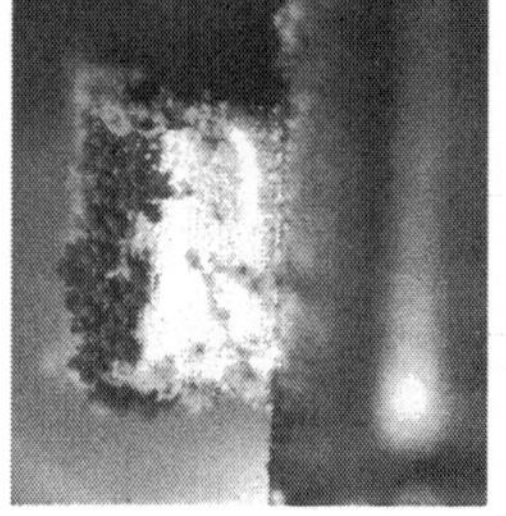

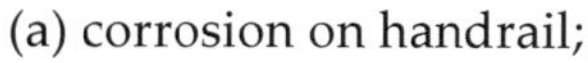

(a) corrosion on handrail; (b) circuit board showing corrosion.

Results of Corrosion

The effects of corrosive attack on process control equipment at wastewater plants are apparent. In the proceeding sections, the odor causing gases and the corrosion causing gases have been linked as one in the same. Following this, actual corrosion on circuit boards was presented. These make a strong case that the corrosion is real. Moreover, the results of this type of corrosion can affect the processes themselves through false signals to control equipment, breaking of control circuits, and ultimately circuit board/control equipment failure. A manufacturer of electronic equipment has also done studies of environmental effects on circuit boards. The company utilized the type of monitoring described later in this paper to evaluate the environment. The length of time until a failure and type of failure for different environments was recorded. As is shown in Table II, mild environments prolonged failures for long periods of time, whereas aggressive environments caused failures within months.

Table. Corrosion Failure in Reactive Environments (Typical)

Environment	Time to Failure	Failure Recorded
Mild	1 – 3 years	intermittent edge card failure
Moderate	6 – 8 months	intermittent edge card failure
Severe	4 – 6 months	corrosion across printed circuit board traces

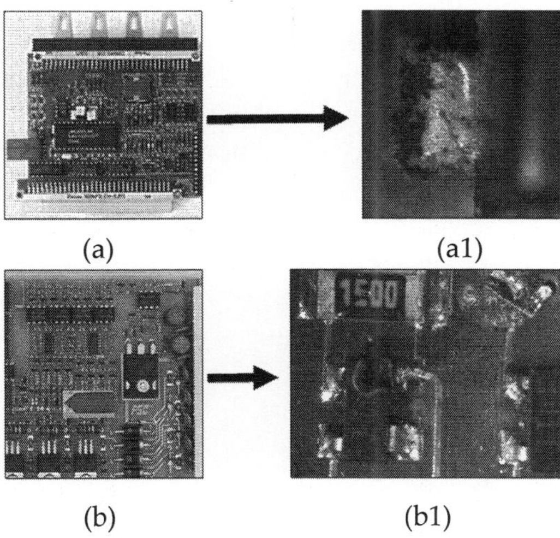

(a) (a1)

(b) (b1)

Fig. (a, b) failed circuit boards with areas of corrosion indicated; (a1) magnified view showing corrosion growth on connector; (b1) magnified view showing component failure due to corrosion.

ARE ODORS CORROSIVE? EVALUATING THE ENVIRONMENT

Odorous gases can make for very aggressive environments with respect to electronic equipment corrosion, but how might one evaluate the environment to gauge this aggressiveness? Direct gas monitoring is one option and requires the use of single gas or multi-gas analyzers/monitors to measure the concentration of specific target gases such as sulfides and chlorine. Some control equipment manufacturers base their warranties on maintaining specific concentration limits of individual gases.

However, when multiple gases are present, there can be synergistic effects that produce more corrosion that might be expected from the individual gases. There may also be some gases not specified or monitored that could contribute to the overall corrosion potential of the environment. Therefore direct gas monitoring is not the preferred method to evaluate the aggressiveness of the environment towards electronic control equipment. As Noon (1987) wrote concerning corrosion and reliability of electronics, "One of the best ways to determine the corrosive aggressiveness of the use environment is to examine reaction by-products of commonly used electronic metals."

Reactivity Monitoring

Corrosion monitoring, or more accurately in the case of electronic equipment, "reactivity monitoring" is an analytical technique that measures the reaction by-products on the metallic components commonly used electronic devices. This monitoring technique uses copper and silver test strips, or coupons, placed in the environment for a period of time. The reaction by-products on these coupons are analyzed most commonly by electrolytic (cathodic) reduction to provide an indication of the types of contamination present (e.g., sulfur and/or chlorine) and the relative amounts present in the environment. Two types of reactivity monitors are shown below Figure 3a shows a copper coupon that is typically analyzed via electrolytic reduction to determined corrosion film chemistry and total film thickness(measured in angstroms (Å, 10-10 meter). Figure 3b shows a quartz crystal microbalance (QCM) sensor used for real-time corrosion monitoring and measures corrosion by mass gain which is converted to film thickness and can be correlated to the results obtained by coupon monitoring. Both techniques provide a measure of film growth over time.

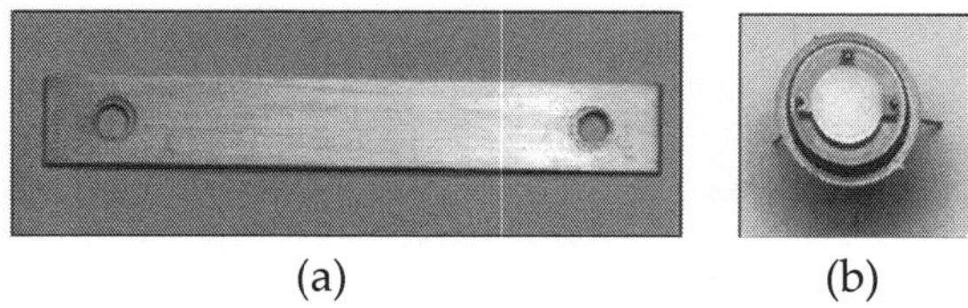

(a) (b)

Fig. (a) copper coupon used to measure corrosion by electrolytic reduction; (b) copper QCM used to measure corrosion by weight gain.

A Standard for Environmental Evaluation

As mentioned above, reactivity monitoring can be used to measures the aggressiveness of an environment towards electronic equipment and devices by measuring the reaction products formed metal coupons/sensors. Reactivity monitoring is used as the basis of determining environmental classifications for process control equipment by the International Society for Automation

(ISA) Standard 71.04-1985: Environmental Conditions for Process Measurement and Control Systems: Airborne Contaminants (ISA, 1985). This standard provides the practical information on the reliability of process control equipment for various environmental classifications or severity levels. These levels are based on the amount of corrosion formed over time normalized to angstroms per thirty days. The least aggressive environment would exhibit <300 Å/30 days and is described as Class G1 - Mild. The most aggressive classification – Class GX - Severe – would exhibit =2,000 Å/30 days. Most equipment warranties are based on maintaining a G1 severity level where "corrosion is not a factor in determining equipment reliability" (ISA, 1985).

Whereas the standard was originally written using copper corrosion as the sole determinant of severity levels and equipment reliability, new environmental laws and regulations have caused changes in the manufacturing of electronics that have resulted in the standard being revised to require the use of both copper and silver to determine severity levels. Specifically, laws prohibiting the use of lead in electronic equipment has caused current electronic equipment to be more sensitive to the environment such that locations that would have been classified as G1(Mild) before these regulations came into effect are now showing classifications of G2 (Moderate) or higher. The revised classification scheme is shown in Table.

Table. ISA Environmental Classification Scheme (Muller, et. al, 2012)

Severity Level	Description	Corrosion Å/30 days		Effect On Electronic Equipment Reliability
		Copper	Silver	
G1	Mild	<300	<200	Corrosion is not a factor.
G2	Moderate	<1,000	<1,000	Corrosion may be a factor.
G3	Harsh	<2,000	<2,000	High probability of corrosive attack. Environmental controls or specially designed and packaged equipment should be used.
GX	Severe	e"2,000	e"2,000	Only specially designed and packaged equipment would be expected to survive without environmental controls.

Corrosion classification coupons (CCCs) employing both copper and silver coupons are excellent tools for environmental monitoring, but they cannot provide the real-time data many plant operators require for the protection of their electronic equipment. They are the most widely used reactivity monitors, based on low cost and the ability to provide information on the types of contamination present in the local environment, but real-time monitoring is gaining in its application and is the preferred technique for many facilities. The most accurate method employs copper and silver-plated quartz crystal microbalances (QCMs) to provide the information necessary to maintain equipment warranties and provide for continuous and profitable operation. These reactivity monitors can be connected directly to the process control system to provide the real-time data necessary to prevent equipment damage and costly shutdowns. Both of these reactivity monitoring techniques provide the data required to develop accurate environmental classifications which can then be used to develop preventative maintenance programs for the protection of process control equipment. The QCM technology can be used to pinpoint corrosion sources and track corrosion events so that modifications can be made to prevent and eliminate such events. Both techniques can also be used to confirm that the control equipment is operating in an environment safe from corrosive attack.

Field Data

Many plant sites employ continuous corrosion monitoring to gauge the impact of the ambient (local) environment on computer process control equipment and other electronic equipment. Areas monitored include control rooms, motor control centers, instrument rooms, office areas, and other areas containing sensitive electronics. Initial monitoring is used to determine the current severity class and to determine the steps necessary to provide the level of protection required for the electronics in use. Figure 4 shows corrosion monitoring data from 30 sites across the U.S. with the percentage of sites in each ISA severity class. Note the percentage of sites classified as GX - Severe in which electronic equipment is not expected to survive.

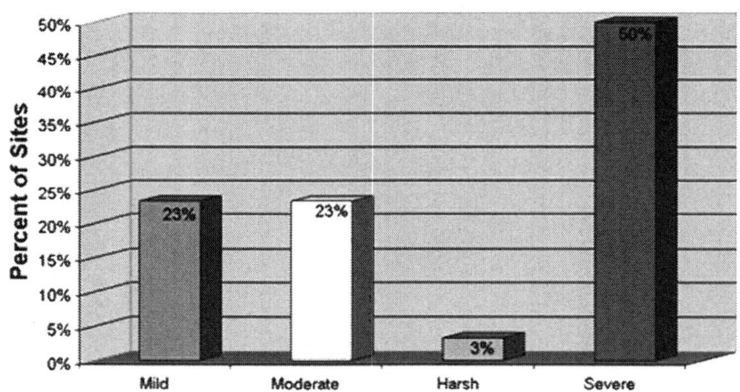

Fig. Summary of Sites evaluated with Coupons

CONTROL OF CORROSION

Corrosion control requires a change in either the metal or the environment. The first approach, changing the metal, is expensive. Also, highly alloyed materials, which are very resistant to general corrosion, are more prone to failure by localized corrosion mechanisms such as stress corrosion cracking. The second approach, changing the environment, is a widely used, practical method of preventing corrosion. In aqueous systems, there are three ways to effect a change in environment to inhibit corrosion:

- Form a protective film of calcium carbonate on the metal surface using the natural calcium and alkalinity in the water
- Remove the corrosive oxygen from the water, either by mechanical or chemical deaeration
- Add corrosion inhibitors

CALCIUM CARBONATE PROTECTIVE SCALE

The Langelier Saturation Index (LSI) is a useful tool for predicting the tendency of a water to deposit or dissolve calcium carbonate (see Chapter 25 for a thorough discussion of LSI). A uniform coating of calcium carbonate, deposited on the metal surfaces, physically segregates the metal from the corrosive environment. To develop the positive LSI required to deposit calcium carbonate, it is usually necessary to adjust the pH, alkalinity, or calcium content of the water. Soda ash, caustic soda, or lime (calcium hydroxide) may be used for this adjustment. Lime is usually the most economical alkali because it raises the calcium content as well as the alkalinity and pH. Theoretically, controlled deposition of calcium carbonate scale can provide a film thick enough to protect, yet thin enough to allow adequate heat transfer. However, low-temperature areas do not permit the development of sufficient scale for corrosion protection, and excessive scale forms in high-temperature areas and interferes with heat transfer. Therefore, this approach is not used for industrial cooling systems. Controlled calcium carbonate deposition has been used successfully in some waterworks distribution systems where substantial temperature increases are not encountered.

Mechanical and Chemical Deaeration

The corrosive qualities of water can be reduced by deaeration. Vacuum deaeration has been used successfully in once-through cooling systems. Where all oxygen is not removed, catalyzed sodium sulfite can be used to remove the remaining oxygen. The sulfite reaction with dissolved oxygen is:

$$Na_2SO_3 + 1/2\ O_2 = Na_2SO_4$$

sodium oxygen sodium sulfite sulfate

The use of catalyzed sodium sulfite for chemical deaeration requires 8 parts of catalyzed sodium sulfite for each part of dissolved oxygen. In certain systems where vacuum deaeration is already used, the application of catalyzed sodium sulfite may be economically justified for removal of the remaining oxygen. The use of sodium sulfite may also be applicable to some closed loop cooling systems. In open recirculating cooling systems, continual replenishment of oxygen as the water passes over the cooling tower makes deaeration impractical.

Corrosion Inhibitors

A corrosion inhibitor is any substance which effectively decreases the corrosion rate when added to an environment. An inhibitor can be identified most accurately in relation to its function: removal of the corrosive substance, passivation, precipitation, or adsorption.

- Deaeration (mechanical or chemical) removes the corrosive substance-oxygen.
- Passivating (anodic) inhibitors form a pro-tective oxide film on the metal surface. They are the best inhibitors because they can be used in economical concentrations, and their protective films are tenacious and tend to be rapidly repaired if damaged (see Figure 24-8).
- Precipitating (cathodic) inhibitors are simply chemicals which form insoluble precipitates that can coat and protect the surface. Precipitated films are not as tenacious as passive films and take longer to repair after a system upset.

A discussion of each of these inhibitors follows, preceded by an overview of the role of polarization in corrosion. Polarization. Figure 24-9 is a schematic diagram showing common potential vs. corrosion current plots. The log current is the rate of electrochemical reaction, and the plots show how the rate of anodic and cathodic reactions change as a function of surface potential. In Figure 24-9(a), uninhibited corrosion is occurring. The corrosion potential, Ecorr, and the corrosion current, Icorr, are defined by the point at which the rate of the anodic reaction equals the rate of the cathodic reaction. Icorr is the actual rate of metal dissolution. Figure 24-9(b) shows the condition after an anodic inhibitor has been applied. The rate of the anodic reaction has been decreased. This causes a decrease in Icorr accompanied by a shift in Ecorr to a more positive (anodic) potential. Figure 24-9(c) shows the effect of a cathodic inhibitor. Here, the rate of the cathodic reaction has been decreased, accompanied again by a decrease in Icorr, but this time the shift in Ecorr is in the negative (cathodic) direction.

Passivation Inhibitors. Examples of passivators (anodic inhibitors) include chromate, nitrite, molybdate, and orthophosphate. All are oxidizers and promote passivation by increasing the electrical potential of the iron. Chromate and nitrite do not require oxygen and thus can be the most effective. Chromate is an excellent aqueous corrosion inhibitor, particularly from a cost perspective. However, due to health and environmental con-cerns, use of chromate has decreased significantly and will probably be outlawed in the near future. Nitrite is also an effective inhibitor, but in open systems it tends to be oxidized to nitrate. Both molybdate and orthophosphate are excellent passivators in the presence of oxygen. Molybdate can be a very effective inhibitor, especially when combined with other chemicals. Its main drawback is its high cost. Orthophosphate is not really an oxidizer per se, but becomes one in the presence of oxygen. If iron is put into a phosphate solution without oxygen present, the corrosion potential remains active and the corrosion rate is not reduced. However, if oxygen is present, the corrosion potential increases in the noble direction and the corrosion rate decreases significantly. A negative attribute of orthophosphate is its tendency to precipitate with calcium hardness found in natural waters. In recent years, deposit control agents that prevent this deposition have been developed. Due to its relatively low cost, orthophosphate is widely used as an industrial corrosion inhibitor.

Precipitating Inhibitors. As discussed earlier, the localized pH at the cathode of the corrosion cell is elevated due to the generation of hydroxide ions. Precipitating inhibitors form complexes which are insoluble at this high pH (1-2 pH units above bulk water), but whose deposition can be controlled at the bulk water pH (typically 7-9 pH). A good example is zinc, which can precipitate as hydroxide, carbonate, or phosphate. Calcium carbonate and calcium orthophosphate are also precipitating inhibitors. Orthophosphate thus exhibits a dual mechanism, acting as both an anodic passivator and a cathodic precipitator.

Copper Corrosion Inhibitors. The most effective corrosion inhibitors for copper and its alloys are the aromatic triazoles, such as benzotriazole (BZT) and tolyltriazole (TTA). These compounds bond directly with cuprous oxide (Cu2O) at the metal surface, forming a "chemisorbed" film. The plane of the triazole lies parallel to the metal surface; thus, each molecule covers a relatively large surface area. The exact mechanism of inhibition is unknown. Various studies indicate anodic inhibition, cathodic inhibition, or a combination of the two. Other studies indicate the formation of an insulating layer between the water surface and the metal surface. A very recent study supports the idea of an electronic stabilization mechanism. The protective cuprous oxide layer is prevented from oxidizing to the nonprotective cupric oxide. This is an anodic mechanism. However, the triazole film exhibits some cathodic properties as well. In addition to bonding with the metal surface, triazoles bond with copper ions in solution. Thus, dissolved copper represents a "demand" for triazole, which must be satisfied before surface filming can occur. Although the surface demand for triazole filming is generally negligible, copper corrosion products can consume a considerable amount of treatment chemical. Excessive chlorination will deactivate the triazoles and significantly increase copper corrosion rates. Due to all of these factors, treatment with triazoles is a complex process.

Adsorption Inhibitors. Adsorption inhibitors must have polar properties in order to be adsorbed and block the surface against further adsorption. Typically, they are organic compounds containing nitrogen groups, such as amines, and organic compounds containing sulfur or hydroxyl groups. The size, orientation, shape, and electrical charge distribution of the molecules are all important factors. Often, these molecules are surfactants and have dual functionality. They contain a hydrophilic group, which adsorbs onto the metal surface, and an opposing hydrophobic group, which prevents further wetting of the metal. Glycine derivatives and aliphatic sulfonates are examples of compounds which can function in this way. The use of these inhibitors in cooling systems is usually limited by their biodegradability and their toxicity toward fish. In addition, they can form thick, oily surface films, which may severely retard heat transfer.

Silicates. For many years, silicates have been used to inhibit aqueous corrosion, particularly in potable water systems. Probably due to the complexity of silicate chemistry, their mechanism of inhibition has not yet been firmly established. They are nonoxidizing and require oxygen to inhibit corrosion, so they are not passivators in the classical sense. Yet they do not form visible precipitates on the metal surface. They appear to inhibit by an adsorption mechanism. It is thought that silica and iron corrosion products interact. However, recent work indicates that this interaction may not be necessary. Silicates are slow-acting inhibitors; in some cases, 2 or 3 weeks may be required to establish protection fully. It is believed that the polysilicate ions or colloidal silica are the active species and these are formed very slowly from monosilicic acid, which is the predominant species in water at the pH levels maintained in cooling systems.

Effect of Conductivity, pH, and Dissolved Oxygen

Figures 24-10 through 24-12 show the effects of several operating parameters on the corrosion tendency in aqueous systems. As shown in Figure 24-10, corrosion rate increases with conductivity.

Figure 24-11 shows the effect of pH on the corrosion of iron. Within the acid range (pH <4), the iron oxide film is continually dissolved. In cooling water, the potential for calcium carbonate precipitation increases with higher pH and alkalinity; thus the corrosion rate decreases slightly as pH is increased from 4 to 10. Above pH 10, iron becomes increasingly passive.

Figure 24-12 shows the effect of oxygen concentration on corrosion at different temperatures. As discussed previously, oxygen is the main driving force for corrosion of steel in cooling water. The increase in corrosion with temperature at a given oxygen concentration is due to more rapid oxygen diffusion occurring at higher temperatures.

Practical Considerations

The success of cooling water corrosion inhibitor programs is affected by the following factors:

- Water Characteristics. Calcium, alkalinity, and pH levels in water are important factors for reasons already cited. Further discussions are covered in the chapters on once-through and open recirculating systems.
- Design Considerations. Low water velocity, which occurs in shell-side cooling, increases deposition. This factor must be addressed in the design of the system.

- Microbiological Control. An effective microbiological control program is necessary to prevent severe fouling problems. Fouling caused by uncontrolled biological growth can contribute to corrosion by one or more mechanisms.
- System Control. Even the best treatment technology available will fail without a reasonable level of control. Therefore, careful consideration must be given to system control-the accuracy with which the pH, inhibitor levels, and other water character-istics are maintained.
- Pretreatment. Grease and/or corrosion products from previous treatment programs should be cleaned out, and the system should be treated with a high level of a good inhibitor before normal operation.
- Contamination. Contamination can also be a problem. Sulfide, ammonia, and hydrocarbons are among the most severe contaminants. Sulfide is corrosive to steel and copper alloys. Ammonia is corrosive to Admiralty and promotes biological growth. Hydrocarbons promote fouling and biological growth.

In the determination of treatment levels, solubility data is important. The Langelier Saturation Index, which defines the solubility of calcium carbonate, is commonly used. Solubility data for calcium orthophosphate and zinc orthophosphate may be needed if the treatment contains phosphate and zinc.

Monitoring

Every cooling water system should include a method of monitoring corrosion in the system. Tools commonly used for this purpose include metal corrosion coupons, instantaneous corrosion rate meters, and heated surfaces such as test heat exchangers and the Betz MonitAll® apparatus. Data obtained from these devices can be used to optimize an inhibitor treatment program to maintain the plant equipment in the best possible condition. When heat transfer data cannot be obtained on operating exchangers, monitoring devices can be useful for evaluating the success of a treatment program without a plant shutdown. Corrosion Coupons. Preweighed metal coupons are still widely used as a reliable method for monitoring corrosion in cooling systems. Coupon weight loss provides a quantitative measure of the corrosion rate, and the visual appearance of the coupon provides an assessment of the type of corrosion and the amount of deposition in the system. In addition, measurement of pit depths on the coupon can indicate the severity of the pitting. Coupons should be installed properly in a corrosion coupon bypass rack with continuous, controlled water flow past the coupons. The metallurgy should match that of the system. One disadvantage of coupons is their lack of heat transfer, resulting in a lower temperature than that of the actual heat exchanger tubes. In addition, only a time-weighted average corrosion rate is obtained.

Corrosion Rate Meters. Additional corrosion monitoring tools have been developed by various instrument manufacturers and water treatment companies. Instantaneous corrosion rate meters can measure the corrosion rate at any given point in time. Instrument methods fall into two general categories: electrical resistance and linear polarization. With either technique, corrosion measurements are made quickly without removal of the sensing device. The electrical resistance method is based on measuring the increase in the electrical resistance of a test electrode as it becomes thinner due to corrosion. This method is desirable because the probes can be installed in both aqueous and nonaqueous streams. However, the electrical resistance method also has its disadvantages: conductive deposits forming on the probe can create misleading results, temperature fluctuations must be compensated for, and pitting character-istics cannot be determined accurately.

The method based on linear polarization at low applied potentials provides instantaneous corrosion rate data that can be read directly from the instrument face in actual corrosion rate units (mils per year). Systems using two or three electrodes are available. This method offers the maximum in performance, simplicity, and reliability. Corrosion rate meters can be used to assess changes in the corrosion rate as a function of time. They are able to respond to sudden changes in system conditions, such as acid spills, chlorine levels, and inhibitor treatment levels. Coupled with recording devices, they can be powerful tools in diagnosing the causes of corrosion or optimizing inhibitor treatment programs (see Figure 24-13).

Test Heat Exchangers. Test heat exchangers are small exchangers that can be set up to simulate operating conditions in the plant. They provide a convenient way to evaluate corrosion and fouling tendencies on heat transfer surfaces and to measure changes in heat transfer efficiency. A typical design uses cooling water on the tube side and condensing steam as a heat source on the shell side. If the test heat exchanger is insulated, a meaningful "U" (overall heat transfer coefficient) can be calculated.

BETZ MonitAll® Apparatus. The BETZ MonitAll (see Figure 24-14) is designed to measure corrosion and deposition under heat transfer conditions. Cooling water flows over a heated tube section within a glass shell. The specimen tube section is slid onto an electrical heater probe. Thermocouples measure bulk water temperature and tubeside "skin" temperature. The heat flux and flow velocity can be varied to simulate plant conditions. The tubes are available in various metallurgies and are preweighed for corrosion rate determination. The tube is visible through the glass enclosure, allowing direct observation of corrosion and scaling tendencies. Scaling/fouling can be quantified through temperature and flow measurements.

BETZ COSMOS™

Cooling System Monitoring Station. The Cosmos is a portable data acquisition station that monitors key parameters of a cooling system. The piping and instrumentation cabinet includes flow, pH, and conductivity sensors as well as a corrosion coupon rack, a corrosion rate probe, and a MonitAll unit. Data from all of these devices is fed into the data acquisition system (see Figure 24-15). The accumulated data can be printed directly by the built-in printer or can be downloaded to a personal computer for spreadsheet analysis.

CORROSION CONTROL IN OIL AND GAS PIPELINES

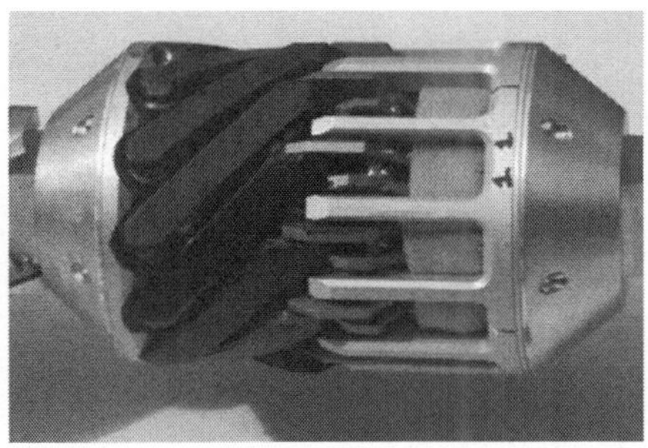

Fig. RFEC system for inspection of unpiggable pipelines.

In the United States, the annual cost associated with corrosion damage of structural components is greater than the combined annual cost of natural disasters, including hurricanes, storms, floods, fires and earthquakes(1). Similar findings have been made by studies conducted in the United Kingdom, Germany, and Japan. According to the U.S. Department of Transportation Office of Pipeline Safety, internal corrosion caused approximately 15 per cent of all reportable incidents affecting gas transmission pipelines over the past several years, leading to an average cost of $3 million annually in property damage, as well as several fatalities. The need to manage and mitigate corrosion damage has rapidly increased as materials are placed in more extreme environments and pushed beyond their original design life.

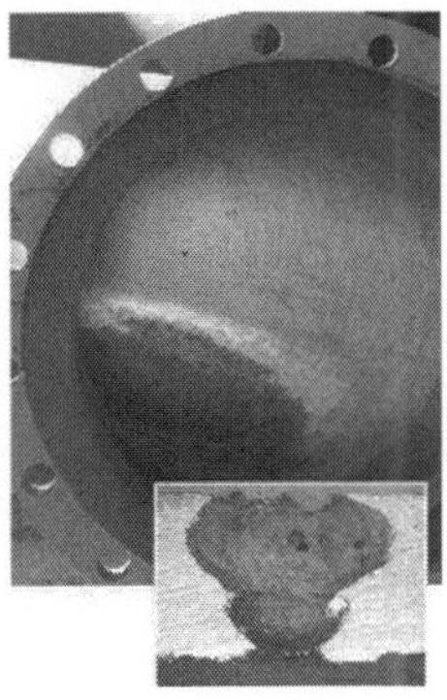

Fig. Localized corrosion in process piping, such as in the stainless steel pipe shown here, can lead to through-wall penetration (inset).

Typical corrosion mechanisms include uniform corrosion, stress corrosion cracking, and pitting corrosion (Figure 1). Corrosion damage and failure are not always considered in the design and construction of many engineered systems. Even if corrosion is considered, unanticipated changes in the environment in which the structure operates can result in unexpected corrosion damage. Moreover, combined effects of corrosion and mechanical damage, and environmentally assisted material damage can result in unexpected failures due to the reduced load carrying capacity of the structure.

Ensuring long-term, cost-effective system integrity requires an integrated approach based on the use of inspection, monitoring, mitigation, forensic evaluation, and prediction. Inspections and monitoring using sensors can provide valuable information regarding past and present exposure conditions but, in general, they do not directly predict remaining life. Carefully validated computer models, on the other hand, can predict remaining life; however, their accuracy is highly dependent on the quality of the computer model and associated inputs. Mitigation (corrosion prevention) methods and forensic evaluations play a key role in materials selection, assessment and design. All of these corrosion-control elements represent long-standing areas of research and development at Southwest Research Institute® (SwRI®).

Pipeline Inspection

A significant portion of many pipeline systems cannot be inspected through traditional methods. Nondestructive evaluation (NDE) and inspection tools are critical to assessing the integrity of pipelines. Traditional NDE methods involve the use of pipeline inspection gauges (PIGs), which travel through the inside of a pipe and detect the presence of mechanical damage or corrosion. Researchers at SwRI have developed an inspection system for inspecting pipelines that cannot accommodate traditional PIGs (Figure 2). This system uses remote field eddy current (RFEC), and was designed for use with the Carnegie Mellon Explorer II Robot. However, this technology can be adapted to other transport mechanisms. The system can expand to inspect 6-8 inch (150-200 mm) diameter lines. The sensor arms retract to accommodate line restrictions, such as elbows, tees and gate valves.

SwRI has also developed a guided-wave inspection technology that can be used to inspect pipelines and other structural components such as tubes, rods, cables and plates. The Magnetostrictive Sensor (MsS) inspection system uses inexpensive ribbon cables and thin magnetostrictive strips that are bonded to the component for inspection. The sensors attached to the pipe can accommodate a range of pipeline diameters, which is a significant advantage of guided-wave inspection systems that use an array of piezoelectric sensors. Because the sensors are low profile and relatively low cost, permanent installation of the sensors to perform structural health monitoring is a practical option.

Fig. RFEC system for inspection of unpiggable pipelines.

Corrosion Fatigue

Corrosion can degrade the mechanical integrity of a material through chemical attack. For example, the presence of hydrogen sulfide (H2S) has been found to reduce the fatigue life of offshore riser materials by approximately a factor of 10, and in the presence of a notch (that acts as an initiation point for corrosion fatigue) the fatigue performance can be decreased by a factor of 100. SwRI has developed customized test facilities for characterizing the performance of pipeline materials in corrosive environments. Figure 3a shows a servohydraulic load frame setup with a custom-designed test cell and redundant H2S containment systems. Full-thickness fatigue specimens (Figure 3b) are machined from riser pipes to preserve through-thickness residual stresses and to capture welds in joined pipe. SwRI recently developed a high pressure, high temperature (HPHT) corrosion fatigue test facility. In this facility the underlying fatigue crack growth behavior of riser materials subject to HPHT H2S (and other aggressive) environments can be quantified (Figure 3c). This unique test facility provides the capability to quantify inter-related corrosion-fatigue mechanisms, and provide data for calibrating and validating corrosion-fatigue computer models.

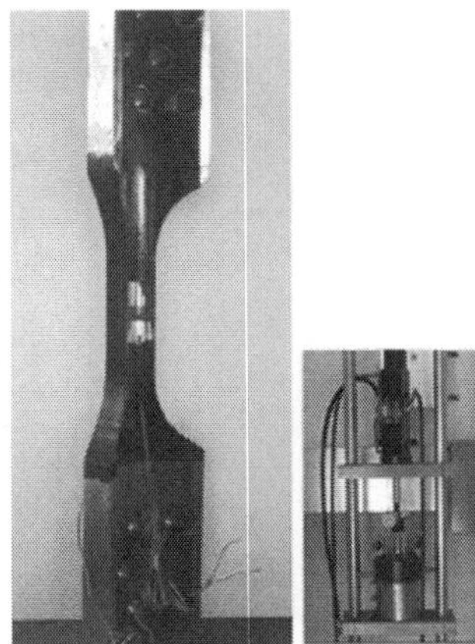

Fig. (a) Servohydraulic load frame with H2S corrosion fatigue test; (b) One-meter-long specimen; (c) high-pressure high-temperature corrosion fatigue test.

Corrosion Exposure Testing

As new materials are developed and environmental conditions change, assessing material performance due to corrosion and stress corrosion cracking is of increasing importance. SwRI has a well-established corrosion testing facility to perform HPHT testing in extremely aggressive environments. In most cases, the testing environment consists of a simulated process or reasonable worst-case scenario. These include determining the effects of H2S, CO2, oxygen, and microbiological organisms on corrosion/cracking of pipeline materials. Testing conforms to NACE, ASTM, API, or ISO standards and test materials are analyzed for mass loss, localized corrosion or stress corrosion cracking (SSC)/sulfide stress cracking (SSC). SwRI staffers are highly experienced in designing, constructing and operating specialty tests to mimic a specific operation that does not conform to standardized methods. One such capability is performing the environmental exposure on the API 16C – Flexible Choke and Kill Systems, which evaluates the effects of gas permeation, gas decompression and test fluid exposure at the rated temperature (Figure 4).

Fig. Photograph of the API 16C – Flexible Choke and Kill line testing.

Corrosion Prediction

Computer modeling is useful to help understand the mechanisms of internal corrosion, external corrosion and stress corrosion cracking, and to predict corrosion damage, failure and the most likely location of corrosion in oil and gas pipelines. These predictions can help support the development of practical guidelines to assist the pipeline industry in mitigating existing, or preventing future, corrosion failures. A four-step, tiered approach (Figure 5) is used by SwRI. The first step is to develop comprehensive fundamental models, which forms the foundation of the approach. Validation of the model against field and laboratory data is performed in the second step to ensure the correct physics are embedded in the model. To simplify use of the model in practical applications, the third step is to develop simplified models. During this step, rate-controlling variables or groups of variables are identified. These simplified models contain only the necessary physics and the values of the corresponding inputs to predict the performance of the system. The end goal of the overall modeling approach is step 4, development of guidelines for practical applications of the model. The tiered modeling approach has been successfully used for several recent applications:

- Predicting corrosion in coating disbonded regions with and without the effect of flow.
- Development of the dry gas internal corrosion direct assessment (ICDA) standard, NACE SP 0206-2006.
- Prediction of the most probable corrosion locations in a long underground pipeline due to variability in elevation, flow characteristics and materials.
- Prediction of the most likely conditions for internal corrosion due to variability in operations, gas quality upsets, and water intrusions.

Corrosion Sensing And Monitoring

While ICDA models can provide general guidelines to identify when internal inspections should occur, environmental and material uncertainties can lead to situations where excavation is performed unnecessarily, or water exists but is not predicted. In either case, costs associated with inspection or failure can be significant. To address this, sensing and monitoring technologies have been developed to enable remote interrogation of the internal corrosion of pipelines. The wireless mobile sensor, shown in Figure 6, travels inside a gas pipeline detecting the presence of water. The system communicates through a distributed wireless sensor network. The sensor body is an injection-molded polymer designed to survive high hydrostatic forces and impact on the pipeline walls while traveling along the pipe. This program has evolved using internal IR&D funding from both SwRI and Aginova, Inc.

The multielectrode array sensor (MAS) probe is ideally suited for monitoring corrosion rates in process streams. Multiple discrete elements or electrodes are used to replicate the material of interest. The MAS probe measures corrosion rates by assessing the current flow between coupled electrodes. The electrodes can be manufactured from a wide range of alloys and product forms. SwRI has used this method to monitor the corrosion of a variety of materials. The wireless mobile sensor and the MAS probe sensor are just two examples of corrosion sensing and monitoring technologies. SwRI has developed a suite of corrosion sensing and monitoring devices. Significant inspection and repair costs can be avoided with the use of tools such as these.

Deposition Coatings

The deposition of material coatings can be effectively employed to protect surfaces of components from wear, erosion and corrosion. A variety of coatings have been studied including metals, ceramics and polymers. A number of deposition techniques have also been developed. One example is magnetron sputtering, where 20-30 μm thick Al-Ce-Co coatings are deposited on Al alloys and 1018 carbon steel, which is sufficient for most applications where corrosion and erosion are possible. A cross-section Al clad and Al-Ce-Co deposition coating. Microstructural analyses show that under certain deposition conditions, amorphous/nano-crystalline structures are obtained, which show superior corrosion resistance in electrochemical tests. Diamond-like carbon (DLC) coatings can be produced using the plasma immersion ion deposition (PIID) process. The coatings are very hard and dense and can be applied to many components for increased wear and erosion resistance. In fact, SwRI recently developed a technique for applying DLC coatings to the inner surface of pipeline segments.

Ti-Si-C-N based nanocomposite coatings deposited using the Plasma Enhanced Magnetron Sputtering (PEMS) process have shown high-hardness (>40GPa) and superior erosion and wear resistance. PEMS was originally developed for use on gas turbine compressor blades and vanes and steam turbine blades against solid particle erosion and liquid droplet erosion, and won an R&D 100 award in 2009. Laboratory tests have shown that the erosion resistance of these coatings can increase the lifetime by a few to more than 100 times as compared to uncoated substrates. The corrosion resistance for Ti-Si-C-N coated samples has been shown to be comparable to or better than the uncoated Ti-6Al-4V substrate, which already exhibits excellent corrosion resistance. SwRI has developed a suite of deposition-coating solutions for addressing a range of erosion, corrosion and wear issues. As noted, large-scale production of corrosion-resistant coatings using vacuum deposition techniques is possible (e.g. Al-Co-Ce coatings and DLC coatings). For more severe environments, vacuum-deposited Ti-Si-C-N nanocomposite coatings have been successfully used to protect important components from erosion, abrasion and corrosion damage.

Forensic Evaluations

Although a comprehensive corrosion-control program based on

inspection, monitoring and model predictions can be an effective means for controlling pipeline corrosion, unexpected events or undocumented changes in operating conditions can still lead to premature pipeline failure. When these occur, it is essential to perform a thorough forensic evaluation of the failure to determine the failure mechanism and its root cause. By identifying the root cause of the failure, the pipeline operator will know if this resulted from an event or operating condition outside of the general conditions included in the corrosion-control program. Steps can then be identified to mitigate future failures by eliminating recurrence of the event. If such an event is not identified as the root cause of failure, the results of the evaluation can be instrumental in identifying necessary changes to the corrosion-control program. Additionally, destructive evaluations, which are a routine part of a forensic evaluation, can be a valuable tool for validating the effectiveness of a corrosion-control program.

Conclusion

Is your odor corrosive? The question should now make sense. The gases identified as odorous at wastewater treatment plants and those that aggressively attack electronic process control equipment are essentially the same. They cause corrosion of electrical circuits at very low concentrations, causing false alarms, intermittent signals, and eventual failure of components and devices. Evaluating the environments electronic process control equipment environments is extremely important to keep facilities operational and to protect the equipment itself. The environmental monitoring technique that links directly to the control equipment is reactivity monitoring, which provides a measure of the severity level which can in turn help develop a necessary plan for the protection of electronics. When the environment is continuously monitored, action can be taken to prevent failures before the fact. Such monitoring can also confirm that the process control equipment is in a safe environment.

4

Electrochemistry of Corrosion

ELECTROCHEMISTRY

Electrochemistry is the area of chemistry dealing with the interconversion of electrical energy and chemical energy. There are many applications of this in every day life. Batteries, control of corrosion, metallurgy and electrolysis are just a few examples of the applications of electrochemistry. This handout will look at the basic principles of electrochemistry and show some of its applications. Electrochemistry always involves an oxidation-reduction process. Recall that oxidation involves the loss of electrons by a substance and that reduction involves the gain of electrons by a substance. You can't have one process without the other. The substance that is reduced is the oxidizing agent. This makes sense, because how you define an oxidizing agent – a compound that is able to oxidize another substance by removing electrons from that substance. Since the oxidizing agent is gaining the electrons that it pulls off the substance it is oxidizing, the oxidizing agent becomes reduced during the process. Similarly, the substance that is oxidized is the reducing agent. A reducing agent is a compound that will reduce another substance by giving away its own electrons to the substance it is reducing. Thus, since the reducing agent is losing those electrons, it becomes oxidized during the process. It is important to be able to determine the oxidation states of the compounds involved in an electrochemical process so as to identify which substance is oxidized and which substance is reduced. If you need help with this, please refer to the handout "Determining Oxidation States" found on this website.

In general, metals tend to make good reducing agents because they can only be oxidized. The reducing ability of the metal is given by the activity series. The more active metal is able to reduce the less active metal cation. This activity series is:

Li>K>Ca>Na>Mg>Al>Zn>Cr>Fe>Ni>Sn>Pb>H2>Cu>Hg>Ag>Pt>Au

most active   least active

So for example, magnesium metal is able to reduce copper(II) ions in solution to form magnesium ions and copper metal:

$$Mg(s) + Cu^{2+}(aq) \rightarrow Mg^{2+}(aq) + Cu(s)$$

Notice how the very reactive alkali metals are the most active. This makes perfect sense. The alkali metals are very easily oxidized, thus readily want give away its electrons to reduce another substance. On the other hand, metals like copper, gold and silver are among the least active. Due to this property, these metals are often used for coinage and jewelry. You certainly wouldn't want your coins or jewelry to rust away with time! Other more active metals such as magnesium and zinc are used to protect more active metals, such as iron. Magnesium disks are bolted to ocean-going ships to protect their hulls from corrosion while traversing the ocean. The magnesium corrodes more easily, thus keeps the iron in the hull from rusting. This process is called "cathodic protection".

GALVANIC CELLS

A common application of electrochemistry is the galvanic cell, commonly known as a battery. A galvanic cell has the capability of producing a voltage and was invented by Luigi Galvani and Alessandro Volta. A schematic of such a cell is shown below:

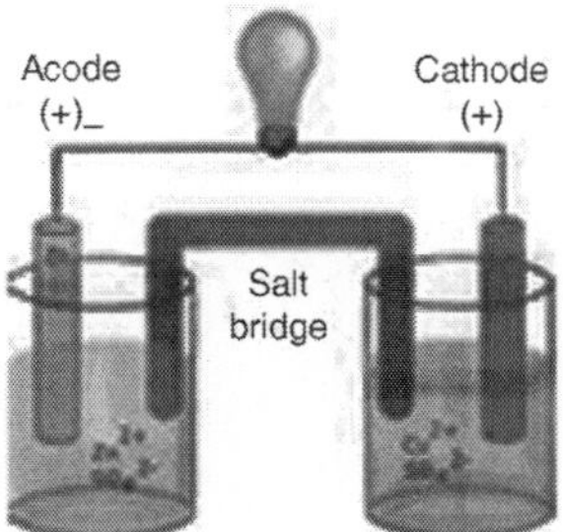

The schematic shows how a zinc anode connected by a salt bridge to a copper cathode will convert chemical energy to electrical energy and allow the light bulb to light up. Lets look at the electrochemistry involved. For any electrochemical cell, oxidation always occurs at the anode and reduction always occurs at the cathode. An easy way to remember this is to keep the vowels together and the consonants together (oxidation at the anode; reduction at the cathode). Also, when drawing a galvanic cell, most people use the convention where the anode is on the left and the cathode is on the right. Use the mnemonic: ABC to remember this convention (Anode/Bridge/Cathode). Since drawing a galvanic cell isn't always convenient, a line notation, called cell notation, is often used to describe a galvanic cell. The cell notation for the above cell would be:

$$Zn(s) \mid Zn^{2+}(aq), 1M; SO_4^{2-}(1M) \mid\mid Cu^{2+}(aq), 1M; SO_4^{2-}(1M) \mid Cu(s)$$

One begins at the solid zinc anode. A single vertical line indicates a phase change from solid zinc anode to aqueous zinc sulfate solution in the anode compartment. The double vertical line indicates the salt bridge connecting the anode compartment to the cathode compartment. Next, one describes the copper(II) sulfate solution in the cathode compartment. Once again, a single vertical line indicates a phase change from aqueous copper(II) solution in the cathode compartment to the solid copper cathode. The cell described above is often referred to as a standard cell. The parameters for a standard cell require the species in solution to have a concentration of 1 M, the partial pressure of gases (or more appropriately, the fugacity) to be 1 atm and the temperature to be 25°C (298K). Using these standard conditions, we can describe a standard electrode potential, having a symbol, E°, in volts and make it possible to calculate the electrical potential of the cell. The superscript, °, denotes standard conditions. You possible have seen this notation prior to this with standard thermodynamic parameters, such as standard enthalpy change, ?H°. The standard electrode potentials are usually listed as reduction potentials and are relative to a standard hydrogen electrode (SHE) having a 0 volt potential as a reference point. There are two methods for calculating the cell potential, E°cell. I will describe both methods and you can choose the method that works best for you.

Method 1

This is probably the most common method. Since all the electrode potentials are listed as reduction potentials, the cell potential is calculated directly form the table using the equation:

$$E^\circ_{cell} = E^\circ_{cathode} - E^\circ_{anode}$$

Let's find the cell potential for the galvanic cell we diagramed earlier. First, let's write out the half cell reactions. One half cell reaction will show the oxidation and the other half cell reaction will show the reduction. Recall oxidation occurs at the anode and reduction at the cathode.

$$\text{anode (oxidation) } Zn(s) \rightarrow Zn^{2+}(aq) + 2e^-$$

$$\text{cathode (reduction) } Cu^{2+}(aq) + 2e^- \rightarrow Cu(s)$$

$$\text{Cell reaction } Zn(s) + Cu^{2+}(aq) \rightarrow Zn^{2+}(aq) + Cu(s)$$

(notice that sulfate is not include, since it is a spectator ion)

Go to any standard textbook, chemical reference book, or a reliable internet site and find a table of standard reduction potentials.

The reduction potential for $Zn^{2+}(aq) + 2e^- \rightarrow Zn(s)$ is –0.76 V and the reduction potential for $Cu^{2+}(aq) + 2e^- \rightarrow Cu(s)$ is 0.34 V. Using these reduction potentials, calculate the cell potential:

$$E^\circ_{cell} = E^\circ_{cathode} - E^\circ_{anode} = 0.34\text{ V} - (-0.76\text{ V}) = +1.10\text{V}$$

Method 2

This is an alternative method that uses more of a "logic" approach. Once again, the oxidation and reduction half cell reactions are used. The sign on the reduction potential for the oxidation half reaction is changed to make it become an actual oxidation potential. In the problem we have been working, the actual half reaction for zinc is an oxidation: $Zn(s) \rightarrow Zn^{2+}(aq) + 2e^-$. The half cell potential for the oxidation would be + 0.76 V. The cell potential becomes the sum of the reduction potential and the oxidation potential:

$$E^\circ_{cell} = E^\circ_{red} + E^\circ_{ox}$$

Thus in our problem:

$$E^\circ_{cell} = E^\circ_{red} + E^\circ_{ox} = 0.34\text{V} + 0.76\text{ V} = +1.10\text{V}$$

An $E^\circ > 0$ means the cell will be spontaneous (produce a voltage). An $E^\circ < 0$ means the cell must be supplied that voltage from an external power supply in order for the cell reaction to run in the direction it is written. If anything, that cell would want to run in the reverse direction. Often these nonspontaneous cells are called electrolytic cells. We will see a mathematical equation that relates cell potential and spontaneity as well as look at an example of an electrolytic process later in this handout. What happens when the cell is not under standard conditions? The Nernst equation will have to be applied under nonstandard conditions:

$$E_{cell} = E^\circ_{cell} - \frac{RT}{nF}\ln Q$$

where Ecell = nonstandard cell potential in volts
E°cell = standard cell potential in volts
R = constant, 8.314 J/mol.K
T = temperature in Kelvin
n = moles of electrons transferred
F = Faraday's constant = 96485 C/mol e^-
note: C represents the coulombs (unit of charge)
Q = the reaction quotient expression

Alternative representations of the Nernst Equation may be used if the temperature is standard temperature (298K) and one has a choice of natural logarithm (ln) or common logarithm, base 10 (log):

$$E_{cell} = E^\circ_{cell} - \frac{0.0257\text{ V}}{\text{n}}\ln Q \text{ or } E_{cell} = E^\circ_{cell} - \frac{0.0592\text{ V}}{\text{n}}\log Q$$

Let's modify our previous cell as follows:

$Zn(s)$ | $Zn^{2+}(aq)$, 3.00M; SO_4^{2-}(3.00M) || $Cu^{2+}(aq)$, 0.0015M; SO_4^{2-} (0.0015M) | Cu(s). We already have E°_{cell} = +1.10V. Let the temperature be standard (298K). The Nernst equation expression becomes:

$$E_{cell} = E^\circ_{cell} - \frac{0.0257\text{ V}}{n}\ln\frac{[Zn^{2+}]}{[Cu^{2+}]} = +1.10\text{ V} - \frac{0.0257\text{ V}}{2\text{ mol e}^-}\ln\frac{3.00\text{ M}}{0.0015\text{ M}}$$

Evaluating the above expression gives:

$$E_{cell} = +1.10\text{ V} - 0.098\text{ V} = 1.00\text{ V}$$

Cell potentials have many other uses. As mentioned earlier, the cell potential indicates whether the cell reaction is spontaneous or nonspontaneous. If you recall, spontaneity for a process is often associated with its Gibbs free energy change, ΔG. There is a mathematical relationship between cell potential and this Gibbs free energy change:

$$\Delta G^\circ = -\ nFE^\circ \text{ under standard conditions}$$

$$\Delta G = -\ nFE \text{ under nonstandard conditions}$$

where ΔG = change in Gibbs free energy in Joules
n = moles of electrons transferred
F = Faraday's constant = 98485 C/mol e^-
E = cell potential

Using the units of Faraday's constant, coulombs/mol e- and volts for cell potential, the Gibbs free energy units would work out to coulomb-volts. Interestingly enough, the units, coulomb-volts, are equivalent to Joules!

Let's calculate the Gibbs free energy change for the nonstandard cell we described in the previous problem:

$$\Delta G = -\ nFE = -\ (2 \text{ mol } e^-)(96485 \text{ C/mol } e^-)(1.00 \text{ V}) = -\ 1.94 \times 10^5 \text{ J}$$

Let's quickly review some thermodynamics to see how other information may be obtained from the Gibbs free energy change.

If you recall from thermodynamics:

$$\Delta G = \Delta G^\circ + RT \ln Q$$

At equilibrium, ΔG = 0, thus Q becomes K_{eq} and the expression simplifies to:

$$\Delta G^\circ = -\ RT \ln K_{eq}$$

Now that the Gibbs free energy is known, it can be related to the equilibrium constant, as well as the cell potential:

$$\Delta G^\circ = -\ nFE^\circ = -\ RT \ln K_{eq}$$

So if we wanted to know the equilibrium constant value for our standard cell, it can be easily calculated:

$$\Delta G^\circ = -RT \ln K_{eq}$$

$$-\ 1.94 \times 105 \text{ J} = -\ (8.314 \text{ J/mol·K})(298 \text{ K}) \ln K_{eq}$$

Solving for $\ln K_{eq}$:

$$\ln K_{eq} = \frac{-1.94 \times 10^5 \text{ J/mol}}{-(8.314 \text{ J/mol} \cdot \text{K})(298 \text{ K})} = 78.3$$

Take the antilog to find K_{eq}:

$$K_{eq} = e^{78.3} = 1.01 \times 10^{34}$$

The large value for K_{eq} indicates the reaction will proceed significantly towards products.

ELECTROLYSIS

The electrolysis process is often used to extract and purify metals from their molten salt form. A current is applied to the molten salt for a period of time. Here is a quick review of the physics before we go further. Current is measured in amperes, A. An ampere is equivalent to the coulombs of charge per second, C/s. In other words:

$$1 \text{ ampere} = \frac{1 \text{ coulomb}}{1 \text{ second}}$$

Let's suppose you are trying to purify aluminum from a melt containing aluminum cations. If a current of 5.00 A is used for 2 hours, how many grams of aluminum metal can be extracted?

For every mole of Al^{3+} cations, 3 moles of electrons are gained to form 1 mole of aluminum metal: $Al^{3+} + 3\ e^- \rightarrow Al$

The rest is getting the units to work out!

$$2 \text{ hr} \times \frac{3600 \text{ s}}{1 \text{ hr}} \times \frac{5.00 \text{ C}}{1 \text{ s}} \times \frac{1 \text{ mol e}^-}{96485 \text{ C}} \times \frac{1 \text{ mol Al}}{3 \text{ mol e}^-} \times \frac{26.98 \text{ g Al}}{1 \text{ mol Al}} = 3.36 \text{ g Al}$$

ELECTROCHEMICAL CELLS

If a copper strip is placed in a solution of copper ions, one of the following reactions may occur

$$Cu^{2+} + 2e^- \rightarrow Cu$$

$$Cu \rightarrow Cu^{2+} + 2e^-$$

The electrical potential that would be developed by these reactions prevents their continuation. These reactions are called half-reactions or half-cell reactions. There is no direct way to measure the electrical potential (electromotive force, emf) of a half-cell reaction. Similarly, a zinc strip in a solution of zinc ions has the possible reactions]

$$Zn^{2+} + 2e^- \rightarrow Zn$$

$$Zn \rightarrow Zn^{2+} + 2e^-$$

But these also are prevented from occurring by the electrical potential that would build up. If the metal electrodes (copper and zinc) in the two solutions are connected by a wire, and if the solutions are electrically connected by perhaps a porous membrane or a bridge that minimizes mixing of the solutions, a flow of electrons will move from one electrode, where the reaction is

$$M_1 \rightarrow M1^{n+} + ne^-$$

To the other electrode, where the reaction is

$$M_2^{n+} + ne^- \rightarrow M_2$$

In this case, the zinc metal goes into solution as zinc ions and the copper ions plated out. The overall cell reaction is

$$Zn + Cu^{2+} \rightarrow Zn^{2+} + Cu$$

The electromotive force for such a cell, which is written as below, can be measured.

$$Zn \mid ZnSO_4 \mid\mid CuSO_4 \mid Cu$$

By convention, all half-cell emf's are compared to the emf of the standard hydrogen electrode. The standard hydrogen electrode is defined as a platinum electrode covered with platinum black that is in contact with hydrogen gas at 1 atmosphere pressure and a 1 molar solution of hydronium ions (actually, it is defined for unit activity). The hydrogen electrode half-cell reaction is

$$2H^+ + 2e^- H_2 \qquad Eo = 0.00 \text{ V (defined)}$$

By definition, the standard reduction potential of the hydrogen half-cell is 0.00 V at standard state. The emf of a half-cell, with respect to the standard hydrogen electrode, is called the reduction potential. Standard reduction potentials, $E°$, are for 1 molar solution. Consequently, the difference between the reduction potentials of two half-cells is the emf they would develop if connected together as a cell. The emf for the Zn-Cu cell described would be

$$E_{cello} = E_{cathodeo} \text{ (Cu)} - E_{anodeo} \text{ (Zn)} = 0.34\text{V} - (-0.77\text{V}) = 1.10 \text{ V}$$

if the solutions are 1.0 molar.

If a cell reaction can be written

$$aA + bB \rightarrow cC + dD$$

then the emf of the cell can be expressed in the form of the following equation, developed by Nernst:

$$E = E° - (2.3RT / nF) \log ([C]^c [D]^d/[A]^a [B]^b)$$

If all the concentrations are 1M, then the logarithmic term becomes zero and

$$E = E°$$

which is the reason for choosing 1 M concentration as the standard condition. If the system is at equilibrium, then

$$\frac{[C]^c[D]^d}{[A]^a[B]^b}$$

and the emf developed by such a cell at equilibrium must be zero. Therefore,

$$0 = E° - (2.3R\,T/nF)\log K_{eq}$$

or

$$\log K_{eq} = (nF\,E°) / (2.3\,RT)$$

In these equations, *F* is the Faraday, *n* is the number of electrons transferred in the oxidation-reduction step, and *R* is the gas constant in units of electrical work.

MEASURING EMF

To measure the emf of a galvanic cell, a sensitive meter is needed, but it is important that the meter not draw a significant amount of current. If the current produced by the cell to be measured is large, the cell will become polarized, and the emf will be decreased. Many solid state voltmeters have sufficiently high impedance that they can be used to measure the emf of a cell accurately.

PROCEDURE:

Part 1: The Daniell cell

Place 0.1 M $ZnSO_4$ solution in a 100-mL beaker, and 0.1M $CuSO_4$ solution in another beaker of the same size. The liquid levels should be the same. In the zinc solution place a clean strip of zinc, and in the copper solution, a clean strip of copper. The zinc strip may be cleaned by placing it in a beaker of about 2M HCl (made by diluting one volume of 6M HCl with two volumes of water). The copper strip can be cleaned by placing it in a beaker of 2M HNO_3 (made by diluting one volume of 6M HNO_3 with two volumes of water). Rinse the electrodes with deionized water thoroughly before using them.

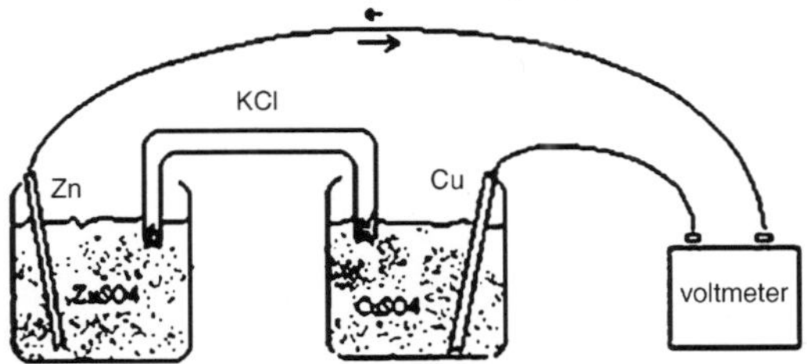

Attach the wires to the electrodes, connecting them to the voltmeter or potentiometer. (On the potentiometer the black terminal is negative, and the red terminal is positive. Connect the wire from the zinc electrode to the black terminal on the potentiometer, and the copper electrode to the red terminal.) The cells are connected with a U-tube filled with 0.5 M potassium chloride and closed with cotton stuffed into the ends. See Fig. 1. Short lengths of plastic or Tygon tubing may be used instead of glass U-tubes. No large air bubbles should be in the U-tube, or the electrical resistance will be high. Electrons will flow from the zinc electrode to the copper electrode because of the reactions occurring in the half-cells.

$$Zn \rightarrow Zn^{2+} + 2e^-$$

$$Cu^{2+} + 2e^- \rightarrow Cu$$

The overall reaction for the cell is

$$Cu^{2+} + Zn \rightarrow Cu + Zn^{2+}$$

The cell may be represented as

$$Zn \mid 0.1M\ ZnSO_4 \mid\mid 0.5M\ KCl \mid\mid 0.1M\ CuSO_4 \mid Cu$$

Measure the voltage of this cell. Also, measure the voltage of a Daniell cell with 0.01 M $ZnSO_4$ and 0.1 M $CuSO_4$, and with 0.1 M $ZnSO_4$ and 0.01 M $CuSO_4$.

Part 2: Solubility product of cupric hydroxide

If an electrochemical cell can be built in which the half-cell reactions are

$$Cu^{2+} + 2e^- \rightarrow Cu$$

$$Cu + 2OH^- \rightarrow Cu(OH)_2\ (s) + 2e^-$$

the overall cell reaction is the reverse of the solubility product of cupric hydroxide

$$Cu^{2+} + 2OH^- \rightleftharpoons Cu(OH)_2\ (s)$$

Such a cell can be written as

$$Cu \mid OH- \mid Cu(OH)_2 \mid\mid KCl \mid\mid Cu^{2+} \mid Cu$$

From the Nernst equation,

$$E = E^\circ - 2.3RT \frac{\log}{2F} \frac{1}{[Cu^{2+}][OH^-]^2}$$

$$E = E^\circ + 2.3RT \frac{\log}{2F} [Cu^{2+}][OH^-]^2$$

$$E^\circ = E - 2.3RT \frac{\log}{2F} [Cu^{2+}][OH^-]^2$$

Consequently, if we can measure the emf, E, of such a cell, we can calculate E°, if we know the concentration of cupric ion in one half-cell and the concentration of hydroxide ion in the other half-cell. In the introductory discussion, we have shown from the Nernst equation that the following relationship exists

$$\log K_{eq} = (nF\ E^\circ) / (2.3\ RT)$$

So if Eo is obtained, it can be converted into a value of the solubility product. The factor 2.3RT/F is equal to 0.0592 V at 25oC. Construct a cell of two beakers. In a 100 or 150 mL beaker, place 0.10 M $CuSO_4$ solution and a clean copper strip. The copper strip can be cleaned in dilute nitric acid, but it must be washed well with distilled water. Until it is put into the copper sulfate solution, it should be kept in dilute HCI, but rinsed with deionized water before use. In a clean beaker, add the same height of 0.10 M potassium hydroxide solution. Add a copper electrode to this KOH solution. Connect wires to the electrodes and the appropriate terminals of the voltmeter or potentiometer. The KOH side is the negative electrode, because copper goes into the oxidized form (Cu^{2+}) because of the low concentration of Cu^{2+} in the KOH solution. Now add the salt bridge, which is a U-tube filled with KCI and closed off by wads of cotton in each end. The cotton wads must be soaked. Measure the emf of the cell as soon as possible. The cell is polarized very easily. To get around this difficulty, clean the electrodes again, rinse, and return them to the solutions. Re-measure the voltage. Also measure the emf of the cell which uses 0.10 M Copper sulfate and 1.0 M potassium hydroxide. Calculate the solubility product, K_{sp}, using the equations above. Literature values range from 10^{-14} to 10^{-20}, the latter being considered the best value. Taking the standard reduction potential ***E***° of the Cu-Cu^{2+} electrode to be 0.34 V, evaluate ***E***° for the Cu-$Cu(OH)_2$, OH– electrode.

Part 3: Formation constant of tetraamminecopper(II) cation

Similarly, if we can construct a cell having the half-reactions

$$Cu^{2+} + 2e^- \rightarrow Cu$$

$$Cu + 4NH_3 \Leftrightarrow Cu(NH_3)_4^{2+} + 2e^-$$

the overall cell reaction is the formation equilibrium for tetraamminecopper(II) cation

$$Cu^{2+} + 4NH_3 \rightarrow Cu(NH_3)_4^{2+}$$

Such a cell can be written

$$Cu \mid NH_3, Cu(NH_3)_4^{2+} \mid\mid KCl \mid\mid Cu^{2+} \; Cu$$

The Nernst equation can be written

$$E = E^\circ - 2.3RT \frac{\log}{2F} \frac{\left[Cu(NH_3)_4^{2+}\right]}{\left[Cu^{2+}\right]\left[NH_3\right]^4}$$

$$E^\circ = E + 2.3RT \frac{\log}{2F} \frac{\left[Cu(NH_3)_4^{2+}\right]}{\left[Cu^{2+}\right]\left[NH_3\right]^4}$$

and the formation constant, Kf, is related by

$$\log K_f = (2F\,E^\circ) / (2.3\,RT)$$

Clean the copper strips in 6M nitric acid. To a 100-ml beaker labeled #1, add 50 mL of 1.0M aqueous ammonia and 0.25 mL (this small volume may be measured using a calibrated medicine dropper, probably about 5 drops) of 0.1M cupric sulfate solution. Mix well and add a rinsed copper electrode. To another 100-mL beaker labeled #2, add 50 mL of 0.1M cupric sulfate and another rinsed copper electrode. Connect the electrodes to the voltmeter. Which electrode is negative? Make the electrical connection by adding the salt bridge. Determine the cell potential. Using the calibrated medicine dropper, add 2.25 ml (probably about 45 drops) more of 0.1M cupric sulfate to the beaker containing cupric ammine complex, mix well, and determine the voltage of the cell. Assuming that the reaction of the 0.25 mL of 0.1M cupric sulfate solution and 1 M aqueous ammonia goes to completion

$$Cu^{2+} + 4NH_3 \rightarrow Cu(NH_3)4^{2+}$$

Calculate the concentration of cupric ammine complex in the first cell. Using this value for $[Cu(NH_3)_4{}^{2+}]$, $[Cu^{2+}] = 0.10$ M, and $[NH_3] = 1.0$ M, calculate E^o from the preceding equation, and then calculate K_f. Do the same for the second cell containing 2.25 mL 0.1M cupric sulfate.

ELECTROCHEMICAL OXIDATION

Water has a number of unique properties that are essential to life and that determine its environmental chemical behaviour. Many of these properties are due to water's polar molecular structure and its ability to form hydrogen bonds. Water also has the highest dielectric constant of any common liquid, a maximum density as a liquid at 40 C and a higher heat capacity than any other liquid except amonnia. But most of the important chemical phenomena associated with water do not occur in solution but rather through interaction of solutes in water with other phases. For example the oxidation -reduction reactions catalyzed by bacteria occur in bacterial cells. Many organic hazardous wastes are carried through water as emulsions of very small particles suspended in water. Some hazardous wastes are deposited in sediments in bodies of water from which they may later enter the water through chemical or physical processes and cause severe pollution effects which need to be eliminateed as much as possible.

The wastewaters are usualy oxidated with ozone, which is a powerful oxidant but the total organic carbon removal was no more than 30 per cent and the same results were obtained using hydrogen peroxide in the presence of Fe^{+2} as a catalyst. In general by a chemical oxidation the organic pollutants are almost completely eliminated but the removal of total organic carbon still remains a problem. These are the reasons why the electrooxidation of the hazardous pollutants has been the subject of extensive studies during recent years. The electrochemical method of depollution presents many important advantages because it does not need auxiliary chemicals, it is applicable on a large range of pollutants and does not need high pressures and temperatures.

There has been new research on the electrochemical oxidation of organic compounds from the wastewaters due to its great efficiency and the rigorous control which it allows. The electrochemical oxidation of pollutants from wastewaters has been studied using anodes made from different materials. This is attributed to the oxidation of the adsorbed organic compounds to carbon dioxide. It has also been proven that under the same conditions the electrooxidation index obtained on the SnO2 anode was higher than on the Pt anode, indicating a higher degree of phenol oxidation.

WASTEWATER TREATMENT

As indicated above, industrial wastewater contains a vast array of pollutants: insoluble, colloidal, and particulate forms, both inorganic and organic. In addition, the required effluent standards are also diverse, varying with the industrial and pollutant class. Consequently, there can be no standard design for industrial water pollution control; rather, each site requires a customized design to achieve optimum performance. However, each of the many proven processes for industrial waste treatment is able to remove more than one type of pollutant and is in general applicable to more than one industry. Generally, a combination of several processes is utilized to achieve the degree of treatment required at the least cost. Much of the experience and data from wastewater treatment has been gained from municipal treatment plants. Industrial liquid waste is similar to wastewater but differs in significant ways. Thus, typical design parameters and standards developed for municipal wastewater operations must not be blindly utilized for industrial wastewater. It is best to run laboratory and small pilot tests with the specific industrial wastewater as part of the design process. It is most important to understand the temporal variations in industrial wastewater strength flow, and waste components and their effect on the performance of various treatment processes. Industry personnel, in an effort to reduce cost, often neglect laboratory and pilot studies and depend on waste characteristics from similar plants. This strategy often results in failure, delay, and increased costs. Careful studies on the actual waste at a plant site cannot be overemphasized.

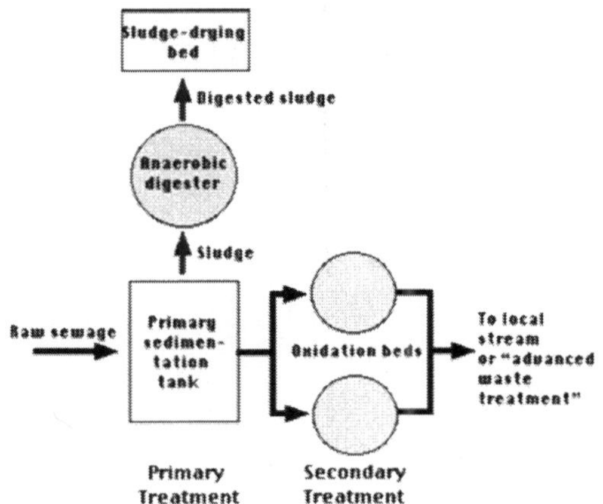

Fig. Traditional overall treatment of wastewater

Organic Hazard in Asia

Life-threatening poisons such as DDT, aldrin, chlordane, dieldrin and heptachlor -all of which are either severely restricted or banned in most countries- continue to be manufactured, stored, used and traded freely in South Asia, according to an investigative report released by the international environmental group Greenpeace. The report titled "Toxic Legacies; Poisoned Futures: Persistent Organic Pollutants in Asia" reveals a story of potentially widespread contamination caused by irresponsible corporate behavior, shortsighted lending agencies and misguided government policies.

"Asia faces a frightening scenario of historic, current and potential poisoning by the most dangerous variety of persistent poisons. This situation is a result of existing stockpiles of obsolete PCBs, the continuing production of organophenols and other chemical pesticides and the unmitigated expansion of dirty chlorine-based industries in the region," Focusing on a class of poisonous chemicals called Persistent Organic Pollutants or POPs, which are now targeted for elimination by ongoing international negotiations under the United Nations Environment Program (UNEP), Greenpeace investigations conducted between April and August 2001 in seven Asian countries, including Bangladesh, India, Nepal and Pakistan, revealed that stocks of 5000 metric tons or more of obsolete pesticides, including POP chemicals, are stored in extremely hazardous conditions in more than a thousand sites in Pakistan and Nepal. A sizeable portion of these pesticides are reported to have arrived as part of aid packages from Western countries, and almost all the pesticides were exported by developed nations and India to Pakistan and Nepal.

Chemical corporations whose products were identified in stockpiles in Pakistan and Nepal by Greenpeace investigators include: Bayer and Hoechst (Germany); DuPont, Dow Chemicals, Diamond Shamrock and Velsicol (USA); Shell (Netherlands); Sumitomo Chemical and Takeda Chemical (Japan); Rhone Poulenc (France); Sandoz (Switzerland); ICI (UK); Bharat Pulverising Mills (India). India is among the three remaining known manufacturers of DDT (10,000 mt capacity) in the world, the other two being Mexico and China; India exports nearly 800,000 kilograms of POP pesticides including aldrin, DDT, BHC, and chlordane to a long list of countries, including countries where their usage is banned. Exports of pesticides that could be branded POPs in the near future such as endosulfan, sodiumpentachlorophenate, 2,4–D, and lindane total more than two million tons. Some pesticides such as aldrin are illegal to manufacture in India. In Pakistan, India, Nepal and Bangladesh, locally banned or severely restricted pesticides are freely available. Greenpeace found DDT, BHC, Dieldrin and Heptachlor openly sold in vegetable markets in Karachi. Hardware stores in New Delhi stock the deadly pesticide aldrin, whose registration was withdrawn more than two years ago.

POPs are a class of synthetic toxic chemicals that cause severe and long-term effects on wildlife, ecosystems and human health. POPs have been implicated in the rising incidence of certain cancers (e.g. breast, prostate, endometriosis, etc.), reproductive deficits such as infertility and sex-linked disorders, declining sperm counts, fetal malformations, neurobehavioral impairment, and immune system dysfunction. Because of major threats to human health, the UNEP process has shortlisted an initial twelve substances for elimination which include organochlorine pesticides (DDT, chlordane, mirex, hexachlorobenzene, endrin, aldrin, toxaphene, heptachlor), industrial chemicals like cancer-causing PCBs (polychlorinated biphenyls), and the super-toxic dioxins and furans. In line with the emerging requirements of the UNEP POPs process, Greenpeace also urges governments in the region to take action now by taking inventory of all sources of POPs in their countries and preventing the expansion of POP-producing technologies such as incinerators, PVC manufacturing, pesticide production facilities, and pulp and paper mills using chlorine bleaching processes.

> "It is unfortunate that while governments in the region are still grappling for ways to dispose of their stockpiles of obsolete imported pesticides, the continuing production and trade of these chemicals goes on unabated. This could only lead to an endless cycle of poisoning whose unwitting and eventual victims are communities and future generations," "Governments should aim for an eventual phase-out of such polluting practices and push for international cooperation in developing viable and sustainable non-chemical alternatives."

"While governments in the region are responsible for taking action against POPs pollution, the liabilities associated with such action must always fall on the polluter–the corporations and international lending agencies- and not on the citizens who have long-endured the polluter–the corporation- and international consequences of toxic pollution," added Jayaraman.

Advances in Wastewater Treatment

Several effective, affordable, and environmentally sound waste disposal options are available; others are within scientific reach. Separating waste categories can go a long way in addressing waste problems. By some estimates, intensive reuse and recycling systems could take care of nearly 80 percent of municipal waste. Current data suggest much lower rates of recycling in most locations: 8 percent in the UK, 28 percent in the US, and so on. Conversion to clean production technologies can both help save business dollars and protect the environment. For toxic industrial and medical wastes, non-incineration technologies such as the gas phase thermo-chemical reduction process achieve virtually 100 percent efficiency in POPs destruction and capture all residues and releases. Other emerging technologies include electrochemical oxidation, molten metal technology, solvated electron process, and supercritical water oxidation. Advanced technologies (non-combustion) available for wastewater treatment are listed in table 1.

Why Only Electrochemical Oxidation?

In electrochemical oxidation (EO), an electrochemical cell, operating at 50 to 60°C (120° to 140°F) and atmospheric pressure, is used to generate an oxidizing species (a mediated metal ion) at the anode (the negative electrode) in an acidic solution. As indicated in the flowsheet, this is accomplished by supplying a voltage across two electrodes immersed in an acidic solution containing the oxidizing species in its reduced, more natural state.

What are Electrochemical Cells

Many oxidation-reduction reactions occur spontaneously, giving off energy. An example involves the spontaneous reaction that occurs when zinc metal is placed in a solution of copper ions as described by the net ionic equation shown below.

$$Cu^{+2}\ (aq) + Zn\ (s) \rightarrow Cu(s) + Zn^{+2}\ (aq)$$

The zinc metal slowly "dissolves" as its oxidation produces zinc ions, which enter into solution. At the same time, the copper ions gain electrons and are converted into copper atoms, which coats the zinc metal, or sediments to the bottom of the container.

IMPROTANCE OF REMOVING ORGANIC POLLUTANTS AND CARCINOGENS

Phenol is human poison by inhalation, ingestion, and skin absorption. It is a severe irritant to the eyes, skin and respiratory system. Human mutation data have been reported for phenol. Also it is a questionable carcinogen in animals and a suspected carcinogen in humans, although the data are inconclusive. Phenol is a general protoplasmic poison that is corrosive to any living tissue it contacts. Phenol is highly soluble in water. Concentrations of 1000 milligrams and more will mix with a cubic decimeter of water. About 26.3 per cent of phenol will eventually end up in air, approximately 73.3 per cent in water, and about 0.2 per cent in terrestrial soil and aquatic sediments. It is used to manufacture various phenolic and epoxy resins, for refining lubricating oils, a fuel-oil sludge inhibitor, and as a reagent in chemical analysis. It is also used in the production or manufacture of a large variety of aromatic compounds including fertilizers, illuminating gases, coke, explosives, lampblack, paints, paint removers, asbestos goods, wood preservatives, textiles, perfumes, bakelite, rubber, and other plastics. Many industrial processes generate wastewater flows with a high concentration of phenols and other related compounds. These compounds are known to betoxic, even at low concentrations, and their treatment is very important.

The removal of phenols from wastewaters is, therefore, an important problem and electrochemical oxidation technologies offer the prospect of relatively simple equipment, environmental friendliness, and the possibility of high-energy efficiency by using Electrochemical oxidation techniques.

Benefits of Electrochemical Over Others

The advantages of biological treatments are very well known, but also their limitations either for a high COD (Chemical Oxygen Demand) value or the presence of very toxic compounds. The possible presence of inorganic compounds, such as heavy metals, may cause a drop of the bacterial count. On the other hand, the incineration of organic compounds can originate the formation of toxic products that are dragged at the same time by the combustion gases; also, the presence of corrosive agents can cause problems in the stability of the materials of the incinerator. Electrochemical degradation (direct and indirect. Electrochemical oxidation) and electrocatalisys of hazardous waste water have several advantages compared with incineration and biological treatment. Electrochemical treatment is able to treat very toxic wastes.

CORROSION PROTECTION OF PETROLEUM FACILITIES

Internal and external corrosion affects many oil and gas facilities. It is the principal cause of leaks of products and rupture of storage tanks and pipelines, resulting sometimes in catastrophic damages (human damages, pollution of the natural environment, additional costs for repair, prolonged stop of pumping). The most common way to prevent a metal from corroding or delay its corrosion is to provide an impervious coating over it. Organic coatings have long been used to protect metals against corrosion. However, most coatings, such as paints, are not permanently impenetrable, and all coating systems eventually fail, either through existing pinholes in the coating or by diffusion of oxygen and water through it. Therefore, as a second line of defence, anticorrosive agents are added to the coating and also metallic pigments, semiconducting metals oxides and salts were conventionally used at relatively high volume concentrations to achieve the conductivity of coatings.

Growing environmental concerns regarding the use of heavy metals as pigments in anti-corrosion coating formulations led to a new coating strategy using intrinsically conducting polymers (ICPs) as key components. Conducting polymers as a new materials class provide a unique set of new properties. Their specificities are mainly their low density, thermal and chemical stability and the ease of synthesis. Coatings based on these polymers are able to meet high demands and are outperforming even the best conventional anti-corrosion coating systems. Representative of these polymers are such substances as polyacetylene, polypyrrole, polythiophene, poly(p-phenylene) and polyaniline. Polyaniline (PANI) is the most popular and one of the most explored applications of it is based on its good corrosion protective properties in coatings. However, many of the applications of PANI are restricted because of its unprocessability and insolubility in common organic solvents. But the alkyl substituted PANI has better solubility in organic solvents. It has been reported by Trivedi and by Mekki Daouadji that among several alkyls the ortho-ethoxy confers to the polymer the best solubility in organic solvents and the best properties as a corrosion inhibitor for the protection of steel in acidic environment. Our research was motivated by these properties of the poly (ortho-ethoxyaniline) "POEA" and by the higher solubility of this polymer compared to that of polyaniline "PANI". So, we have considered that the solubilisation of POEA could open up a new area for its use in anticorrosion coating formulations.

In the petroleum industry, all the facilities exposed to humidity, all storage tanks, all buried and sea water immersed pipelines suffer considerably by the presence of the corrosion. In this study, we have developed a coating containing a conducting polymer (POEA) as anticorrosive agent in order to substitute the conventional coatings used in this field. Therefore, we have tested and compared the performances against saline corrosion of steel between the developed coating and a conventional one generally used to protect tanks and pipelines. Then, the combination of the measurement of the Open Circuit Potential, the Potentiodynamic Polarisation and the Electrochemical Impedance Spectroscopy allowed us to obtain a good comparison of their efficiency.

By the developed coating, we have expected a high corrosion protection of steel with safety, environmental benefits and cost reduction. This fact is due to the specific properties of the polymer composed coating which can unlike the conventional one, work indefinitely as a redox catalyst by passivating and ennobling the metal maintaining a protective oxide layer on it.

CHEMICAL STRUCTURE OF THE DEVELOPED POLYMER (POEA)

The poly (ortho-ethoxyaniline) (POEA), the conducting polymer resulting from the oxidative polymerisation of the substituted aniline: ortho-ethoxyaniline, has emerged recently as one of the best candidates among conducting polymers , particularly among substituted polyaniline because of its high solubility in various organic solvents and its good properties as corrosion inhibitor. POEA exists in several different oxidation states. Each of the forms can be reversibly interconverted through electron transfer. The base form and the most stable of the polymer is the emeraldine oxidation state.

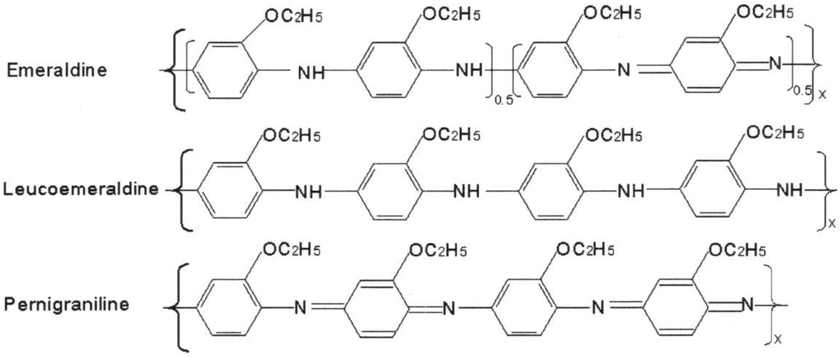

Fig. Poly (ortho-ethoxyaniline) oxydation forms.

The exceptional environment stability of conducting POEA is mainly due to the chemical flexible –NH group present in the polymeric backbone flanked on either side by a phenylene ring. This –NH group has very high chemical flexibility, such as protonation and deprotonation, in addition to the availability of a lone pair of electrons that can easily take part in surface adsorption phenomenon.

Theory of POEA Protection

Corrosion occurs in the presence of water, salts and oxygen; during which iron or steel undergo an electrochemical process in which different locations of the iron surface act as electrodes. At the local anode, iron is oxidised to soluble Fe^{2+} and Fe^{3+} ions:

$$n\text{Fe(s)} \rightarrow m\text{Fe}^{2+}\text{(aq)} + (n-m)\text{Fe}^{3+}\text{(aq)} + 3(n-m)\text{e}^{-}$$

At the local cathode, hydroxide ions are formed:

$$\text{O}_2\text{(aq)} + 2\text{H}_2\text{O(l)} + 4\text{e}^{-} \rightarrow 4\text{OH}^{-}\text{(aq)}$$

Rust, which forms subsequently, is composed of FeO, Fe_2O_3, Fe_3O_4 and other mixed oxides, $Fe(OH)_x$, and Fe^{x+} salts (chlorides, sulfates *etc*). This highly irreproducible and irregular composition opens up new surfaces for self-catalytic growth – mainly Fe^{3+} salts acting as rust formation catalysts. Rust that forms this way does not adhere well on the iron surface.

In the presence of the conducting polymer (POEA) coating, a completely different galvanic process occurs, in which POEA replaces iron as the cathode due to its metallic property, as it is situated like PANI slightly less noble than silver in the galvanic series. This arrangement is relatively evenly distributed over the whole surface. It involves Fe-oxidation by POEA (the more noble metal, Emeraldine salt ES), which is thereby reduced to Leucoemeraldine base (LE); further oxidation of Fe (II) to Fe (III) and reoxidation of LE to POEA (ES) via the Emeraldine base EB occur both by oxygen; and Fe_2O_3 deposition by resulting OH^-. This scheme shows furthermore, that POEA coating is different from conventional coatings in that it does not protect simply by offering a physical barrier but it acts as a redox catalyst, and the full catalytic cycle (ES LE EB and back to ES) will take place providing a continuous protection of metal as long as the mechanical integrity of the polymer film remains intact. By POEA coating, a remarkable corrosion potential shift ('ennobling') and an iron oxide layer formation ('passivation') together lead to a significant anticorrosion effect.

COATING CHARACTERIZATION METHODS

DC Electrochemical Monitoring Techniques: Potentiodynamic Methods

Corrosion on metals occurs at a reaction rate determined by opposing electrochemical reaction equilibria established between the metal and the electrolyte solution. One reaction is the anodic reaction, in which the metal is oxidized, releasing electrons from its surface. The other is the cathodic reaction, in which solution species like O_2 or H^+, or even the protective coatings and oxides films that cover the metal, are reduced, attracting electrons from the metal. The potentiodynamic technique is used to examine the passivation of a metal or allow in an electrochemical system. The behaviour of the potentiodynamic polarisation curves for metal covered with anticorrosion coatings tell us whether the metal is cathodically protected (sacrificial anode principle) or anodically protected (passivation principle). All experimental details and principles can be found in ASTM Standard G3. The hypothetical potentiodynamic polarisation curve for a material. The extrapolation of the linear portions of both cathodic and anodic branchs on a potential versus log (current) plot to their point of intersection is given by Tafel equations and can be combined into the Stern-Geary [8] equation,

$$i = i_{corr}\left\{\exp\left[\frac{2.303(E - E_{corr})}{\beta_a}\right] - \exp\left[\frac{2.303(E - E_{corr})}{\beta_c}\right]\right\}$$

Where i is the measured cell current in amperes icorr is the corrosion current in amperes, which is a measure of the corrosion rate

E_{corr} is the corrosion potential in volts

E is the applied electrode potential in volts

β_a and β_c are the cathodic and anodic Tafel beta coefficients in volts/decade

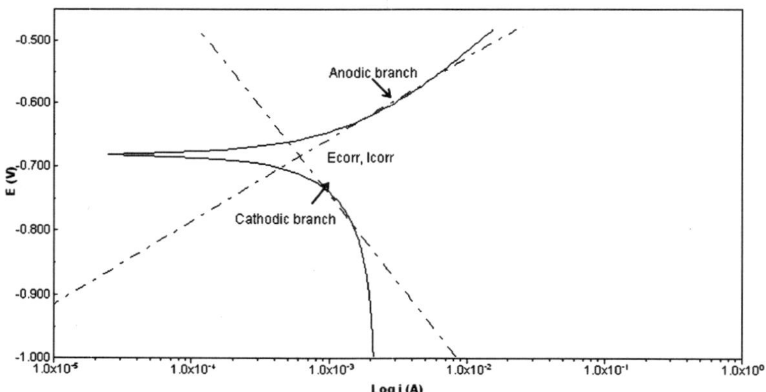

Fig. Hypothetical potentiodynamic polarisation curve for a material.

AC Monitoring Techniques: Electrochemical Impedance Spectroscopy

In the Electrochemical Impedance Spectroscopy (EIS) measurement, an alternating voltage is applied to the metal, and the impedance, Z, is measured. In an EIS experiment, the impedance is measured for a number of frequencies that span a range from 0.001 Hertz to 100 K–Hertz. The result is a complex number which is a mathematical representation of two associated quantities, the real and imaginary components of electrical impedance and it is expressed by:

$$Z(w) = \mathrm{Re}\,Z + j\,\mathrm{Im}\,Z$$

EIS is useful in the evaluation of coatings, the elucidation of transport phenomena in electrochemical systems, and the determination of corrosion mechanisms and rates. If Δ *E* is the perturbation brought to the impressed potential between the working electrode and the reference electrode, the theoretical expression of the total Faradaic impedance Z_f is :

$$\frac{1}{Z_f} = \left(\frac{\partial I}{\partial E}\right)_{\theta i,Cj} + \sum_i \left(\frac{\partial I}{\partial \theta_i}\right)_{Cj} \frac{\Delta\theta_i}{\Delta E} + \sum_j \left(\frac{\partial I}{\partial C_j}\right)_{\theta i} \frac{\Delta C_j}{\Delta E}$$

Where

θ_i is the covering factor of the *i* specie,

C_j is the concentration of the *j* specie.

Bode and Nyquist plots are the most common data output formats, and an example of the Nyquist plot is shown in Fig.3. At very high frequency, the imaginary component, Z'', disappears, leaving only the solution resistance, Rs. At very low frequency, Z'' again disappears, leaving a sum of Rs and the Faradaic reaction resistance or polarisation resistance, R_p, which is inversely proportional to the corrosion rate.

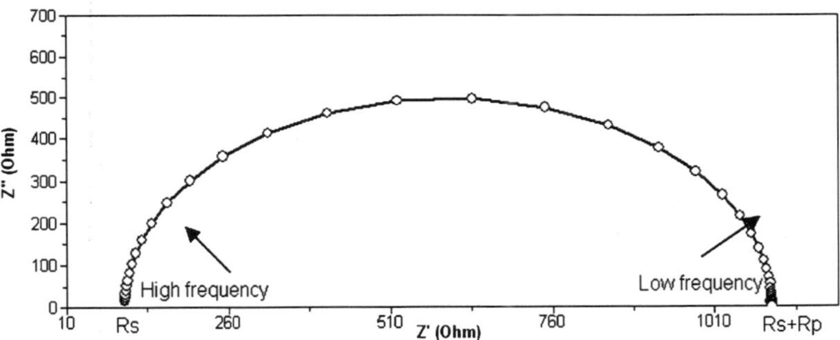

Fig. Nyquist plot for a simple electrochemical system

Evaluation of Anti-Corrosion Coating Performances

In this study, samples of uncoated and coated carbon steel (with coating thickness of about 45 μm) were tested in a severe environment (3.5 per cent in weight NaCl aqueous solution) using the electrochemical methods described in section 4. Then, both conventional and POEA coatings were evaluated for their anti-corrosion performances for the protection of carbon steel. The corrosion cell consisted of a three – electrode system. A stainless steel electrode and a saturated calomel electrode (SCE) were used as the counter and reference electrodes, respectively. All experiments were carried out using a potentiostet/ galvanostat model Autolab, type PGSTAT 20. Before testing, samples were immersed into the solution and the open circuit potential (OCP) was monitored until a constant value was reached. The first test, which we performed with Potentiodynamic Polarisation, tells us about the protection mechanism of steel. Indeed, Fig.4 clearly shows that contrary to the conventional coating, the POEA coating ennobles the metal by passivation effect. So, compared to the uncoated steel, a shift in the corrosion potential to more positive value (ΔE = 160 mV) occurs when the sample surface is covered by the POEA coating and a shift to more negative value (ΔE = 125 mV) when it is covered by the conventional one. The second test, which has been performed with Electrochemical Impedance Spectroscopy, indicates the barrier properties of coating. We want a system that does not allow ions (salts and corrosion products) to migrate through the coating. We can quantitatively describe this tendency by the coating resistance (R_c). From the Nyquist plots (Fig.5), we can see the difference existing between the semicircles of the impedance diagrams for the studied samples immersed during 1 hour in the saline solution. By fitting curves, we obtain a coating resistance value of the POEA coating higher than that of the conventional one, and as a direct consequence, an increase of the metal protection against corrosion by a factor of 10. In order to confirm this result, the studied samples were immersed during 7 days in the same solution. EIS measurements were periodically monitored and the results obtained from the Nyquist plots are giver in table1.

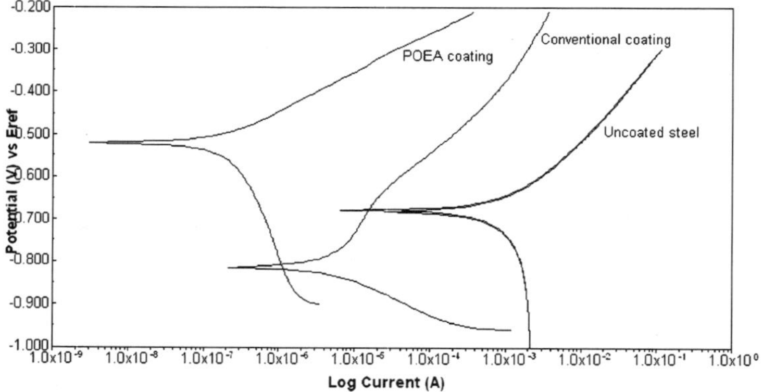

Fig. Potentiodynamic polarisation (Tafel plots) for uncoated and coated steel in saline environment.

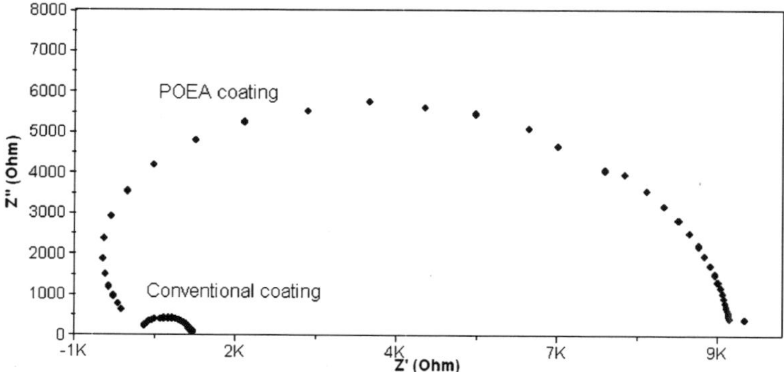

Fig. Impedance diagrams (Nyquist plots) for the studied samples (coated steel) immersed during 1 hour in saline environment.

Table: Resistance Values for the Studied Samples (Coated Steel) in Saline Environment (3.5 per cent wt NaCl).

Coatings	Coating Resistances Rc (Ohm)					
	1 hour	1 day	2 days	3 days	4 days	7 days
Conventional coating	820	500	300	210	180	150
POEA coating	9 890	9 100	7 400	5 600	5 000	4 600

The coating resistance values given in table 1 indicate that the conventional coating fails faster than the POEA coating. This fact is clearly shown in Fig.6, where the ratio of coating resistances ($R_{POEA} : R_{conv}$.) Increase with immersion time in saline environment.

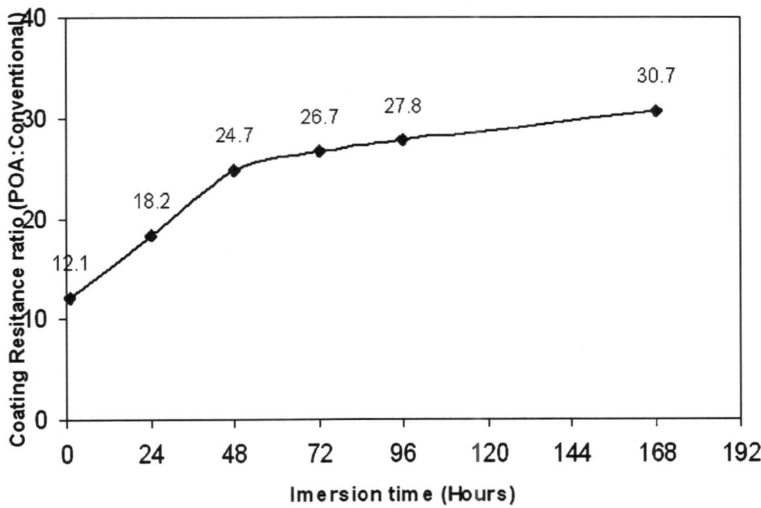

Fig.:Degradation rate comparison of the studied coatings in saline environment.

Conclusion

Our results clearly indicate that a crucial factor that determined whether coatings systems provide a maximum of corrosion protection was the choice of the corrosion inhibitor used in the coating formulations. Indeed, the coating developed in this study and which is based on a conducting polymer (POEA) as corrosion inhibitor is outperforming the conventional epoxy coating used in the petroleum industry. Due to the intrinsically properties of this polymer, coatings based on it are able to provide, with safety working, a high corrosion protection of any equipment or pipeline. Moreover, both chemical and physical properties of this polymer and the ease of its synthesis, allow us to preserve the natural environment. It should be noted, that this developed coating can also lead to a significant cost reduction for oil and gas industry because of the new ennobling and passivation technology which is only feasible with well advanced conducting polymer (POEA) in especially composed coating.

5

Corrosion Discovery of Resistant Existing Beams

INTRODUCTION

In recent years, the field of structural health monitoring (SHM) has been the subject of intensive investigation because of its feasibility and practical importance in damage detection. The main purpose in SHM is to detect damage at its earliest possible stage to prevent severe deterioration and reduce subsequent repair costs. SHM has been used successfully in mechanical, aerospace, and civil engineering structures. To obtain the best results in conventional damage detection methods such as ultrasonic pulse waves, schmidt rebound hammer and half-cell potential which are visual and localized methods, location of damage should be primarily accessible and determined by an in situ inspection. For overcoming this limitation, there has been an emerging trend towards the use of vibration characteristics of structures to detect damage. The basic concept of this method is that modal parameters (notably: natural frequencies, mode shapes and modal damping) are functions of physical and mechanical properties of the structures (mass, damping and stiffness). Therefore, changes in these properties will cause changes in modal parameters.

One of the dynamic properties which has been used by many researchers is, shift in natural frequencies. In this method, modal frequency is obtained by special equipment in situ and then it is compared with mathematically predicted frequency of intact structure. Shifts in modal frequency prove the existence of damage but no more information about the location and severity of damage can be extracted. Tests conducted on I-40 Bridge demonstrated that severe damage in a localized section of the bridge causes a minor shift in modal frequency. In fact, modal frequencies are related to global properties and do not provide any information about local changes in structural characteristics. Mode shape is another dynamic property which has been used by researchers. West was probably the first who used mode shapes to predict the location of damage in structures. Pandey et al. [4] presented a method which uses mode shape curvature as an indicator of damage for evaluation of damaged and undamaged structures. It should be noted that most of these methods need a baseline obtained from the undamaged structure which is the main deficiency, since in most of the practical cases only the properties of damaged structure are available.

One of the first attempts for detecting structural damage without baseline information is gapped smoothing method propounded by Ratclif et al. Wavelet analysis is an efficient tool in signal processing in vibration analyses and recently has been used in the damage detection in the SHM field. Staszewski presents a summary of recent advances and applications of wavelet analysis for damage detection. Kijewski and Kareem evaluated the feasibility and practicality of wavelet transforms for damage detection and identication of a structure in civil engineering. Continuous and discrete wavelet transforms are the two main types of transforms for structure analysis. The main advantage of the continuous wavelet transform (CWT) is its ability to elucidate real-time information in time (or space) and scale (or frequency) with adaptive windows.

An application of wavelet theory in spatial domain is detection of crack in structures which was proposed by Liew and Wang. The crack location is indicated by a peak in the profile of the wavelets coefficients along the length of the beam. Wang and Deng, 1999 proposed that the wavelet transform can be directly applied to spatially distributed structural response signals, such as surface pro?le, displacement, strain or acceleration measurements. Several studies used CWT of the fundamental mode shape exponent to detect the location and extent of damage in a beam Hong et al. , Gentile and Messina, Douka et al. and Wu and Wang.

Reinforcement corrosion in RC structures is an electro chemical reaction which can cause severe reduction in serviceability and ultimate capacity of the structures. Corrosion products cause extensive volumetric increase and therefore lead to splitting in concrete and meanwhile reduce reinforcement cross sectional area and bonding strength. Although corrosion causes severe damages in structures particularly bridges, using corrosion detection techniques based on health monitoring approaches can efficiently prevent these damages. Therefore, the present study aims to develop a new corrosion detection method based on wavelet analysis to detect early stage corrosion without baseline information. Discrete and continuous wavelet transforms were used in this study. Minimum number of vanishing moments of wavelets is also considered.

APPLICATION OF WAVELET TRANSFORM IN DAMAGE DETECTION

Wavelet transform is a powerful tool for image and signal processing. In recent years, some studies have proven the efficacy of this theory as a damage detection and localization tool (e.g. [9-13]). Wavelet transform has the ability to detect fluctuations in a signal and therefore can be used to detect fluctuations in a damaged structure profile like mode shape or its derivatives. In this section, application of continuous and discrete wavelet transform in damage detection is presented, respectively.

Continuous Wavelet Transform (CWT)

A wavelet is an oscillatory, real- or complex-valued function $\psi(x) \in L^2(R)$ .This function has zero average and finite length. Function $\psi(x)$ is a mother wavelet and $L^2(R)$ denotes the Hilbert space function which is measurable, square-integrable and one-dimensional. This function is used to create a family of wavelet as follows:

$$\psi_{u,s}(x) = \frac{1}{\sqrt{s}} \psi\left(\frac{x-u}{s}\right)$$

where, u and s are real translation and dilation parameters, respectively. Translation parameter determines the location of wavelet window along the space axis and dilation parameter represents the width of wavelet window. Therefore, contrary to Fourier transform, information related to space and frequency is kept simultaneously. Therefore, CWT for a given signal $f(x)$ is defined as follows:

$$Wf(u,s) = \frac{1}{\sqrt{s}} \int_{-\infty}^{+\infty} f(x)\psi\left(\frac{x-u}{s}\right)$$

where $Wf(u,s)$ gives the wavelet coefficient of a given signal and its values range between 0 and 1. This coefficient determines the level of similarity between signal and corresponding wavelet function at the vicinity of the point u with scale s. Value 1 denotes the absolute similarity (presence of fluctuations in signal) and value 0 (signal is smooth at vicinity of point u) represents no similarity. Therefore, a peak or sudden change in wavelet coefficient can show the presence of fluctuations in response to the damaged structure. It should be noted that application of wavelet transforms with appropriate number of vanishing moments plays an important role in detection of signal singularities. A wavelet with n vanishing moments should satisfy Eq. (below):

$$\int_{-\infty}^{+\infty} x^k \psi(x) d(x) = 0, \; k = 0,1,2,3,\ldots,n-1$$

Wavelet with n vanishing moments ensures zero or very small value for wavelet coefficient because it is orthogonal to polynomials up to degree n–1. In the case of mode shapes which are similar to polynomial degree 4 wavelets with at least 4 vanishing moments is required for a robust damage detection procedure.

Discrete Wavelet Transform (DWT)

In practical signal processing, a discrete version of wavelet transform is mostly employed by choosing discrete dilation parameter s and the translation parameter u [15]. In fact, continuous transform includes redundant information which is not necessary for signal processing. In general, the procedure becomes much more efficient and the involved computational efforts will be greatly reduced if dyadic values of u and s are used as follows:

$$s = 2^j, \; u = 2^i k \quad j,k \in Z$$

where Z is the set of positive integers and discrete wavelet function is written as follows (Eq. below):

$$\psi_{j,k}(x) = 2^{\frac{j}{2}} \psi\left(2^j x - k\right)$$

Wavelet expansion function (*F* (*x*)) and the coefficients of the wavelet expansion, $\alpha_{j,k}$ are defined according to Eqs. (below):

$$F(x) = \sum_j \sum_k \alpha_{j,k} \psi_{j,k}(x)$$

$$\alpha_{j,k} = \int_{-\infty}^{+\infty} F(x) \overline{\psi}_{j,k}(x) d(x)$$

where the $\overline{\psi_{j,k}}(x)$ is the complex conjugate of the function $\psi(x)$. Finally, using a DWT, a signal can be decomposed into its approximation and detail coefficients at the J_{th}–level as Eqs. (below), respectively:

$$A_J = \sum_{j>J} D_J$$

$$D_J = \sum_{k \in Z} \alpha_{J,k} \psi_{J,k}(x)$$

reconstructed the two components is presented in Eq. (below):

$$F(x) = A_J + \sum_{j \le J} D_J$$

In the field of damage detection method, wavelet approximation coefficient conveys the intact and sound structure information, while the wavelet detail coefficient contains the information of the damaged or noisy parts of the whole structure. Presence of a peak in detail coefficient can represent the location of a defect if there is not any noise (like wind or vehicle vibration) in the system. In the presence of noise, decomposition should be done into various levels by selecting different dyadic scales to separate the effects of noise and damage. In this paper, both CWT and DWT are used and suitable mother wavelets and minimum number of vanishing moments for optimum detection of corrosion is evaluated.

MATERIALS MODELING

The applied models for concrete, reinforcing steel and steel-to-concrete bond behavior are introduced in this section. These models have the ability to take into account the corrosion effects on material degradation.

Concrete Modeling

The idea of coupled mechanical-environmental damage model developed by Saeta et al. is used in this study to represent the concrete behavior under static as well as dynamic loading, including environmental degradation. Two different parameters of d_t and d_c for tension and compression damage are used as mechanical damage modifiers in the damaged plasticity model which are represented according to Eqs (below):

$$\sigma_t = (1-d_t)E_0(\varepsilon_t - \tilde{\varepsilon}_t^{pl})$$

$$\sigma_c = (1-d_c)E_0(\varepsilon_c - \tilde{\varepsilon}_c^{pl})$$

where E_0 is the initial or undamaged elastic stiffness, the subscripts t and c refer to tension and compression, respectively. $\tilde{\varepsilon}_t^{pl}$ and $\tilde{\varepsilon}_c^{pl}$ are the equivalent plastic strains. σ_t and σ_c denote tension and compression strength, respectively. ε_t and ε_c represent tension and compression strains, respectively.

Considering tension stiffening in stress-strain relation, the post failure behavior of concrete under tension (cracked concrete) and the effects of reinforcing steel-concrete interaction can be modeled easily. As discussed by, for reinforced concrete it is accepted to assume that tension in concrete becomes zero linearly at a total strain of about 10 times the cracking strain(ε_{ot}^{el}) as it is shown in Fig. 1.a. It is also possible to model concrete behavior outside the elastic range by defining tabular data of compressive stress as a function of inelastic strain $\tilde{\varepsilon}_c^{pl}$, as it is presented in Fig. 1b.

$$\tilde{\varepsilon}_c^{pl} = \varepsilon_c - \varepsilon_{oc}^{el}$$

where

$$\varepsilon_{oc}^{el} = \frac{\sigma_{co}}{E_0}$$

and σ_{co} denotes maximum compression stress within the elastic zone.

A yield function proposed by Lubliner et al. is employed to account for different evolutions of strength under tension and compression. In terms of effective stresses the yield function takes the mathematical form as follows (Eq. below):

$$F(\bar{\sigma}, \tilde{\varepsilon}^{PL}) = \frac{1}{1-\alpha}(\bar{q} - 3\alpha\bar{p} + \beta(\tilde{e}^{PL}) < \hat{\bar{\sigma}}_{\max} > -\gamma < -\hat{\bar{\sigma}}_{\max} >) - \bar{\sigma}_c(\tilde{\varepsilon}^{PL})$$

where α and β are dimensionless material constants; $\bar{p}$ is the effective hydrostatic pressure; $\bar{q}$ is the Mises equivalent effective stress; $\bar{s}$ is the deviatoric part of the effective stress tensor and $\hat{\bar{\sigma}}_{max}$ is the algebraically maximum eigenvalue of $\bar{\sigma}$ and the parameter γ takes the value of 3. The flow potential G chosen for this model is the Drucker-Prager hyperbolic function as:

$$G = \sqrt{\left(\in \sigma_{to} \tan\psi\right)^2 + \bar{q}^{-2}} - \bar{p}\tan\psi$$

where ψ is the dilation angle measured in the *p–q* plane at high confining pressure; σ_{to} is the uniaxial tensile stress at failure; and $\in$ is a parameter, referred to as the eccentricity, that defines the rate at which the function approaches the asymptote (the flow potential tends to a straight line as the eccentricity tends to zero). This flow potential, which is continuous and smooth, ensures that the flow direction is defined uniquely.

To account for the mechanical effects of corrosion in concrete structures, a coupled chemical–mechanical damage model has been developed, by introducing an additional internal variable, called 'environmental damage d_{chem} in the stress-strain relationship. Substituting this variable into the Eqs. (above), they can be re-written as follows:

$$\sigma_t = (1 - d_{chem})(1 - d_t)E_0(\varepsilon_t - \tilde{\varepsilon}_t^{pl})$$

$$\sigma_t = (1 - d_{chem})(1 - d_c)E_0(\varepsilon_c - \tilde{\varepsilon}_c^{pl})$$

The environmental damage parameter dchem is a function of development degree of the chemical reaction and the relative residual strength of the material. The environmental damage parameter is assumed, to be represented by an increasing function with time. Despite the other damage parameter, It has the same effect in tension and compression.

Reinforced Concrete Modeling

Within the framework of finite element method, to model the reinforcing steel in concrete, two different approaches can be used: "smeared approach" in which steel is smeared into a layer between concrete layers; the second one is a more detailed and called "detailed approach" in which steel and concrete are modeled as distinct elements. The first method, which is frequently used for analyzing the global behavior of RC structural elements, is not directly applicable when a corrosion phenomenon is considered. Actually such an approach cannot easily simulate the loss of bond between steel and concrete, which is a fundamental issue in corrosion problems. Among different possibilities allowed by the "detailed approach", an elastic constitutive relationship with isotropic hardening has been adopted for reinforcing steel. A linear elastic material model is used to model reinforced concrete. The total stress is defined from the total elastic strain as follows :

$$\sigma = D^{el}\varepsilon^{el}$$

where σ is the total stress, or Cauchy (true) stress in finite-strain problems, D^{el} is the fourth-order elasticity tensor, and eel as the total elastic strain. As discussed in, corrosion effects on cross sectional area and ductility of reinforcement are both considered and reduction in yield and ultimate strength are neglected. Reduction in cross section is applied based on equation proposed by Berto et al. as follows:

$$As = \frac{Ns^{\pi}[D_0 - nx]^2}{4}$$

where N_s is the number of steel bars of the same diameter, D_0 is the original diameter of steel bars, n takes into account the possibility of a one-sided or two-sided corrosion attack and x is the corrosion depth in the reinforcement.

In this phase of the research, the effect of corrosion on steel bar ductility is introduced in the model by adopting the approach recently proposed in. According to this approach the steel ultimate strain ε'_{su} can be considered as linearly dependent on the reduction of the bar cross-section, by defining α_{pit} as (Eq. below):

$$\alpha_{pit} = \frac{\Delta A_{pit}}{A_{rebar}}$$

where ΔA_{pit} is the area reduction because of corrosion and A_{rebar} is the nominal reinforcement cross-section, Therefore it is assumed that (Eq. below) :

$$\varepsilon'_{su} = \left(\varepsilon_{su} - \varepsilon_{su}\right)\left(1 - \frac{\alpha_{pit}}{\alpha_{pit}^{\max}}\right)$$

Bond Modeling

The bonding strength at the steel-concrete interface plays an important role in defining the behavior of RC structures, both to ensure the safety level and prepare adequate structural ductility (e.g.). Therefore, an efficient modeling of bond behavior is a major issue to prevent relative displacements between steel and concrete. This model should include not only the effect of bond strength value, τ_{max}, but also definition of the shape of the bonding curve similar to concrete stress – strain relationship. Furthermore, it should consider the corrosion level. To achieve this aim, following the approach described in section 3.1 for environmental damage in concrete, a corrosion bond damage parameter, d_{bond}, was introduced, similar to the variable d_{chem} of Eq. (17) and defined as a monotonically increasing function of the corrosion level, with values ranging between 0 and 1. Hence, the relationship between parameters τ and s can be mathematically represented as Eq. (below):

$$\tau_{\max} = \left(1 - d_{bond}\right)G\gamma$$

where G is the elastic shear modulus; *s* is relative slip between concrete and reinforcement; γ = s/t is the shear strain and *t* is element thickness. Consequently, by adopting the unique parameter d_{bond}, the proposed model can describe the reduction of bond strength as well as the peak stress value versus elevation of corrosion levels, as illustrated in the τ–s curve.

NUMERICAL ANALYSES

To demonstrate and prove the capability of the proposed method, a simply supported reinforced concrete beam of 1000 mm length and 50 mm x 100 mm cross sectional area was modeled for damage detection. To obtain Concrete Damaged Plasticity (CDP) parameters, experimental results of JANKOWIAK, as summarized in Table 1, were used. Bond model parameters were obtained by experimental results provided by and are shown in Table.2. Maximum value of 20 percent was considered for environmental damage parameter, dchem based on the findings of. For the FE analysis, solid (continuum) elements with 8 integration points (CD8R), 5 mm of mesh size were used to ensure accurate solutions. An interface layer was modeled between concrete and reinforcement elements and bonding properties were attributed to this layer. Field investigation has proven that corrosion starts from a small part of the reinforced and propagates to other parts gradually [19]. Therefore, this idea was used in modeling of damage with different corrosion percentages in reinforced damaged zone as shown in Fig. 3. Damaged zone was divided into 3 sub zones: side zone, middle zone and central zone. Central, middle, and side zones were modeled with highest, moderate, and lowest corrosion levels, respectively. 15 scenarios were considered for corrosion detection as displayed in Table 3. In each case, static deflection profile of the beam under uniform load of 2.5 N/m or mode shape profile related to its free vibration was extracted numerically.

Then, these profiles were analyzed by wavelet transform to detect corrosion. Scenario number 1 was defined to verify the correctness and precision of the finite element modeling by considering an undamaged simply supported beam. Concrete, steel and bonding degradation were considered individually in scenarios number 2, 3, and 4, respectively. It should be mentioned that in other damaged scenarios, these effects were considered simultaneously. Scenarios numbers 5 to 9 were developed with single damage location and different corrosion percentage in damaged zone to investigate the effectiveness of wavelet analysis in damage detection. Then, these five scenarios were employed to define a relation between corrosion percentage and wavelet coefficient. Capability of the proposed method in detection of multiple damages was assessed with scenarios numbers 10 to 14 which consist of two damage locations at one third and two thirds of the beam length with the same corrosion level and damage length in each location. Finally, to verify the ability of the proposed relation in detection of corrosion severity, scenario number 15 was evaluated.

Results

In this section, the results of numerical studies are presented and discussed. At first, to verify the accuracy of the finite element modeling presented in previous section, an undamaged beam (scenario 1) was analyzed and its deflection profile as well as its first mode shape was calculated and compared with the result of basic theoretical formulations of structural mechanics. As it can be seen from these figures, they are in good agreement with each other. Corrosion was modeled taking into account the role and behaviors of concrete, reinforcing metal and bonding degradation. To investigate the sensitivity of this damage detection technique to these deteriorations individually, the first mode shape with respect to the scenarios numbers 2, 3, and 4 were analyzed with discrete wavelet transform and in each case, similar peaks in wavelet coefficients profile was observed. It should be noted that, scenario number 3 (taking into account only the reinforcement deterioration) had the highest peak. Therefore, corrosion can be efficiently modeled via reinforcement degradation which is possible with less computational efforts. However, to provide accurate results, corrosion was modeled in scenarios 5 to 15 taking into account all the three degradation effects simultaneously. Corrosion propagation could be prevented if it is detected at its early stage. Therefore, static deflection profile of damaged beam with 2 percent corrosion (scenario number 9) was analyzed to demonstrate the capability of presented method in detection of corrosion in its early stage. Continuous and discrete wavelet transforms were used in this step and both showed high efficiency in the damage detection. It can be easily seen that in the vicinity of corrosion zone, there is a peak in wavelet coefficients that can be used as an indicator for damage existence and location. Therefore, it can be concluded that both CWT and DWT have the ability to detect corrosion in RC beams. However, discrete wavelet transform was used for the rest of the study because of its simplicity and reduced amount of data, it can provide.

The first mode shape of the scenario number 9 was also analyzed through different mother wavelets with the same number of vanishing moments like the studies conducted by Bior, Coiflet, Symlet, and Daubechies. Our findings were in consistence with the previous studies showing high efficiency of these mother wavelets in corrosion detection. It should be noted that, Daubechies mother wavelet is more sensitive to damage and leads to wavelet coefficients of greater value. Therefore, in this paper Daubechies mother wavelet is mostly used to investigate other capabilities of wavelet transform in the field of damage detection. The damaged beam can be used in corrosion detection, but using static profile in corrosion detection is more suitable because it can be obtained very easily in practice. The minimum number of vanishing moments should be determined for an accurate corrosion detection procedure in damaged structures. Therefore, Daubechies mother wavelet with different number of vanishing moments was used for the wavelet analysis of mode shapes. Scenario number 9 was used in this section and the obtained results are summarized. It can be concluded that in the case of corrosion detection in simply supported RC beams, wavelets with at least 4 vanishing moments should be used for accurate corrosion detection. Other Mother wavelets like Symlet, Coiflet and Bior were also used and the same results about the number of vanishing moments were observed. Next, scenarios numbers 5 to 9 were used to demonstrate the capability of wavelet analysis in developing a relationship between wavelet coefficients and corrosion percentage. In each case, maximum value of continuous wavelet coefficient of damaged zone is calculated by utilizing Coiflet mother wavelet, as summarized in Fig. 9 For practical applications, a linear trend-line is best fitted to these data, as presented in Eq. (below):

$$CP = 3.33 \times 10^5 CW_{coif} + 666.6$$

where CW denotes maximum of continuous wavelet coefficients in damaged zone, the subscript coif implies the Coiflet mother wavelet, which was used for developing of this equation and CP represents maximum corrosion percentage of damaged zone. Using other mother wavelets is also possible, but the constant values of the equation may change. Similar procedure was implemented with DWT and Eq. (below) was obtained as follows:

$$CP = 3 \times 10^8 DW_{coif} + 2670$$

where DW_{coif} denotes maximum of discrete wavelet coefficients of damaged zone. Therefore, implementing Eqs. in practical conditions, corrosion severity can be determined by performing wavelet analysis on mode shape or static profile of the damaged beam. To verify the reliability and accurateness of the proposed equation, scenario number 15 with 7 percent of corrosion expansion was analyzed and then, its wavelet coefficient was used as an input for the presented equations. Based on Eqs., the corresponding corrosion percentages of 7.2 and 7.22 percent respectively, which are reasonable. Capability of the presented method to detect multiple damages was evaluated, considering scenarios numbers 10 to 14 (with two damaged locations). The three first mode shapes were extracted and analyzed. The high efficiency of the presented method for detecting multiple damages in each mode shape can be concluded. In cases where corrosion is occurred in more than one location. In these cases, the reinforcing beam should be divided into several parts in a manner that each part includes just one corrosion zone. Then, corrosion severity can be investigated in each part separately. It should be noted that, extracting mode shapes higher than mode of order three is possible with lots of difficulties in practice, so damage detection by using lower mode shape is more suitable. Therefore, in this paper just first three mode shapes has been investigated and reasonable results are obtained. Peak of wavelet coefficients in mode 3 has greater values than mode 1 and 2, thus at the presence of e there is noise in the system because of the vehicle motion or wind, using mode shape 3 is more efficient because it is more sensitive to damage.

6

Pitting Corrosion in Occurrence of Oxidants and Inhibitors

INTRODUCTION

We showed that for the design of an open cooling circuit, the choice of the carbon proportion of the steel should take into account the chloride quantity of the medium in which it will be exposed (Boucherit, 2005). Even if the steel does not present local attacks after several years (Boucherit, 2006), preventive measures must be considered because localised corrosion occurs rapidly at a microscopic level, with an insidious manner and can be even stimulated by numerous thermohydraulic factors. A preventive procedure on a regular basis consists of incorporating inhibitors directly in the coolant. During several decades, chromate was the best option for its cost/effectiveness. However, the toxicity caused by its hexavalent form (Wise, 2006) encouraged research for non-toxic and more environmentally acceptable corrosion inhibitors. The interest went spontaneously on molybdate and tungstate due to chemical similarities in the group VI of the periodic table.

Many studies related to molybdate and tungstate inhibition were published during the last decade. It was reported that in acidic medium, molybdate is more efficient than chromate (Refaey, 2005). Like dichromate, molybdate suppresses pit nucleation but inversely its polymerisation increases the pH, whereas dichromate decreases the pH due to the oxidation/reduction reaction (Zhao, 2002). Another comparative study between oxyanions showed that under static conditions the order of inhibition performance is $WO_4^{2} \rightarrow MoO_4^{2} \rightarrow VO_3-$ and this order is altered under electrode rotation $WO_4^{2} \rightarrow VO_3 \rightarrow MoO_42^-$ (Martini, 2000). Tungstate seems to be more interesting than molybdate because of its applicability under a broader pH range. Nevertheless, molybdate and tungstate lack the oxidising properties of chromate. To enhance their efficiency, they must be used in presence of dissolved oxygen (Azambuja, 2003) or additional oxidant agent like iodate (Shibli, 2005). Yet, the use of an oxidant agent as a co-inhibitor is dangerous because pitting can occur at low concentration (Luo, 2000).

The performance of an inhibitor is usually judged by its capacity to shift anodically the pitting potential. This one is taken from voltammograms and corresponds to the potential where the anodic current exceeds a threshold (Kaneko, 2002). In reality, this threshold is imposed by the sensitivity of the electrochemical equipment. It is generally of the order of 1 μA, which is relatively important. Because pits are initiated at lower potentials and there is a broad region where metastable pits exist. To affirm detecting the first currents generated by pits initiation, it is necessary to be able to record charges as low as 10^{-13} C (Maurice, 2001). Some attempts to access weaker currents are published. They are based either on the study of microelectrodes (Kobayashi, 2000), or on the amplification of the anodic current (Burstein, 2004), or directly on the provocation of just one dissolution site (Verhoff, 2000). The investigation of the exact limit of the material resistance becomes more complicated when it comes to observe the influence caused by weak variations in the chemical composition of the solution. In addition, the knowledge of the phenomena taking place at the material surface is not enough to predict performance, because the pitting initiation depends on multiple factors related to the solution and the material properties as well as the exposition conditions.

The difficulty in obtaining experimentally the pitting potential, and the complexity of its expression by multiple parameters, justify the application of a mathematical model, which would make it possible to predict this potential starting from measurable parameters. The neural networks are relatively recent mathematical tools applied with enthusiasm in various fields including corrosion problems (Pleune, 2000), (Diaz, 2006). Their application for pitting potential prediction is possible since a great number of experiments are available.

PITTING CORROSION

Pitting corrosion, or pitting, is a form of extremely localized corrosion that leads to the creation of small holes in the metal. The driving power for pitting corrosion is the depassivation of a small area, which becomes anodic while an unknown but potentially vast area becomes cathodic, leading to very localized galvanic corrosion. The corrosion penetrates the mass of the metal, with limited diffusion of ions. The mechanism of pitting corrosion is probably the same as crevice corrosion.

MECHANISM

It is supposed by some that gravitation causes downward-oriented concentration gradient of the dissolved ions in the hole caused by the corrosion, as the concentrated solution is denser. This however is unlikely. The more conventional explanation is that the acidity inside the pit is maintained by the spatial separation of the cathodic and anodic half-reactions, which creates a potential gradient and electromigration of aggressive anions into the pit. This kind of corrosion is extremely insidious, as it causes little loss of material with small effect on its surface, while it damages the deep structures of the metal. The pits on the surface are often obscured by corrosion products. Pitting can be initiated by a small surface defect, being a scratch or a local change in composition, or a damage to protective coating. Polished surfaces display higher resistance to pitting.

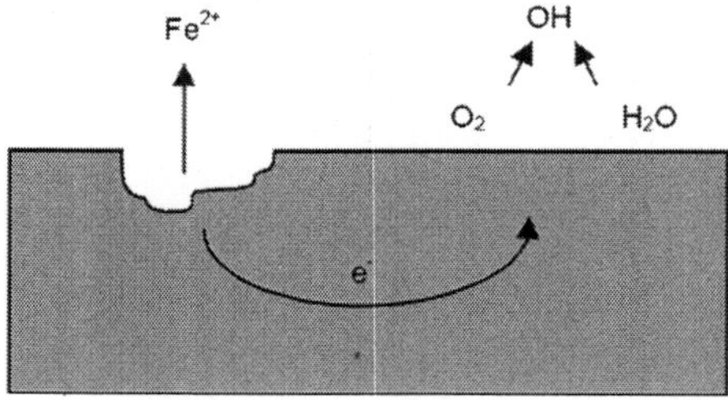

Fig. Diagram showing a mechanism of localized corrosion developing on metal in a solution containing oxygen

Susceptible Alloys

Alloys most susceptible to pitting corrosion are usually the ones where corrosion resistance is caused by a passivation layer: stainless steels, nickel alloys, aluminum alloys. Metals that are susceptible to uniform corrosion in turn do not tend to suffer from pitting. Thus, a regular carbon steel will corrode uniformly in sea water, while stainless steel will pit. Additions of about 2 per cent of molybdenum increases pitting resistance of stainless steels.

Environment

The presence of chlorides, e.g. in sea water, significantly aggravates the conditions for formation and growth of the pits through an autocatalytic process. The pits become loaded with positive metal ions through anodic dissociation. The Cl^- ions become concentrated in the pits for charge neutrality and encourage the reaction of positive metal ions with water to form a hydroxide corrosion product and H^+ ions. Now, the pits are weakly acidic, which accelerates the process. Besides chlorides, other anions implicated in pitting include thiosulfates ($S_2O_3^{2-}$), fluorides and iodides. Stagnant water conditions favour pitting. Thiosulfates are particularly aggressive species and are formed by partial oxidation of pyrite, or partial reduction of sulfate. Thiosulfates are a concern for corrosion in many industries: sulfide ores processing, oil wells and pipelines transporting soured oils, Kraft paper production plants, photographic industry, methionine and lysine factories. Corrosion inhibitors, when present in sufficient amount, will provide protection against pitting. However, too low level of them can aggravate pitting by forming local anodes.

Examples

A single pit in a critical point can cause a great deal of damage. One example is the explosion in Guadalajara, Mexico on April 22, 1992, when gasoline fumes accumulated in sewers destroyed kilometers of streets. The vapors originated from a leak of gasoline through a single hole formed by corrosion between a steel gasoline pipe and a zinc-plated water pipe. Firearms can also suffer from pitting, most notably in the bore of the barrel when corrosive ammunition is used and the barrel is not cleaned soon afterward. Deformities in the bore caused by pitting can greatly reduce the firearms accuracy. To prevent pitting in firearm bores, most modern firearms have a bore lined with chromium. Pitting corrosion can also help initiate stress corrosion cracking, as happened when a single eyebar on the Silver Bridge, West Virginia failed, killing 46 people on the bridge in December, 1967.[clarification needed]

EXPERIMENTAL

All the experimental results were obtained from electrochemical measurements on a carbon steel with an exposed surface of 0,1 cm^2. The steel composition and the electrode preparation, as well as the description of the equipment used were detailed in a previous paper (Boucherit, 2005). We carried out a program of 260 experiments. Each experiment consists of tracing the voltammogram of the steel in a solution by sweeping the potential from – 0,8 to 1 V/ECS at 1 mV/s. The solutions contain a pitting agent, Cl^-, an inhibitor, MoO_4^{2-} or $WO4^{2-}$ and an oxidant, IO_3–. Figure 1, shows the anions concentrations distribution. Tungstate and molybdate were studied separately. The study of each inhibitor required 140 experiments. By deducing the 20 cases where the inhibitor concentration is null, the total number of experiments considered in this work is 260.

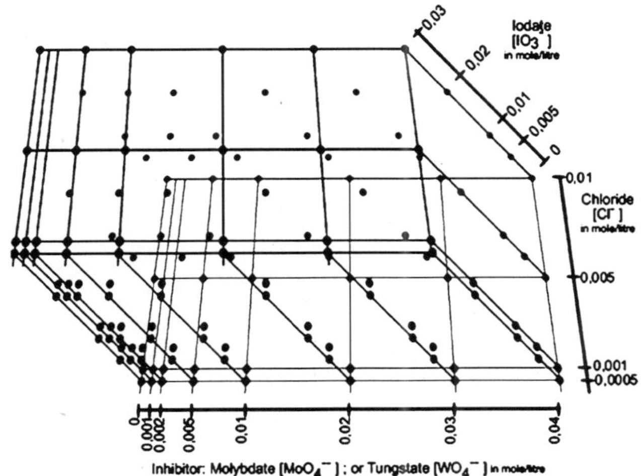

Fig. Anions concentrations distribution

At the beginning of each experiment, the ionic conductivity and the pH of the solution are measured. At the end the pitting potential is taken from the voltammogram. We defined the pitting potential as the potential where the anodic current reaches 5 μA. The prediction of the pitting potential by neural networks method was carried out using MATLAB v6.5 and nntool v4.0.

Results

Figure 2 shows voltammograms obtained for various chloride concentrations in the absence of any inhibitor or oxidant. The sensitivity of our equipment enables us to measure current of 1 μA. By zooming, it appears that the chloride concentration does not influence the pitting potential but affects only the slope of the current.

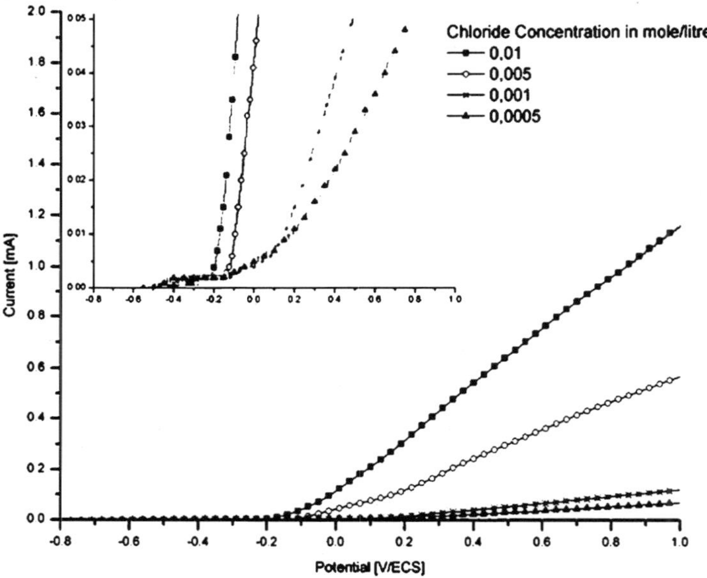

Fig. Voltammograms of the steel in water containing various chloride concentrations

If the variation of the chloride concentration does not affect the pitting potential, this one is however very influenced by the presence of inhibitors or oxidants. On figure 3, we collected voltammograms obtained with a low chloride concentration: 0,0005 mole/litre. Parts (a) and (b) correspond respectively to the presence of molybdate and tungstate with a fixed concentration of 0,001 mole/litre. The influence of iodate is thus evaluated through the variation of its concentration. All the solutions have a pH of about 6,3. The variation of the iodate concentration does not affect the pH. In the absence of iodate, the increase in the anodic current around 0 V/ECS is done in a continuous way. However, in the case of tungstate, the increase is considerably attenuated. Taking into account the pH value and the Pourbaix diagram, this potential would correspond to iron dissolution in Fe^{2+}. By increasing the iodate concentration, a current peak appears revealing a passivation process. Iodate, with its oxidizing capacity, should oxidize the

Fe^{2+} ions to Fe^{3+} supporting passive film formation. The pitting resistance increases with iodate concentration.

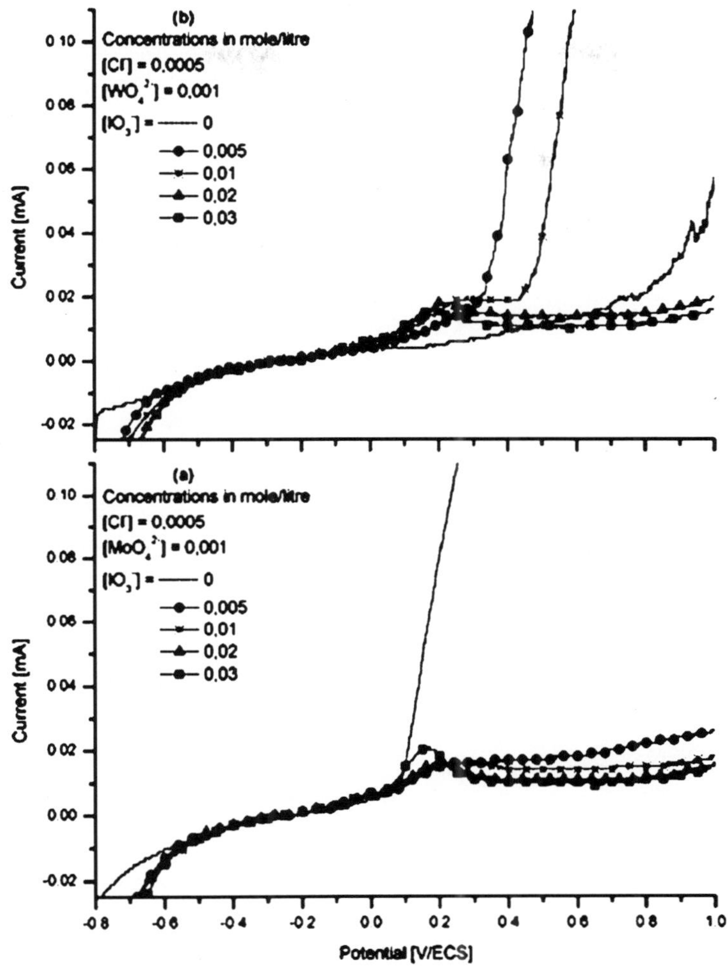

Fig. Effect of the iodate concentration in presence of small chloride and inhibitor quantities

The capacity of iodate to improve the pitting potential is not always granted. For some inhibitor concentrations the addition of iodate can even be prejudicial. This is clearly visible on figure 4, where we gathered pitting potentials for solutions containing a fixed chloride concentration of 0,001 moles/litre and by varying the inhibitor concentration: molybdate and tungstate. The figure makes it possible to see, by following a curve, the iodate concentration influence on the pitting potential while keeping constant the pitting agent and the inhibitor concentrations. In the absence of inhibitors, it is clear that the pitting potential increases with the iodate concentration. But with a molybdate concentration equal to 0,01 mole/litre, progressive addition of iodate reduces the steel resistance, whereas in the absence of iodate, the steel remains resistant until the higher potential limit. These observations are confirmed with tungstate.

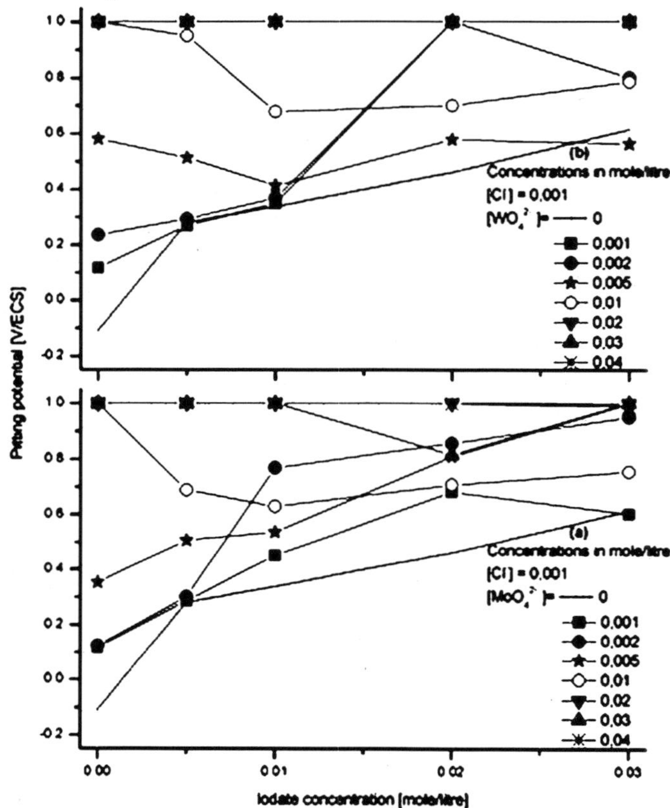

Fig. Effect of iodate concentration on the inhibitor performance

The performance evaluation, by considering the pitting potential, such as defined for figure 4, causes some practical problems. Indeed, the current threshold corresponding to the rupture such as we fixed it (5 μA) is lower than the current peaks values observed on figure 3 and which correspond not to pitting but rather to dissolution followed by passivation. Reading the pitting potential when it takes place around 0, 2 V/ECS is often not very accurate. On the other hand, the current evolution varies greatly after pitting and does not depend on the pitting potential value. In some cases we observe a pitting followed by a weak increase in current and in other cases the pitting is delayed but the current slope following is quite steep.

Another method to estimate the steel performance consists of taking the voltammograms, fixing a threshold current and choosing a performance potential. Then, the steel is considered resistant if its potential reaches the performance potential with an anodic current not exceeding the fixed threshold current. The performance thus defined, is represented on figure 5. The threshold current "is" is fixed at 25 μA which is higher than the current peaks observed in figure 3. The performance potential corresponds to the anodic limit of the sweeping interval: 1 V/ECS. If the anodic current does not exceed "is" at the end of the sweeping, the experience is indicated by an open circle. Otherwise, it is represented by a small square. On the horizontal axis, we represented the inhibitor and the oxidant predominance, defined by the relation: [inhibitor]-[iodate])/([inhibitor]+[iodate]). On the vertical axis we represented the inhibitor and the oxidant concentrations sum divided by the chloride concentration: [inhibitor]+[iodate]/[chloride]. The advantage of these figures is that they allow visualizing synergy and antagonism fields between the oxidant and the inhibitor. By drawing a horizontal line, we sweep the cases where the oxidant and inhibitor concentrations are variable but their sum is maintained constant. Thus at the left we are in presence only of the oxidant, whereas at the right just the inhibitor is present. All the experiments with the various chloride concentrations are represented. Each graph on the figure gathers the data of 140 experiments.

The obtained results are slightly different for the two inhibitors. In absence of oxidant, the steel is resistant if the inhibitor concentration exceeds 10 times that of chloride. But, without the inhibitor the oxidant concentration must be 60 times higher than that of chloride to have a similar result. By drawing a horizontal line, and starting from the right side of the figures, one notes that the substitution of the inhibitor by oxidant reduces the performance. Whereas, from the left side, one realizes that the substitution of oxidant by the inhibitor improves initially the performance before reducing it when the concentration ratio [inhibitor]/[iodate] increases. The performance improvement with a weak [inhibitor]/[iodate] ratio is particularly interesting because it affects weakly the pH. It is an interesting option especially for cooling circuits where it is important to avoid high pH values and the scale deposition in heat exchangers.

On the basis of our experimental data, we notice that the anions concentration influence on the pitting potential is rather regular. It would be interesting to have a model which enables to find the pitting potential from experimental data and to predict this potential from supposed data. For similar studies, the neural networks were applied with a certain degree of satisfaction (Parthiban, 2005). The main advantage of the neural networks rises from their capacity to model phenomena without having knowledge of the mechanisms involved. This corresponds to our case, where the pitting potential cannot be directly expressed by the pH, the conductivity and the anions concentrations. Nevertheless the accurateness of neural networks modelling requires some considerations: a significant number of experiments; a large field of experimental parameters variation because extrapolated situations are not always modelled successfully; and finally an appropriate parameter initialisation to the network input.

The two inhibitors were studied separately. In each case, 140 experiments were considered. An output database was represented by a vector of dimension (1,140) corresponding to the pitting potentials. An input database was modelled by a rectangular matrix of dimension (8,140). The 8 input parameters are related to the experimental parameters combined between them according to their interrelation and also to facilitate the model convergence. Each experiment is thus described by: the conductivity; the pH; the logarithm of the chloride concentration; the concentration ratio [inhibitor]/[chloride]; the concentration ratio [iodate]/[chloride]; the difference between the inhibitor and the oxidant concentrations [inhibitor]-[iodate] which takes into account the antagonism effect between anions concentrations at the material surface; a Boolean value ([iodate]=0) which allows to distinguish the experiments where the oxidant was used. Finally, a very important parameter was introduced to discern the cases where no pits occurred. This Boolean parameter takes value 1 in the absence of pits and 0 otherwise. The databases were then divided in two parts: a training base of 93 experiments and a test base of 47 experiments. The network architecture optimization consists on the determination of the hidden neurons number, the hidden and the output layers activation functions and the training algorithm. By taking the case of molybdate, optimized results are presented in table 1. The representation of the network is given by figure 6.

Table 1. The Neural Network Structure

Input layer	Hidden layer		Output layer		
neurons number	neurons number	Activation function	Neurons number	activation function	training algorithm
8	14	Hyperbolic tangent`	1	Hyperbolic tangent	Levenberg –Marquard t

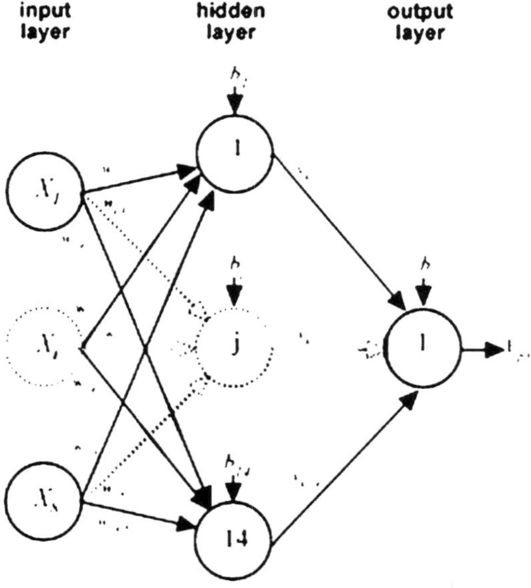

Fig. the neural network representation

In order to facilitate the learning process, the program normalizes all the variables to a zero mean, unitary variance. The symbol *i* refers to one of the 8 input parameters; *j* refers to one of the 14 neurons in the hidden layer; *k* is equal to 1 and corresponds to the unique neuron in the output layer.

The jth neuron of the hidden layer produces a value:

$$V_j = \sum_{i=1}^{8} w_{ji} X_i + b_j$$

The network output value is given by $g(y) = \tanh(Y)$ with

$$Y = \sum_{j=1}^{14} v_{1j} \tanh(V_j) + b$$

The weights w_{ji}, v_{ji} and the biases b_j, b are fixed after the training phase.

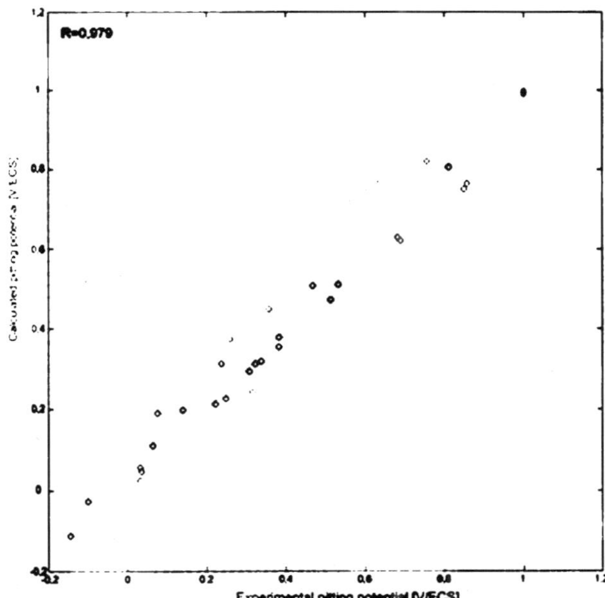

Fig. Calculated pitting potential values vs. those experimentally observed

Figure 7 shows the pitting potentials calculated by the neural network versus the experimental pitting potentials. The regression factor is 0,979. In general, the experimental points follow the diagonal. A perfect modelling would have given a total distribution of these points on the diagonal and a regression factor equal to 1. It is possible to improve the regression factor by increasing the hidden layers number or the neurons number by layer. On the other hand, more experiments or an adequate adjustment of the parameters in the input table would give a better convergence. The important fact in this figure is the disparity noted beyond 0,4 V/ECS. In this potential area, the current evolution follows different forms. In some experiments the pitting is followed by a fast current increase whereas in other experiments the current slope is very low. This situation creates confusion in the automatic pitting potential reading. Thus, through the use of neural networks, it appears clearly that the pitting potential, defined by the potential from which the current exceeds a fixed threshold, is certainly not the best performance indicator.

Discussion

The absence of inhibitors or passivating agents and around a neutral pH, chloride causes localised attacks at the same potential regardless of its concentration. By localizing this potential and the pH in the Pourbaix diagram, this potential should correspond to the oxidation of iron to Fe^{2+}. Therefore, without a passive film, the potential region of metastable pits existence is reduced. Consequently, it is the iron thermodynamic properties which would control pits initiation and not the chloride concentration which has just an influence on pits evolution. That means that a weak chlorine concentration can generate localised attack. In an open cooling circuit exposed to the air, and even if the chlorine concentration is low, it is possible that when a pit is initiated, it can evolve quickly under the influence of various thermo-hydraulic factors. In nuclear facilities subject to strict cooling procedures, the risk of localised corrosion must be maintained as low as possible. Also the use of localised corrosion inhibitors in the coolant is of the highest importance, even if the pitting agents concentration is considered weak (Boucherit, 2006).

Molybdate and tungstate are efficient inhibitors. Their efficiency is comparable. The two inhibitors increase incontestably the pH. It was indeed proven that at pH 6 the MoO42- ion is unstable (Vukasovich, 1986). It converts into paramolybdate ion according to the reaction:

$$7MoO_4^{2-} + 8H^+ \rightarrow Mo7O_{24}^{6-} + 4H_2O$$

When the pH is lower than 4,5 the paramolybdate ion changes to give:

$$Mo_7O_{24}^{6-} + MoO_4^{2-} + 4H^+ \rightarrow Mo_8O_{26}^{4-} + 2H_2O$$

Molybdic acid appears in more acidic medium. All these reactions consume protons and thus tend to increase the pH. Inside a pit, it is known that the material hydrolysis reduces the local pH. The molybdate can then intervene, according to preceding reactions, to increase the pH and to form, in parallel, salts such as $Na_6(Mo_7O_{24}) \times H_2O$ which can precipitate. The iron ions, which dissolve inside the pits, can even contribute to the formation of hetero-poly-complex such as $[FeMo_6O_{24}^{9-}]$ and thus avoid their hydrolysis which could reduce again the pH. We also noticed, that inhibition becomes complete when the concentration ratio [inhibitor]/[chloride] reaches the value 10. Under these conditions the molybdate concentration, at the surface, becomes so important compared to that of chloride. This could lead to a build-up of a protective layer of Fe-Mo (VI) permeable to cations but not to anions. Tungstate should have a similar behaviour to that of molybdate.

Oppositely to molybdate and tungstate, iodate does not affect the pH but influences clearly the pitting potential. Anodic pitting potential shift but also a passivation effect as the iodate concentration increases. The role of iodate would be to oxidize Fe^{2+} ions resulting from the material dissolution in Fe3+. Ferric ions can then form a hydroxide which precipitates to passivate the steel surface and then to delay the pits evolution. The use of iodate is beneficial when we want to limit the pH increase and consequently the scale deposition. Figure 5 shows that it is possible to have a good performance in a solution containing a weak [inhibitor]/[iodate] ratio. However, the addition of iodate is not always advantageous. An inhibitor concentration equal to 0,01 mole/litre, the progressive iodate concentration increase does not improve the performance. This was observed with tungstate and molybdate. This performance reduction is due mainly to the obstacles created by the anions following their migration towards the surface of the material under anodic polarization. For a constant global inhibitor concentration, successive oxidant additions gradually reduce the inhibitor concentration close to the surface. Inhibition by polyanions decreases and the Fe^{2+} ions oxidation by iodate is not sufficient to passivate the entire surface. Under these conditions, the iodate concentration increase is vain. Even if the separate effects of the inhibitor and the oxidant are beneficial, their combination can be useless or even risky if the concentration ratio is not adequate.

The availability of a great number of experiments has made it possible to apply neural networks to predict the pitting potential. The purpose was not the optimization of the model by reducing the difference between computed pitting potential values and those obtained by experiments. The interest of these results is that they underline the inaccuracy to consider the pitting potential as a performance indicator, as it is usually defined. It remains however possible to improve the model by reconsidering specifically the input database.

7

Internal Corrosion in Pipes

INTRODUCTION

WHAT IS CORROSION AND WHY DOES IT OCCUR?

Corrosion is the deterioration of metal that results from a reaction with the environment which changes the iron contained in pipe to iron oxide (rust). For example, if your car develops a rust spot, that is corrosion of metal. The same process can occur in various forms on pipelines. As is the case with your car, there are effective methods for preventing and arresting corrosion damage to pipelines.

Internal corrosion occurs due to chemical attack on the interior surface of the steel pipe from the commodities being transported within the pipeline. In some cases, the corrosive liquids may be contaminants such as water or other chemicals entrained or suspended within the commodity being transported. Typically, either the commodity's quality is controlled, internal coatings are applied, or corrosion inhibitors utilized to prevent internal corrosion. When one or more of these protective measures break down, internal corrosion can occur.

What are the Risks from Internal Corrosion?

Internal corrosion can result in the gradual reduction of the wall thickness of the pipe and a resulting loss of pipe strength. It can occur relatively evenly over an area of the pipe surface (sometimes referred to as "General Corrosion") or in isolated spots on the pipe. This loss of pipe strength could result in leakage or rupture of the pipeline due to internal pressure stresses unless the corrosion is repaired, the affected pipeline section is replaced, or the operating pressure of the pipeline is reduced. Internal pipeline corrosion creates weaknesses at points in the pipe, which in turn makes the pipe more susceptible to third party damage, overpressure events, etc. (i.e., corrosion doesn't necessarily need to cause the leak or rupture itself to increase risk).

Pipeline failurer rates from Corrosion

While data specific to internal corrosion alone is not available, corrosion in all its various forms is one of the most prevalent causes of pipeline spills or failures. For the period 2002 through 2003, incidents attributable to corrosion have represented 25 per cent of the incidents reported to OPS for both Natural Gas Transmission Pipelines and Hazardous Liquid Transmission Pipelines. Over this same period, approximately 1 percent of the incidents reported to OPS for Gas Distribution Pipelines were due to corrosion.

What is Being Done to Prevent/Mitigate Internal Corrosion?

- Modern manufacturing processes for steel pipe and coatings are subject to rigorous fabrication and installation standards and quality control to reduce the occurrence of defects that can lead to corrosion-induced failures.
- Operators attempt to control the moisture and chemical content of the products transported through their pipelines, where possible, to reduce the occurrence of internal corrosion.
- Operators routinely run devices called "cleaning pigs" through their lines to remove accumulations of materials that can lead to internal corrosion.
- Operators introduce corrosion inhibitors into the pipeline to control internal corrosion.
- The Office of Pipeline Safety has implemented new Pipeline Integrity Management ("IM") regulations that require all pipeline operators to inspect and assess all of their pipelines that could affect areas of high consequence such as populated areas or environmentally sensitive areas. The operators are required to inspect and assess their pipelines for integrity issues, such as corrosion, and repair or replace affected pipe.
- By implementing the requirements of the regulations and through responsible maintenance programs, pipeline operators continuously inspect their pipelines for internal corrosion damage.

Internal Corrosion: What More can be Done?

- *Public:* Be aware of pipelines located near you. Be observant for signs of pipeline damage, leakage, or security concerns. Report any concerns you have regarding pipeline safety to the pipeline operator immediately. Always respect the pipeline right-of-way. Do not dig or build on a pipeline right-of-way without first contacting the pipeline operator or your state one-call center.
- *Industry:* Pipeline operators and industry stakeholders can continue to develop and implement improved corrosion detection and prevention technologies. Operators must continue to implement

corrosion protection effectively and strengthen pipeline integrity management programs. Operators must mitigate the effects of corrosion when it is detected.

- *Regulators:* OPS and state regulators must continue to inspect pipeline operators to ensure they effectively implement required integrity management and corrosion control programs to ensure that risks to pipelines are identified and mitigated at the earliest possible time. Better coordination is needed between local permitting agencies and pipeline operators to facilitate expeditious granting of permits when public safety is potentially threatened.

INTERNAL CORROSION IN GAS PIPELINES

Corrosion on the internal wall of a natural gas pipeline can occur when the pipe wall is exposed to water and contaminants in the gas, such as O_2, H2S, CO_2, or chlorides. The nature and extent of the corrosion damage that may occur are functions of the concentration and particular combinations of these various corrosive constituents within the pipe, as well as of the operating conditions of the pipeline. For example, gas velocity and temperature in the pipeline play a significant role in determining if and where corrosion damage may occur. In other words, a particular gas composition may cause corrosion under some operating conditions but not others. Therefore, it would be difficult to develop a precise definition of the term "corrosive gas" that would be universally applicable under all operating conditions.

Corrosion may also be caused or facilitated by the activity of microorganisms living on the pipe wall. Referred to as microbiologically influenced corrosion, or MIC, this type of corrosion can occur when microbes and nutrients are available and where water, corrosion products, deposits, etc., present on the pipe wall provide sites favorable for the colonization of microbes. Microbial activity, in turn, may create concentration cells or produce organic acids or acid-producing gases, making the environment aggressive for carbon steel. The microbes can also metabolize sulfur or sulfur compounds to produce products that are corrosive to steel or that otherwise accelerate the attack on steel.

Internal corrosion in a gas pipeline may be detected by any of several methods, including visual examination of the inside of a pipeline when it is opened, external measurement of the pipe wall thickness with instruments, evaluation of corrosion coupons or probes placed inside the pipeline, or inspection of the pipe with an in-line inspection tool to identify areas of pitting or metal loss. Internal corrosion may be kept under control by establishing appropriate pipeline operating conditions and by using corrosion-mitigation techniques. One method for reducing the potential for internal corrosion to occur is to control the quality of gas entering the pipeline. Also, by periodically sampling and analyzing the gas, liquids, and solids removed from the pipeline to detect the presence and concentration of any corrosive contaminants, including bacteria, as well as to detect evidence of corrosion products, a pipeline operator can determine if detrimental corrosion may be occurring, identify the cause(s) of the corrosion, and develop corrosion control measures.

STATEMENT OF THE PROBLEM:

Most of the pipes that are made of metals are subject to corrosion invariably. A very important problem that is faced by many industries, especially in chemical engineering plants, is measurement of remaining wall thickness in pipes that may be corroded on the inside surface. This is the thickness of the metal left by heavy corrosion, often in the form of pitting. Such corrosion is often not detectable by visual inspection without cutting or disassembling the pipe. If undetected over a period of time, corrosion will weaken walls and possibly lead to dangerous structural failures. This calls for periodic inspection of such structures using a reliable NDT technique. Ultrasonic testing is a widely accepted nondestructive method for performing such inspection. A highly versatile technique called the Phased array ultrasonic technique offers a lot of flexibility in such applications involving scanning of complex geometries apart from reducing the inspection time.

Corrosion Pits

Pitting corrosion is a localized form of corrosion by which cavities or "holes" are produced in the material. Pitting corrosion occurs when discrete areas of a material undergo rapid attack while the vast majority of the surface remains virtually unaffected. This is in sharp contrast to uniform corrosion in which all parts of the exposed surface recede at approximately the same rate. Pitting is considered to be more dangerous than uniform corrosion damage because it is more difficult to detect, predict and design against. Pitting corrosion, which, for example, is almost a common denominator of all types of localized corrosion attack. Apart from the localized loss of thickness, corrosion pits can also be harmful by acting as stress risers. Pits can be either hemispherical or cup-shaped. Fatigue and stress corrosion cracking may initiate at the base of corrosion pits. Pitting is initiated by localized chemical or mechanical damage to the protective coating /oxide film or due to the presence of non-uniformities in the metal structure of the component, e.g. nonmetallic inclusions.

PITTING CORROSION

Pitting corrosion is a form of localized corrosion in which metal loss occurs in the form of holes with cross section small relative to the overall exposed surface. Most of the surface often suffers little or no metal loss. The penetration can be so great that the wall can be completely perforated resulting in leakage. Or, the penetration can stop at a certain depth or stop and then restart. For a component under tensile stress, pits can be initiation sites for cracks, which can then grow at a rapid rate, eventually ending in failure or breaking of the part. Pitting corrosion does not need a well-defined surface heterogeneity to initiate. The sites of pitting, though, usually have microscopic (or smaller) heterogeneities in the passive layer on the surface (e.g. sulfide inclusions in stainless steels). Pits come in a number of geometries. An example of more hemispherical pit is shown in this picture.

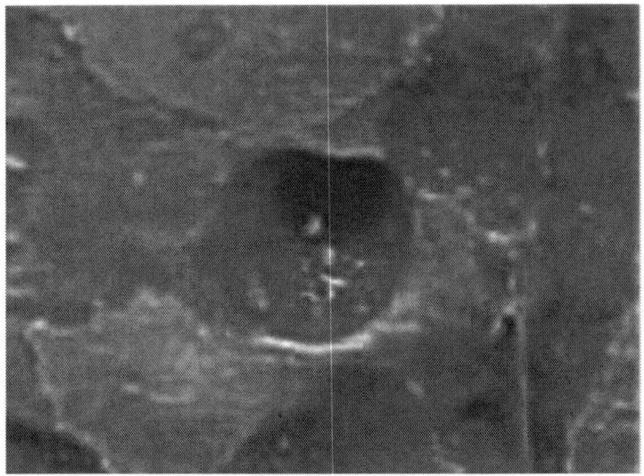

Additional examples are shown in these figures.

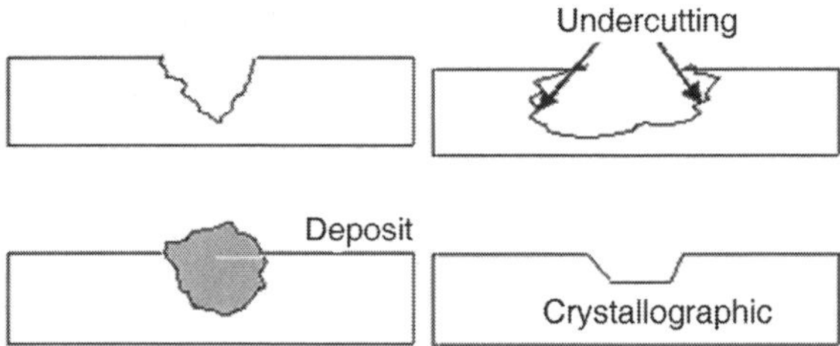

The first figure above shows that not all pits are hemispherical. Often pitting starts and stops, restarts, stops again, etc. Pits can undercut the surface so that the cross section observed at the surface can be far less than the cross section of the entire pit. Sometimes individual pits will be filled with corrosion products. One possible cause is the current density emanating from the pit base sometimes reaching values as high as 1 amp/cm2. The metal salt that enters the immediate pit environment can reach saturation quickly. Further hydrolysis reactions can result in a fairly insoluble but fairly porous corrosion product. This product can depress the pit propagation rate but the metal that is lost remains lost.

General Mechanism Information

One distinct mechanism cannot be invoked to describe pitting on all alloys and in all environments. In fact, disagreement still exists as to the exact mechanism that causes pits to initiate and propagate. But, certain characteristics are common to most types of pitting.

- Most common pitting is associated with halide and halide containing ions such as chloride, bromide, and hypochlorites.

- A large cathode vs. small anode area relationship tends to exist. A large portion of the surface that is not attacked can act as the cathode and only a small region that is attacked can act as the anode.
- Ionic concentrations in the pit and in the bulk fluid are different. Ionic concentrations are much greater within the pit region.
- Hydrolysis reactions involving the metal-containing cations within the pit cause the acidity to increase, i.e. the pH to decrease significantly.
- Sometimes the hydrolysis reaction products can create an autocatalytic effect in which their presence accelerates pit propagation.
- Initiation can occur at discontinuities in the either the passive layer in the surface of the alloy or between the base metal and inclusions.
- Surfaces exposed to stagnant conditions (absence or diminished fluid motion) are often observed to pit more easily than the same surfaces exposed to fluid motion.

Pitting of Iron-Based and Nickel-Based Alloys

Many iron-based and nickel-based alloys rely on a fairly thin metal oxide surface region to impart corrosion resistance to the bulk material. This region or layer is often called a "passive layer" or "passive film". Alloys are not homogeneous. The surface region is not homogeneous. Commercial alloys contain numerous inclusions, second phases, and regions of composition-based heterogeneities. These regions are believed to provide initiation points for pitting in alloys. In addition, pitting can occur in homogeneous alloys depending on the presence of certain species in the environment, e.g. chloride ions. The figure below depicts a propagating pit in an iron or nickel based alloy containing chromium in a chloride containing environment and a pH at which surface passivity is assured.

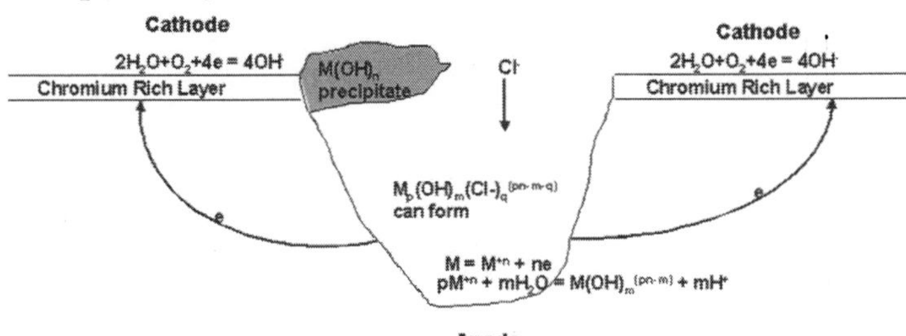

Once initiated, pits can propagate deeper into the alloy. The mechanism has the following characteristics:

- Oxygen reduction proceeds at the surface outside of the pit. The surface area for reduction is far greater than the pit surface area.
- The alloy consituents dissolve within the pit via oxidation. Chromium, nickel, and iron can all dissolve depending on the pH within the pit.

- The metal ions formed react with water (hydrolyze) to form hydroxides and to liberate hydrogen ions. This liberation of hydrogen ions decreases the pH which can fall to values as low as 2 even if the outside environment is at a pH of 7 to 9.
- The pH is maintained at a a low value at the bottom of the pit further accelerating dissolution.
- The counter ion, chloride, increases in concentration within the pit to maintain charge neutrality.
- A potential difference exists between the base of the pit and outside metal. This potential difference is maintained by the various electrochemical reactions.
- A porous metal hydroxide can form at the pit mouth. This solid, when it forms, allows electrolyte contact between the pit and the environment.
- Oxygen reduction at the alloy surface completes the electrochemical cell.
- The counter ion concentration (e.g. chloride ion) can actually increase the activity of the hydrogen ion so that the measured pH can reach zero.
- Sometimes, the metal ion concentration at the pit base can become so large that a metal salt layer precipitates at the bottom of the pit.

Pitting of Aluminum Alloys

A generalized picture of the propagation of a pit in aluminum in aerated solutions containing chloride ions is shown in the figure below. The pH of the environment is assumed to be in the range of about 5 to 9.

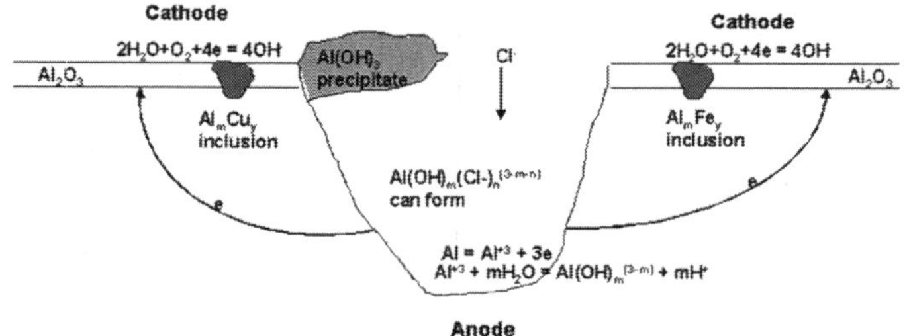

Though the exact mechanism of aluminum pit initiation depends on the alloy type, some general characteristics of the process can be summarized as follows.

- Microflaws exist in the aluminum oxide layers that provide passivity to the alloy. The surface is very likely a hydroxide form of aluminum.
- Chloride adsorption in the microflaws might aid in pit initiation.
- Alloys of aluminum contain intermetallic compounds can form dissimilar metal junctions at the surface. For example, 2000 series aluminum contain copper aluminide inclusions and 3000 series

aluminum contain iron-aluminide inclusions. The copper aluminide inclusion may decompose to redeposit copper on the surface.

- A potential difference is created between the "unflawed" aluminum surface and that created by pre-existing flaws, chloride adsorption in flaws, or the intermetallic inclusions.
- As a result of the potential difference, oxygen reduction on the "unflawed" surface or on the intermetallic compound drives the anodic dissolution in the region of the flaw or on the aluminum surface adjacent to the intermetallic compound.

The result is a micropit. Some of the micropits repassivate. Some propagate to larger pits. The above figure shows the process as the pit is propagating. The propagation process has the following characteristics.

- Aluminum dissolution proceeds within the pit especially at the bottom or base of the pit.
- Aluminum ions react with water (hydrolysis) to form aluminum hydroxide cations and hydrogen ions.
- The formation of hydrogen ions decreases the pH relative to the environment outside of the pit further accelerating the dissolution process.
- In a chloride environment, chloride ions increase in concentration to maintain charge neutrality. These ions can react with the aluminum hydroxide ions to form chloride containing adducts.
- Aluminum hydroxide can precipitate at the pit-environment boundary. Sometimes this hydroxide can cover the pit surface but maintain electrolyte contact between the pit and environment.
- Hydrogen ions can be reduced to form hydrogen gas bubbles.
- Oxygen reduction continues on the surface as the cathodic driver. In addition, copper from the copper-aluminide intermetallics at the surface can be reduced further driving the process.

Electrochemical Characteristics

In electrochemical terms, the "critical pitting potential" (sometimes called the "rupture potential") is an electrochemical characteristic that alloys relying on passivity share if they undergo pitting corrosion. This potential is the most negative potential above which pits can initiate and propagate. Assuming measurement artifacts are absent, the value of this potential provides an upper bound value. Control at higher (more anodic or more noble) potentials would destroy passivity and promote pitting. That is, if the corrosion potential is greater than (anodic with respect to) the pitting potential, pitting will initiate. These potentials are often estimated from Cyclic Potentiodynamic Polarization Scans. The cyclic potentiodynamic polarization scan in the figure below shows the relative relationships among potentials. In this particular case, the pitting potential lies about 200 mV anodic with respect to the corrosion potential. Once the potential is forced above that value, the increase in current signifies the rapid breakdown in passivity and the generation of pits. In this case, repassivation of the surface does not occur until the potential decreases to a point below or cathodic with respect to the original corrosion potential. Pits that were initiated when the potential rose above the pitting potential continued to grow until the potential fell cathodic with respect to the corrosion potential.

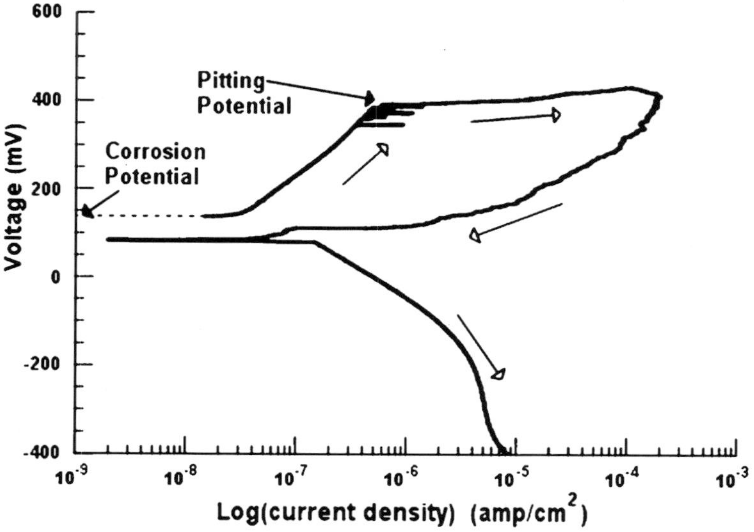

Stress Corrosion Cracking

Stress Corrosion Cracking is a form of Environmental Assisted Cracking (EAC). The term EAC is used to describe all types of cracking in pipeline that is influenced by the environment and stress. In a pipeline environment when water comes in contact with steel there is potential for the minerals and gases to create an initial corrosion site that is acted on by stresses that result in crack growth. Contributing factors to crack growth are residual stresses, temperature, load stress, bending, and local stresses. If there is no stress then crack growth will not occur and the result will be general wall thinning or pitting through corrosion.

THEORY OF OPERATION

Most of the corrosion detection is performed using guided wave ultrasonics which can be used for long range detection of presence of corrosion. This technique is usually not very effective at detecting pitting and measuring true thickness of corroded pipe and pipes with especially with high surface roughness. Thus, it is not recommended for corrosion gauging as the extent of corrosion in the structure cannot be reliably quantified by this method. Hence, the common practice is to detect the corroded region using the guided wave technique and hence find the extent of corrosion using bulk wave techniques. The phased array technique can be effectively used to qualitatively image the corrosion damage and quantify its extent.

Introduction to Phased Array:

The Phased Array concept is based on the use of transducers made up of individual elements that can each be independently driven. To generate a beam, the various probe elements are pulsed at slightly different times. By precisely controlling the delays between the probe elements, beams of various angles, focal distance, and focal spot size can be produced. The illustration shows how a beam can be focused at an angle and a given distance by firing the left elements slightly ahead of the corresponding elements on the right. It is possible to change the angle, focal distance, or focal spot size, simply by changing the timing to the various elements. Beam focusing and simultaneous beam steering-focusing is shown in figures 1(a) and (b) respectively.

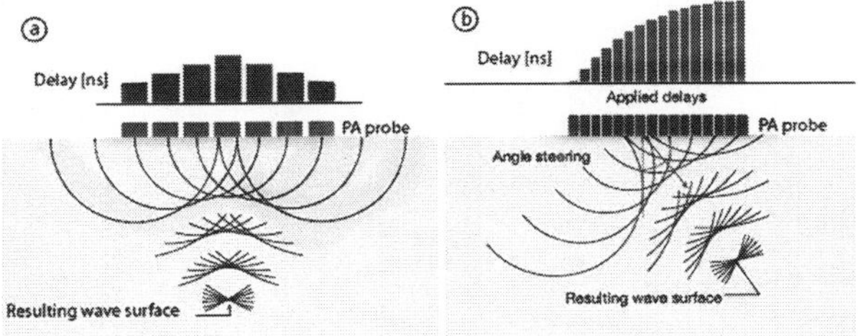

Fig. (a) Beam focusing in a phased array system (b) steered-focused beam in a phased array system

KEEPING PIPELINES SAFE FROM INTERNAL CORROSION

Internal corrosion is a leading cause of pipeline failure -- and one of the most difficult to detect. Pipeline accidents have caused catastrophic injury and destruction, resulting in imposition of integrity management requirements on pipeline operators by the U.S. Dept of Transportation. To aid operator compliance, an efficient, reliable means of monitoring internal corrosion before it causes problems has been developed.

The Problem

Keeping pipelines safe from internal corrosion can be a challenging and expensive business; industry estimates run to several billion dollars annually in the U.S. alone. Panhandle Energy, a subsidiary of Southern Union, operates approximately 18,000 miles of interstate pipelines.

> "In 2006, Panhandle will perform over 300 anomaly investigations," says David McQuilling, a Panhandle senior engineer. "On average, we budget $50,000 per investigation in rural areas, but if you"re in an urban area, $50,000 may not even cover permitting; I"ve heard of operators spending over $250,000 on a single dig."

Internal corrosion can occur when impurities are present within the natural gas, crude, and refined products being transported. Bill Shaw is an rngineering professor at the University of Calgary and director of the Pipeline

Engineering Centre, which studies corrosion and monitoring. "Moisture is the big thing; if you had no moisture, for the most part, you"d be fine," he says. "It mainly mixes with salts, like chlorine, and sulphur compounds."

Corrosion can typically reduce a pipeline wall thickness at a rate of 2-3 mils/year, but it can happen much more quickly in the upstream area of operations. "Gathering systems are really bad; you have large quantities of hydrogen sulphide and brine," says Shaw. Water can chemically react to form very corrosive liquids that collect in low-lying spots. Corrosion rates in these areas may reach several hundred mils per year.

What the Industry Is Doing

To mitigate the potential for incidents related to internal corrosion, the pipeline industry works assiduously to reduce risk. The U.S. Dept of Transport"s Pipeline and Hazardous Materials Safety Administration (PHMSA) notes that 258 natural gas and liquid pipeline accidents (14 per cent of which were caused by internal corrosion) occurred in 2004.

> "It is important to place natural gas pipeline safety data in perspective," says Don Santa, president of the Interstate Natural Gas Association of America. "INGAA"s members operate a network of approximately 200,000 miles of transmission pipeline, and this total does not include all of the intrastate transmission pipelines and LDC-owned pipelines that also are subject to the Dept of Transportation"s integrity management rules. Pipelines are one of the safest modes of transportation."

But when they do fail, they tend to get noticed. Early in the morning of Saturday, August 19, 2000, a natural gas line exploded in southeastern New Mexico near the Pecos River. Eleven nearby campers were killed by the blast and fire. The ensuing fireball was large enough to be seen in Carlsbad, NM, 20 miles to the north. Federal investigators determined that the cause was severe internal corrosion; over 70 per cent of the wall had been eaten away near the rupture.

Pipeline Safety Improvement Act

Partly in response to the tragedy, the Bush Administration enacted the Pipeline Safety Improvement Act (PSIA) in 2002. In order to comply, pipeline operators rely on a range of survey methods supplied by third parties. Prominent companies supplying internal corrosion monitoring equipment include Rohrback Cosasco of the US, Corrocean in Norway, Cormon in the UK, and Caproco in Canada. The international market for corrosion monitoring equipment is in the neighbourhood of $70 million annually, and is growing rapidly due to the fast pace of oil and gas developments, expansion of pipeline networks around the world, and the increasing stringency of regulation.

Monitoring Methods: Intrusive

Generally, internal corrosion monitoring and detection is broken down into three techniques, intrusive, In-Line Inspection (ILI) and non-intrusive. "Intrusive methods include electrical resistance probes and coupons," says Shaw. The coupon is the original form of intrusive corrosion monitoring. It consists of a strip of metal about 3 in. long and 1/8 in. thick, made of material similar to the pipeline. It is weighed, then inserted into an access point and left for at least six months. The operator then removes the coupon and weighs it again to see what percentage is missing. The coupon is inexpensive and reliable, but several factors can outweigh the advantages. First and foremost is access above the area of concern.

> "Corrosion often occurs at low spots, under roads or rivers," says McQuilling. "It''s often impossible to put a coupon in the location where the internal corrosion is occurring. If access to the area is possible, burial depths must be considered, as coupons are commonly installed and retrieved at line pressure. It is impractical to install coupons at pipe depths greater than 10-15 ft below grade. I''ve had situations where pipe was removed and replaced due to internal corrosion but a coupon was not considered due to safety concerns about insertion and extraction of a 27 ft long coupon/ coupon staff assembly at 1,950 psig."

Second is the number of locations where coupons can potentially be installed. "With external pipeline corrosion protection, you''ve got a monitoring point every mile; that''s not a lot -- quite a few assumptions have to be made," says McQuilling. "With internal corrosion, you''ve got one every hundred miles if you''re lucky; it''s like getting eight pieces of a 1,000-piece jigsaw puzzle and asking what it looks like." Third, because coupons are most commonly left in the pipeline over a 6 month period, the operator doesn''t know if the corrosion occurred slowly over the time period, or quickly over a few days due to an unusual event.

Monitoring Methods: In-Line Inspection

In-Line Inspection (ILI) tools, or smart pigs, are intelligent sensing devices that are introduced into the line at a specialized entry point and most commonly conveyed by product flow along the length of the pipe.

> "Pigs have great sensitivity, but there''s a lot of piping that cannot be pigged easily," says McQuilling. "It was common to telescope vintage pipelines from 24-18 in. Also, you can have a vintage 36 in. pipeline with 24 in. valves; you saved a considerable amount of money during construction and the flow restrictions due to undersized valves are negligible, but a pig won''t go through -- it''s expensive to make some lines piggable."
>
> "The preferred methodology is inline inspection with a pig, but certain sections can''*t* use a pig due to age and configuration," agrees Santa. "You then have to dig up and visually inspect them, or hydrostatically test them, which from a practical standpoint is not all that convenient and potentially destructive."

External sensors help avoid such expensive intervention; various devices measure the physical properties of the pipeline through secondary techniques. "You can use acoustics to measure wall thickness or listen to the noise of corrosion," says Shaw. "Vacuum foil techniques measure hydrogen permeation."

Monitoring Methods: Non-Intrusive

The most common non-intrusive device is the ultrasonic monitor. To conduct a survey, the pipeline is dug up, then a portable device is held against the metal. Inside the device, voltage is applied across a piezoelectric crystal to generate an ultrasonic sound wave that propagates through the metal. The time it takes to travel through the metal and back to the transducer is directly proportional to its thickness. The devices are quick, easy to use and inexpensive, and operators do not have to shut off flow or risk breeching the pipeline in order to take a reading. On the other hand, they still have to expend thousands of dollars digging up the pipeline each time they run a test. Also, traditional ultrasonic monitors have a sensitivity range in the order of 5-10 mils accuracy, so if the corrosion is only 3 mils/y, it takes three years to start to see it in a statistically significant manner. Rohrback Cosasco Systems, based in California, is a pioneer in the measurement of corrosion by way of electrical resistance and ultrasound. The company has over 80 employees and 50 representative companies around the world (75 per cent of their business is in the international O and G sector), with annual sales of over $20 million. "We saw the potential for a high-speed, high sensitivity device that could be permanently installed, and began to focus on development in 2003," says CEO Brent Ford.

Another Alternative

The company devised the UltraCorr® wall thickness monitor, a combination of permanently mounted transducers and a portable datalogger.

<="" p="">

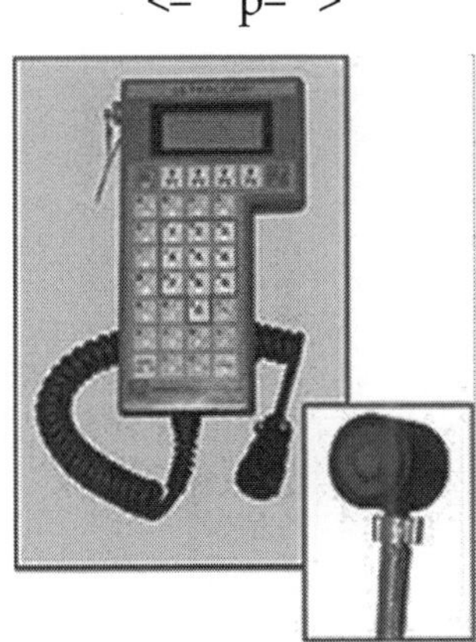

With a permanently installed system, internal corrosion surveys can be incorporated into normal maintenance practice; most pipeline companies have field technicians patrolling the right-of-way on a frequent basis to check cathodic protection systems.

> "The technician takes an UltraCorr Corrosion Data Logger, a handheld device, and loads the transducer information from the test post," says Ford. "It"s easy to use, and only takes seconds per reading." Readings stored on the portable datalogging unit can be uploaded to a PC, where proprietary software is used to organize, store, and graphically display data. Simple plots of thickness versus time are augmented with a cursor-driven corrosion rate calculator permitting detailed event analysis. "Operators know with confidence the condition of their pipe."

Costs for UltraCorr are reasonable. "Depending on the number of transducers required, our device might be $8-$12,000 per dig to install, but you only have to do it once," says Ford. "The handheld device is around $6,500; you only need one per field area." Right from the beginning, Panhandle Energy has been interested in the system. "It"s simple, clean and cheap; you don"t have to blow down the pipe, which really gets expensive, or do a hot tap," says McQuilling. "You just buff off a spot on the pipe and attach it with epoxy, then backfill and set up a post for the transducer wires, and you"re done."

<="" p="">

Since the majority of Panhandle"s system handles clean, dry gas, there is little corrosion along mainlines.

> "Panhandle has numerous locations where we believe the UltraCorr will be beneficial," says McQuilling. "Historically, most of our internal corrosion issues occur at our underground gas storage fields. Panhandle operates six gas storage fields which are depleted gas reservoirs used to store gas. We inject clean, dry, pipeline quality gas, but it often comes back up wet and potentially corrosive. A typical storage field has 30–40 miles of pipe and 70–80 wells, creating numerous opportunities for internal corrosion monitoring."

Although McQuilling is enthusiastic about the system, he is well aware that others may not be so. "Lots of readers will be skeptical; some operators put earlier versions of UltraCorr into operation and the epoxy failed within months," he notes.

> "We had an early prototype with reliability issues," acknowledges Ford. "We spent over a year and $1 million improving reliability and sensitivity before starting field tests. We are confident that the new system will meet expectations."

"They"ve put the latest epoxy through torture to improve it," notes McQuilling. "We"ve had the latest UltraCorr probes in service since October 2005, and so far, it"s worked and worked well. We"ve stuck with them because there"s great potential for a low-cost monitoring system that may eliminate the need for future excavations. With this system, when the Office of Pipeline Safety comes in after seven years and says "You"ve got to re-inspect that portion of pipeline," we hope to say "We"ve done real-time monitoring; here"s the data.""

CORROSION OF PIPES IN DISTRIBUTION SYSTEMS FACT SHEET

The concern that using fluorosilicate additives to fluoridate drinking water causes water system pipes to corrode is not supported by science. At the level recommended by the U.S. Public Health Service for fluoridation of public water supplies (0.7 to 1.2 mg/L, or parts per million), the fluoride ion has little influence on either corrosion or on the amounts of corroded metals released into the water. Fluorosilicates contribute to better water stability with less potential for corrosion, because silica stabilizes the pipe surface.

CAUSES OF CORROSION IN WATER SYSTEM PIPES

Pipes used to distribute drinking water are made of plastic, concrete, or metal (e.g., steel, galvanized steel, ductile iron, copper, or aluminum). Plastic and concrete pipes tend to be resistant to corrosion. Metal pipe corrosion is a continuous and variable process of ion release from the pipe into the water. Under certain environmental conditions, metal pipes can become corroded based on the properties of the pipe, the soil surrounding the pipe, the water properties, and stray electric currents. When metal pipe corrosion occurs, it is a result of the electrochemical electron exchange resulting from the differential galvanic properties between metals, the ionic influences of solutions, aquatic buffering, or the solution pH. For corrosion of metal water pipes to occur, an electrochemical cell must be present. An electrochemical cell can be thought of as a battery, with an electric current between a positive potential (anode) and a negative potential (cathode). The corrosive electrical potential is typically created by differences in the types of chemicals in soil or the surface of the metal pipe.

Galvanic Properties between Dissimilar Metals

All metals have slightly different properties, and galvanic differences are the tendency of one metal to release electrons to another metal. The galvanic series of metals is the hierarchy of which metals will release their electrons to other metals. Metals lower in the galvanic series more negatively charged will sacrifice their electrons to metals higher in the series. An example that many people are familiar with is zinc galvanizing of steel, where the zinc surface coating protects the steel from rusting. The galvanic interaction of different metals has a significant role in pipe corrosion, because many commercial metals are alloys of various metals. Therefore, the interior or exterior surfaces of the pipe can provide locations for an electrochemical cell which can start the process of pipe corrosion.

Influence of Ionic Impurities on Corrosion

Chemical additives are added to water during the water treatment process. More than 40 chemical additives can be used to treat drinking water. Many of these commonly used additives are acidic, such as ferric chloride and aluminum sulfate, which are added to remove turbidity and other particulate matter. Various chlorine disinfectants, also act as acids and have the potential to reduce pH, alkalinity, and buffer intensity. These acidic water treatment additives can interfere with corrosion protection. The amounts of each of these other additives used in water treatment typically are 5 to 10 times the amount of the fluoride additive for fluoridation of drinking water; therefore, their potential effect on the factors affecting water corrosivity is proportionately greater. The fluoride ion interacts weakly with common metals in plumbing materials and the American Water Works Association Research Foundation has reported that fluoride ions contribute to corrosion to the same extent as at the same concentration chloride and sulfate ions. Most of the fluoride interaction will be to form a precipitate that will be incorporated into pipe scale (the deposits on the inside of pipes that are mostly calcium) or removed by routine system flushing. Therefore, the corrosive influence of fluoride in drinking water is not significant compared with other ionic influences.

Lead and Copper in Drinking Water

Lead and copper are rarely detected in most drinking water supplies. However, these metals are a concern to consumers. Because some household plumbing fixtures may contain lead or copper, corrosive waters may leach (pick up) lead and copper from household plumbing pipes after entering a home. This is a greater issue for older houses (i.e., houses built before 1981, if the plumbing system has not been replaced) than for newer houses. The most common reason for water utilities to add corrosion inhibitors is to avoid lead and copper corrosion with older homes, and the second most common reason is to minimize corrosion of pipes in the distribution system.

When waters are naturally corrosive, many substances have a tendency to dissolve in water. Because of this tendency, the U.S. Environmental Protection Agency (EPA) has issued a Lead and Copper Rule that requires all water systems to periodically monitor a set number of samples for lead and copper levels at different locations. This is based on population size and previous tests of lead and copper content. If a certain percentage of the samples exceeds the "action level," the utility system must take corrective actions to control the potential for corrosion in the water system. This often involves the addition of corrosion inhibitors.

Water Properties Influencing Corrosion

Many water quality factors affect corrosion of pipes used in water distribution, including the chemistry and characteristics of the water (e.g., pH, alkalinity, biology), salts and chemicals that are dissolved in the water, and the physical properties of the water (e.g., temperature, gases, solid particles). The tendency of water to be corrosive is controlled principally by monitoring or adjusting the pH, buffer intensity, alkalinity, and concentrations of calcium, magnesium, phosphates, and silicates in the water. Actions by a water system to address these factors can lead to reduced corrosion by reducing the potential for the metal surface to be under the influence of an electrochemical potential.

Waters differ in their resistance to changes in their chemistry. All waters contain divalent metals such as calcium and magnesium that cause water to have properties characterized as hardness and softness. If a water is "hard," it is less likely to "leach" metals from plumbing pipes but often leaves a deposit on the inside of the pipe, while if a water is "soft" it has less of a tendency to leave deposits on the inside of plumbing pipes. If a water is soft, then it has low hardness. Some people in communities with hard water will use water softeners. Water systems adjust the hardness and softness of water because of these tendencies and also for taste considerations.

Alkalinity is a characteristic of water related to hardness. Waters with low hardness, or alkalinity (less than 50 mg/L as calcium carbonate), are more susceptible to the factors affecting corrosion; such systems will typically use additives that can prevent corrosion (corrosion inhibitors) to comply with federal and state regulations.

Corrosion Inhibitors

Chemical additives used for corrosion control include phosphates, silicates, and those affecting the carbonate system equilibrium (amount of carbonate in the system), such as calcium hydroxide, sodium hydroxide, sodium bicarbonate, and sodium carbonate. Corrosion inhibitors are commonly used to address the corrosion influence of acidic water treatment additives. The most common forms of fluoride for approximately 92 per cent of the drinking water that is fluoridated are fluorosilicates, as either fluorosilicic acid or sodium fluorosilicate. Using fluorosilicates to fluoridate drinking water adds silica, a corrosion inhibitor, to the water and increases the silicates available for stabilizing the pipe surface, which contributes to reduced corrosion.

Many Substances with Fluoride have Low Solubility in Water

The water fluoridation additives that are used to increase the fluoride content of water are carefully chosen for their favorable solubility in water. Many divalent metals or heavy metal substances that have an ionic association with fluoride have poor solubility. These include calcium and magnesium cations, as well as many of the heavy metal ions such as nickel and lead. As the pH of the water increases to basic levels, these compounds will precipitate out of the water and be incorporated into a calcium-carbonate scale that will form on the pipe surface.

Soft Waters with Low Buffering

A special case exists when the water source is a high-purity groundwater with little natural buffering. Buffering is the ability of a water to resist pH changes when acids or bases are added to it. Low natural buffering is not typical for community water systems. In such cases, adding acidic chemical additives, such as fluorosilicic acid or sodium fluorosilicate, could potentially result in a slight increase in corrosion because of the influence of the acid additive.

However, the acidity added by such fluoride additives would be less than the acidity introduced from chlorine disinfectants. Any change in water properties is typically addressed by adding a corrosion inhibitor or adjusting the pH. This would be a standard water system practice, since water systems regularly monitor for compliance with the U.S. E.P.A. Lead and Copper Rule and take corrective action, particularly if the regulatory action levels for lead and copper are being approached.

BASIC COMPONENTS OF A PHASED ARRAY SYSTEM

The main components required for a basic scanning system with phased array instruments are presented in Figure 2(a) below.

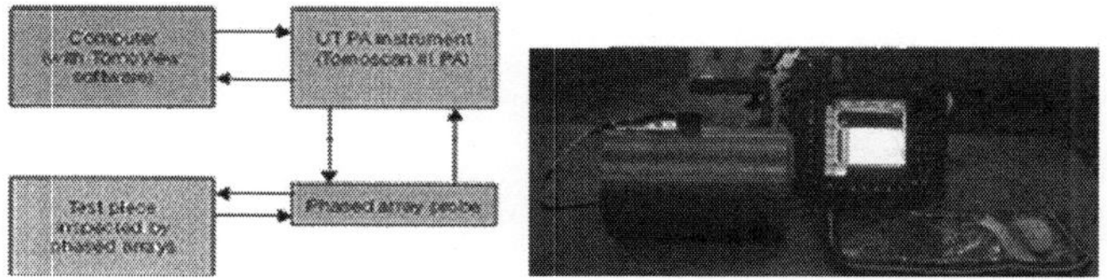

Amplitude Drop Method:

The amplitude drop technique is an easy and fast technique for sizing large and planar defects. It is based on the principle that when half of the ultrasonic energy is not reflected by a defect, the echo is 6 dB less than in the case when the entire beam is reflected. It is then assumed that when half of the beam is returned, the transducer centerline is directly over the edge of the defect. This method also suffers from some disadvantages. Since, the sizing of the defect is based on the amount of reflected ultrasonic energy, planar defects with straight edges only can be accurately sized and hence in most practical cases it only gives an approximation of the defect size. For sizing flat bottom and side drilled holes, usually 2 dB, 3 dB and 6 dB amplitude drop methods are used. Figure 2 (c) shows 1 dB, 2 dB, 3 dB and 6 dB amplitude drop methods to size a 2mm deep side drilled hole. The basic procedure involved in this method is first locating the flaw and maximizing the echo and noting the signal amplitude. Now the probe is translated along the scan axis till a specified amplitude (example 6 dB) drop is observed. The length of the drop along the scan axis is an indication of the size of the defect.

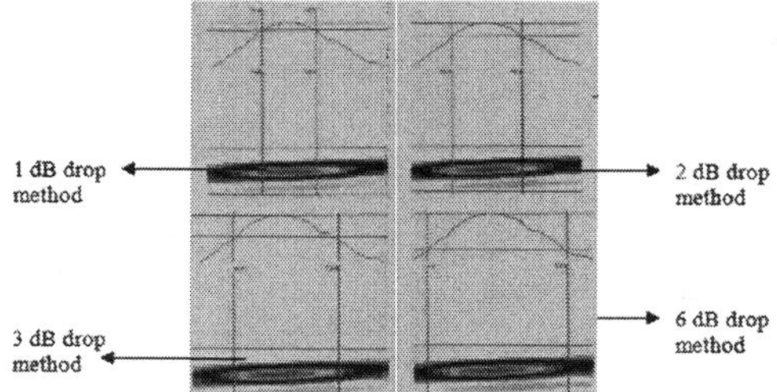

Fig, Various amplitude drop methods to size a 2 mm deep side drilled hole

ISSUES ASSOCIATED WITH CORROSION DETECTION BY ULTRASONICS:

Surface Condition:

The surface condition plays a very important role in the imaging of corrosion damage. This is because signal to noise ratio is directly affected by the quality of the surface. The smoother the surface, the better is the signal to noise ratio. A poor surface finish may interfere with the coupling of sound energy from the transducer into the test material and hence give misleading data which can be interpreted to be the presence of a defect.

location of corrosion pits:

The location of corrosion pits is difficult to control and a wide range of pit morphologies can result when components are corroded, even in a laboratory environment. Additionally, it is difficult to control the location of individual pits.

Specimen Preparation:

A 13 mm aluminium flat specimen was taken. The sides of the specimen were brought to a correct rectangular shape using a shaping machine. It was then given in an end mill cutter and faced to 11.0 mm. The specimen was then taken to a centerless rolling machine and was rolled to the requisite diameter. The specimen was then taken to an EDM and slots of various depths were cut in it, all of 10 mm length and 0.6 mm width. The specimen was then cleaned with kerosene to remove all adhering dirt and was tested using the phased array testing machine. In Al sample all the three notches are rectangular EDM notches of depths 3 mm, 6 mm, and 9 mm. The sample is shown in figure 3(a). The EDM notches on the Aluminium pipe sample was made on the inspection side to simulate same side stress corrosion cracking type defects. The specifications of the EDM notches are given in table 1.

A 11 mm thick 164.5 mm diameter mild steel pipe of 300 mm length was taken and the surface was turned to 10 mm thickness to obtain a uniform and a relatively good surface finish. The specimen was then cut into two halves longitudinally using a gas cutting machine. The edges of the two halves were ground to get a good surface finish. Circular notches of three different depths and sizes were made on the mild steel pipe specimen in a milling machine to simulate corrosion pits. The sample is shown in figure 3(b).The specifications of the circular notches are given in table 2. The notches in the pipe specimen were exposed to concentrated corrosive acids / salt solution for regulated periods of time (total exposure times were varied from 24 to 96 hours.) and the extent of the corrosion was imaged.

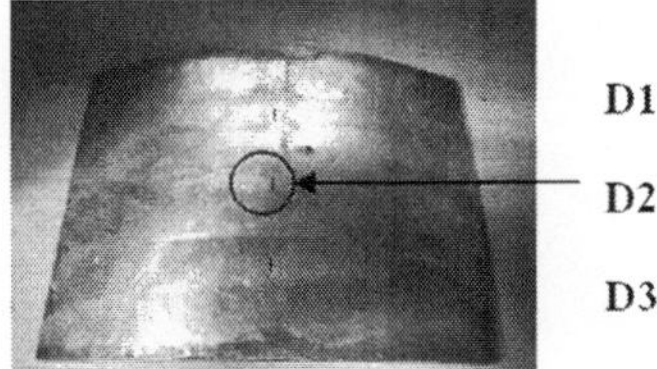

Fig. (a): 11 mm-thick Aluminium Pipe sample with three simulated defects

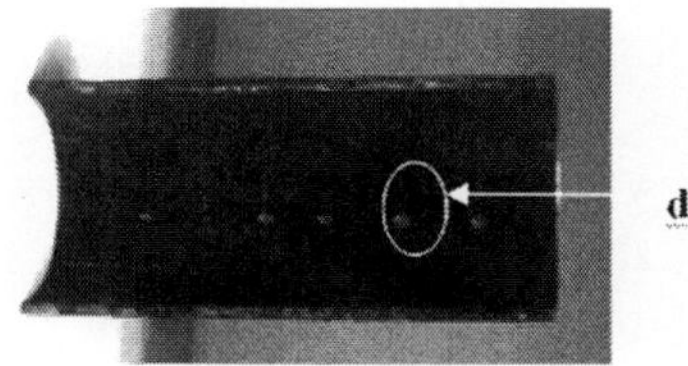

3(b): 10 mm-thick Mild Steel Pipe sample with three simulated defects

Defect Number	Type	Length(mm)	Width(mm)	Actual Depth (mm)
D1	Surface breaking [rectangular]	10	0.6	9
D2	Surface breaking [rectangular]	10	0.6	6
D3	Surface breaking [rectangular]	10	0.6	3

Table. The details of EDM Notches, size and Angles with Respect to Vertical of 11 mm thick Aluminium Pipe Sample

Defect .No.	Defect type	Defect diameter(mm)	Defect depth(mm)
d1	Flat bottom Hole	6	3
d2	Flat bottom Hole	6	6
d3	Flat bottom Hole	8	3
d4	Flat bottom Hole	8	6
d5	Flat bottom Hole	10	3
d6	Flat bottom Hole	10	6

Experiments:

The experimental setup used in this study consists of the R/D Tech Omniscan MX ultrasonic Phased Array system. The data presented in this paper were acquired using 5MHz, 64 elements (46 mm X 15 mm area) probe. This paper discusses the test results obtained for simulated defects such as corrosion pits treated with corrosive acids / salt solution and stress corrosion cracks in a 165 mm diameter 10mm thick mild steel pipe sample. The inspection was carried out by fixing the phased array probe on to an encoder and manually moving over the specimen. The defects were imaged by phasing the elements of the probe to generate a shear wave at an angle of 45 degrees in the linear scan mode. The simulated defects include three corrosion pits of different sizes and depths on the mild steel pipe sample and three EDM notches on a SS pipe to simulate stress corrosion cracks. The B-scan images obtained were then used to determine extent (area) and the depth to which metal loss had occurred.

Results and Discussions:

Experiments were carried out on EDM notches of three different depths to simulate same side stress corrosion cracking and the defects were imaged. The B-scan image of the 11mm thick pipe specimen with a defect is shown in figure 4. The reflection from the defect can be clearly seen in the B-scans. The defect depths were sized using 3 dB and 6 dB amplitude drop method. The defect images of three notches are shown in figure 4 and the sizes of the defects obtained using the two methods and the percentage errors associated with the same have been summarized in table 3 below. It was observed that 6 dB amplitude drop method is more reliable compared to the 3 dB amplitude drop method in sizing simulated same side stress corrosion cracks. Also, the error associated with sizing smaller corrosion cracks is relatively higher compared to sizing larger cracks with respect to amplitude drop method.

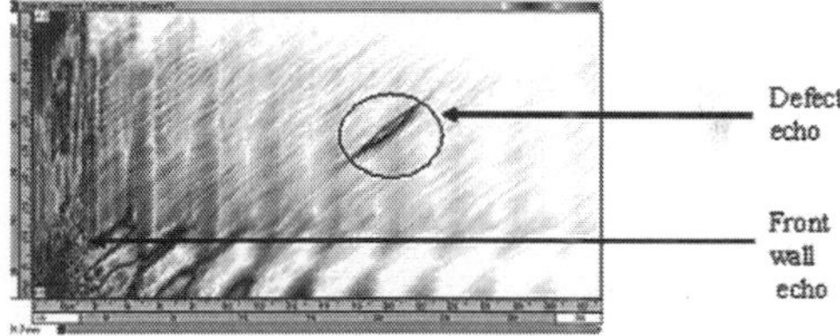

Fig. B-scan of a 3 mm deep EDM notch obtained using a 64 element phased array probe

Table. The Details of Circular Notches, of 10 mm Thick 164.5 mm Mild steel Pipe Sample

Defect Number	Actual Depth (mm)	Defect depth by 3dB amplitude drop method(mm)	Percentage error	Defect depth by 6dB amplitude drop method(mm)	Percentage error
D1	9	8.25	8.33	9.6	–6.67
D2	6	7.5	-25	7.2	–20.00
D3	3	4.25	-41.66	3.7	–23.33

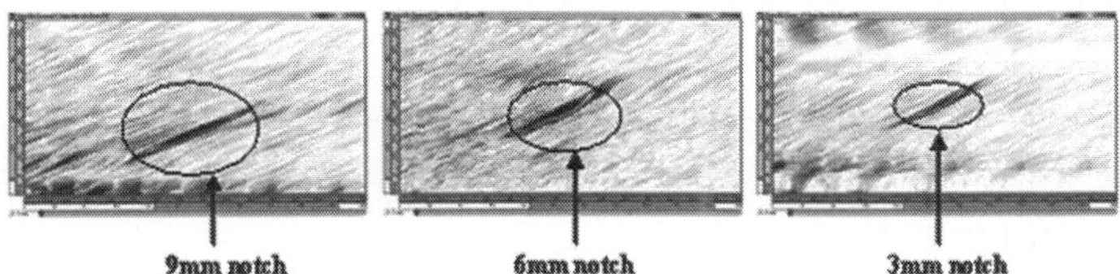

Fig. Defect images of simulated SCC defects on 10 mm thick Aluminium pipe

Table 3: Comparison of the notch depths obtained by time of flight method and amplitude drop method with the actual notch depth of 11 mm thick Aluminium pipe sample. Experiments were carried out on simulated corrosion pits of three different depths The B-scan image of the 10 mm thick mild steel pipe specimen with a defect is shown in figure 6. The reflection from the defect can be clearly seen in the B-scans.

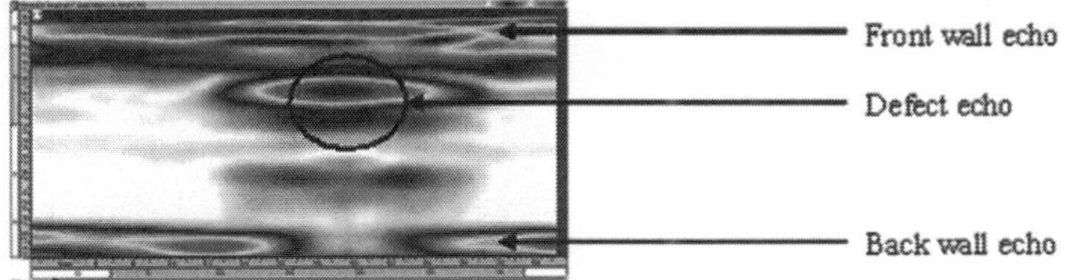

Fig. B-scan of a 6mm deep corrosion pit obtained using a 64 element phased array probe

The defect depths were sized using 2 dB, 3 dB and 6 dB amplitude drop method. The defect images of the six simulated corrosion pits are shown in figure 7 and the sizes of the defects obtained using the two methods have been summarized in table 4.

Table. Comparison of the Notch Depths Obtained by Amplitude Drop Method With the Actual Notch Depth of 10 Mm Thick Mild Steel Pipe Sample

Defect No.	Defect Depth (mm)	Defect depth by 2 dB amplitude drop method (mm)	Defect depth by 3 dB amplitude drop method (mm)	Defect depth by 6 dB amplitude drop method (mm)
d1	3	2.4	3	3.9
d2	6	4.8	5.4	6.5
d3	3	3.4	5.6	5.5
d4	6	4.2	5.4	6.3
d5	3	2.4	3.1	3.3
d6	6	4.2	5.4	6.1

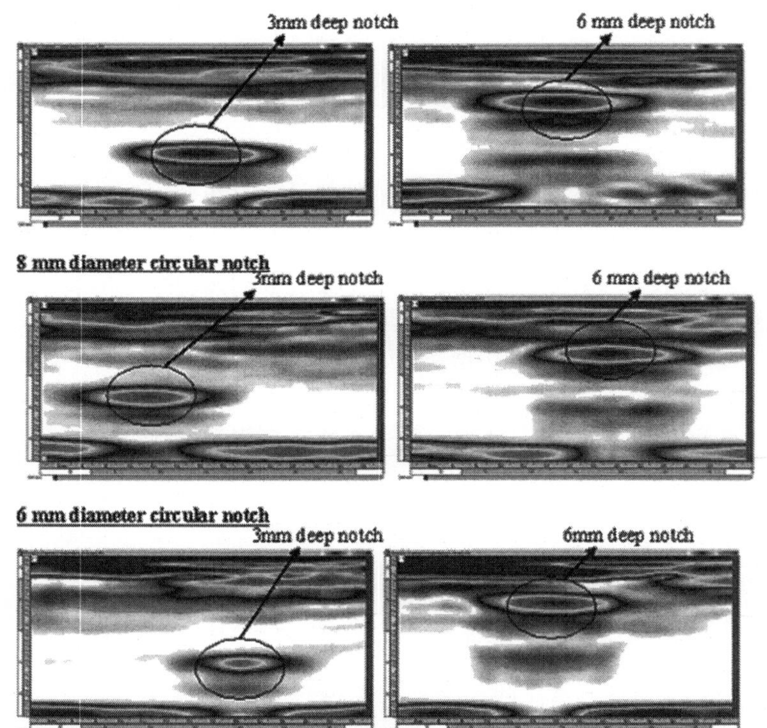

Fig. B-scan of simulated corrosion pits of three different depths and sizes obtained using a 64 element phased array probe

Conclusions:

Corrosion defects like localized corrosion pits and stress corrosion cracks were simulated on pipe samples and were imaged using phased array ultrasonic technique. The sizes of the defects obtained using phased array was sufficiently accurate and hence the technique can be concluded as a reliable technique for measuring metal loss or reduction in wall thickness apart from imaging crack like defects. Among the amplitude drop methods, the 6 dB drop method gave results closer to the actual defect sizes. The error associated with sizing smaller corrosion cracks is relatively higher compared to sizing larger cracks with respect to amplitude drop method.

8

Corrosion in Sewer Drops

INTRODUCTION

In sewer drops a free-falling stream of wastewater flow releases hydrogen sulfide (H2S) and other odorous gasses. These emissions are the cause of public odor complaints coming from areas surrounding sewer drops. The H2S emission from the drop structures initiates chemical processes that can result in rapid, extensive damage to concrete and metal sewer piping and mechanical equipment. The substantial kinetic energy of the falling wastewater can also damage the structure walls by abrasive wear. With the combination of chemical corrosion and mechanical wear, sewer drops are very vulnerable points of conveyance systems. Unfortunately, present technologies for liquid and gas phase treatment require large capital investment and expensive continuous maintenance.

Fig. The Vortex Insert Assembly.

A new and more versatile realization of the Vortex Drop method is the Vortex Insert Assembly (VIA). The VIA, pictured in Figure 1, is a simple, prefabricated insert for existing or new drop structures. Made from relatively inexpensive materials, the VIA installs quickly and converts a typical sewer drop into an energy dissipater and aerator. It creates a slight vacuum, preventing emission of odorous gasses. Air is drawn in and vigorously mixed with the wastewater, oxidizing the hydrogen sulfides. The VIA also solves the problem of abrasive wear on the structure by removing direct contact between the flow and structure walls.

PRINCIPLE OF OPERATION

The VIA installed in a drop structure. The VIA can be made from non-corrosive materials such as PVC, High Density Polyethylene (HDPE), and fiberglass. The Vortex Form at the top has a channel of decreasing radius and a supercritical slope, which creates accelerating spinning flow. No entrance flume or other complex components are needed to create the necessary acceleration. The VIA's Vortex Form is contoured to maximize acceleration over a relatively short distance, and eliminates all means for gas emission.

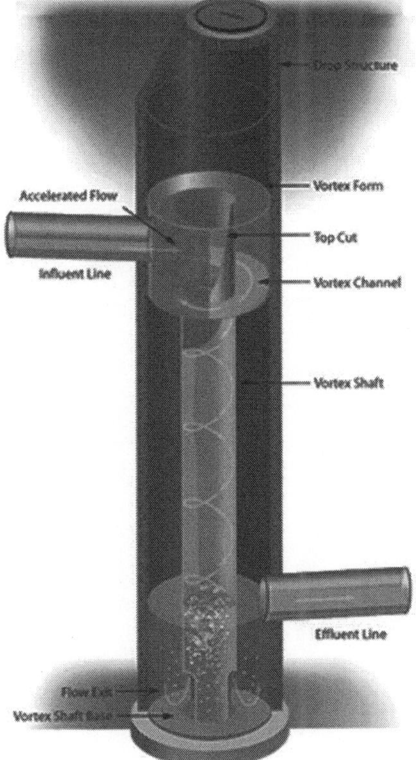

Fig. The Vortex Drop Structure

The flow is directed through a special top cut into a vortex shaft with a much smaller diameter. The top cut serves also as a vortex regulator. It is shaped to mirror the flow velocity profile, ensuring proper vortex in the entire flow range regardless of flow fluctuations. The flow continues spiraling downward in a combined field of gravity and centrifugal forces. Due to acceleration, the sharp reduction in diameter, and significant increase of centrifugal forces, the flow maintains intimate contact with the vortex shaft wall. This creates a stable air core. In the vortex shaft, the flow drags air down creating a slightly negative air pressure above the vortex, preventing gas from escaping up and out of the structure. The air is therefore entrained and mixed in with the wastewater. The most intensive processes of vigorous mixing and aeration occur in the submerged part of the vortex shaft. The wastewater is saturated with oxygen at this point. The dissolved H_2S concentration reduces dramatically.

The flow exits the vortex shaft at the bottom into an energy-dissipating pool. The remaining flow energy is dissipated through mixing and internal friction. A calm flow saturated with air exits the energy dissipating pool to the effluent line. As a result, in the time it takes for the wastewater to pass through the structure, the wastewater is dramatically changed. Its oxygen level is boosted, dissolved H_2S concentration is dropped to immeasurable levels, and the flow is made tranquil. The effect of aeration changes the sewer flow characteristics and reduces potential for corrosion on the long distance of sewer pipe downstream. The Vortex Shaft top cut extends the hydraulic limits of efficient operation and provides a stable spinning flow with air core at the minimum flow that in 10–12 times less than the VIA hydraulic capacity. This is important for the systems with relatively low initial flows, systems having inflow/infiltration problems, and the combined systems where the flow range can be extreme.

Many existing drop structures have the outflow pipe at the base of the structure and do not have pool. For those drop structures the VIA is fabricated with a submerged flow exit. It has an energy dissipating pool as part of the Assembly. The existing structure bottom is to be flat and clean for proper installation. As part of the VIA is a vent that balances the air pressures in the upper and lower pipes and provides air re-circulation to the Vortex Form and into the air core. The VIA is effective in the flow drop height range from 4 ft to 100–110 ft. Each VIA is optimized for its installation using three parameters: inner drop shaft diameter, drop height, and design flow rate. From this, the best Vortex Form and Vortex Shaft dimensions are derived.

The VIA is specifically designed for fast installation. It is pre-fabricated for its destination structure to minimize the on-site labor and installation time. It is recommended that the drop structure walls have a corrosion protective coating applied prior to installation. The process of installation consists of simply lowering the VIA into the open drop structure, and sealing around the influent connection. Should access be needed to the lower lines during the structure's lifetime, the VIA can be designed for easy removal and re-installation. For cases where flow bypassing is especially difficult, the VIA can be designed for insertion directly into falling flow.

EXPERIENCE

The Vortex Drop method has been in use in the Minneapolis/St/ Paul metro area since 1998. The Vortex Drop Structures work on the main interceptor drops, force main discharge, and in the pumping station wet well. The design flow range is up to 24 MGD and drop height is up to 52 ft. New projects incorporating Vortex Drop Method are under way in Hennepin, Anoka, and Washington Counties in Minnesota. The design flow range for each of two VDS in Hennepin County is 8.5 MGD–33 MGD, and 4 MGD–54 MGD for each of three VDS in Washington County. Experiments were conducted at multiple installations of Vortex Drop Structures to measure H2S concentration in the pipes and air quality in areas surrounding the installations. Multiple wastewater samples were taken simultaneously upstream and downstream of two VDS installed in Hennepin Co. Analysis showed a significant decrease of dissolved H_2S and a sharp rise in the dissolved oxygen concentrations downstream of the structures (Table 1). The results of air quality monitoring around the Vortex Drops indicated no H_2S gas emission.

At previously troublesome interceptors, following implementation of the Vortex Drop method, chemical feed for odor has been eliminated. Even more significant savings will be seen over the life of the interceptors since no repairs or rehabilitation due to corrosion are needed. Most importantly, odor complaints from adjacent homeowners have disappeared.

Table. The Average H_2S and Oxygen Dissolved Concentrations.

	Upstream of Vds		Dowstream of Vds	
VDS Installation	**Hydrogen Sulfide, mg/l**	**Oxygen, mg/l**	**Hydrogen Sulfide, mg/l**	**Oxygen, mg/l**
Natchez Ave, Golden Valley	2.88	1.05	0.93	4.57
Hiawatha Ave., Minneapolis	0.38	0.41	0.27	4.28

NEW INSTALLATIONS

As part of a centralization program an existing WWTP in Chaska, Minnesota, was abandoned and an interim sewer pumping station was built in 2000. A severe odor problem in the surrounding residential area was documented. The existing sludge pump chamber in the headwork building was transformed into a wet well housing four identical pumps with flow capacity 2200 gpm (3.2 MGD). The influent flow comes in the existing channel, goes through metering flume, and drops into the new wet well 4-7 ft. A pre-fabricated VIA made from PVC was installed into the wet well (Figure 3, 4) to reduce turbulence and odor in the wet well and in the headwork building.

Fig. The VIA installed into the Chaska Pumping Station Wet Well.

Fig. The working VIA in the pumping station wet well.

The VIA effectively operates over a wide range of flow, from 1.1 MGD up to 8.0 MGD and over a changing drop height from 3,5 ft to 5.5ft. The wastewater in wet well is saturated with small bubbles of air at any flow. The aeration facilitates oxidation of sulfides in wet well and in the long 16000 ft force main. The maintenance personnel noticed reduction in odorous gas emission. There were no public odor complains from surrounding area after the installation. Hunter Water Corporation, the government owned organization providing water supply and sewer services to a population of 500,000 in Newcastle, Australia, built a Maryland-Minmi Transportation Redirection project in 1999-2000. The project included three sewer pumping systems with long force mains. Each of the force mains has a barometric loop at the discharge point to ensure stable operation. The third force main discharge is located at a treatment plant inlet. The Vortex Insert Assembly was installed in each of the barometric loops to reduce emission of H2S gas. The barometric loop at the pumping system 2 discharge with the VIA is shown on Figure. 5. Hunter Water Corporation (HWC) performed hydrogen sulfide gas concentration monitoring above two working Assemblies and carried out liquid sampling in the pumping station wet wells immediately downstream of the Vortex Assemblies No.1 and No.2 (Figure 6). The Vortex Assembly No. 3 effluent flow cannot be isolated from various sources in the treatment plant inlet. Therefore no sampling or gas monitoring data was carried out at this installation.

Fig. The VIA installed into barometric loop.

Fig. The vortex flow in the VIA installed into barometric loop.

According to the HWC data the average hydrogen sulfide gas concentration for one week of continuous monitoring was 1.29 ppm on discharge of the pumping system 1, and 2.27 ppm on discharge of the pumping system 2. For a long force main discharge with a flow drop an average H2S gas concentration is typically at least ten times higher. The dissolved hydrogen sulfide concentrations from samples were 0.34-0.70 mg/l and 4.2-4.5 mg/l downstream of the system 1 and system 2 discharges accordingly. The HWC compared the levels of dissolved H_2S and found them significantly lower than those recorded at non-vortex barometric loops. Compared to similar HWC force main systems with barometric loops, odorous gas emission was significantly lower in the VIA installations. HWC concluded that the VIA is effective in reducing H2S gas emission. VIA implementation allowed reduction of chemical feed dosing rates. Another VIA for the Stockton to Shortland

Wastewater Transfer System in New South Wales, Australia is in fabrication now.

CONTROL SEWER CROWN CORROSION USING THE CROWN SPRAY PROCESS

The County Sanitation Districts of Los Angles County (Districts) serve the wastewater and solid waste management needs of about five million people in Los Angeles County. The role of the Districts is to collect, treat and dispose of wastewater and provide for the disposal and management of solid wastes in a cost effective and environmentally sound manner. Approximately 560 miles of the Districts' 1,300 miles of sewer are made of unlined concrete that is susceptible to sulfide corrosion. In 1989, the Districts developed the sewer crown spray process to neutralize the acid formed by sulfur oxidizing bacteria (SOB) and control sulfide corrosion.

Sulfide Corrosion

The sulfide corrosion of concrete sewers is caused by a two-step biological process. In the first step, sulfur-reducing bacteria residing in the slime layer on the submerged sewer pipe wall reduce sulfate ions to sulfide ions. Through chemical equilibrium, some of the sulfide ions form hydrogen sulfide (H2S), which is volatilized from the liquid phase into the gaseous phase (or sewer head space). In the second step, some of the hydrogen sulfide in the sewer head space dissolves in the moisture present on the unsubmerged portions of the pipe wall. Sulfide oxidizing bacteria (SOB) residing hl the unsubmerged portions of the pipe oxidize the hydrogen sulfide to sulfuric acid. The sulfuric acid is neutralized by the calcium hydroxide binder in the concrete to form calcium sulfate (gypsum). Since gypsum has no structural strength this process weakens the sewer (USEPA, 1985).

Sulfide corrosion of concrete sanitary sewers has become a major problem for the Districts since the reduction in the levels of metals and toxic constituents in sewage due to regulations enacted in the 1970s and 1980s. It is believed that the metals and toxic constituents in the sewage had a toxic effect on the sulfate reducing bacteria and inhibited the sulfide ion generation process (Jin and Sung, 1993). By the mid 1980s, many of the Districts' unlined concrete sewers, which were in fairly good condition in the ndd-1970s, were severely corroded with reinforcing steel visible (Figure 1). In some cases, the crown vf the sewer was completely corroded and the soil surrounding the pipe was exposed.

The Crown Spray Process

The crown spray process was formulated by the Districts in 1986 to directly control the acid formation on the crown of the sewer. The goal was to find a new process for controlling sulfide corrosion by chemical deactivation of the organisms responsible for the conversion of sulfides to sulfuric acid on the crown areas of the sewers (Esfandi, 1986). It was theorized that a high-pH mixture sprayed on the crown of the sewer would deactivate the SOB present, neutralize the acid generated on the crown, and leave a residual alkalinity on the sewer to neutralize the acid produced as the SOB recolonize the sewer.

The goal of the crown spray process is to maintain the crown pH above 4. Core sampling and inspection of many different concrete sewers in the Districts' system from 1973 to 1987 shows that the crown pH is inversely proportional to the log of the corrosion rate (see Figure 2). At a pH of 4 the corrosion rate is 0.025 inch/year. The Districts began the crown spray process experimentally with a 24 per cent sodium hydroxide solution in 1989. Test data showed it took about 60 days for the sewer crown pH to drop to 4 after spray treatment with sodium hydroxide. Therefore crown spray with sodium hydroxide would require six treatments per year (Badia et al., 1991)

In 1993, research began for a chemical that would decrease the number of required treatments, with particular interest in non-hazardous chemicals. The seven chemical mixtures tested were 15 per cent sodium carbonate, 45 per cent potassium carbonate, 12 per cent trisodium phosphate, 12 per cent magnesium hydroxide, 50 per cent magnesium hydroxide, 45 per cent calcium hydroxide and a mixture of 15 per cent sodium carbonate and 7,000 ppm benzalkonium chloride. The 50 per cent magnesium hydroxide and 45 per cent calcium hydroxide were the most effective chemicals evaluated maintaining a pH above 4 for approximately 250 days (Figure 3). Although the 45 per cent calcium hydroxide was effective at maintaining a high pH, it did not perform well in the spray application. It adhered tenaciously to the tank surface and in the chemical pump requiring more frequent maintenance. Since 1994, the Districts has used an approximate 50 per cent magnesium hydroxide slurry, to crown spray its sewers.

Magnesium hydroxide was the most effective chemical evaluated probably because it has the highest alkalinity of all the chemicals tested (1206 gm/l as CaCO3). The primary mechanism for the reduction in concrete loss due to sulfide corrosion in sewers treated with magnesium hydroxide is the addition of sacrificial alkalinity. Microbial enumeration shows that while magnesium hydroxide initially reduces the population of SOB, the SOB recover within a couple of months, yet the pit remains high for several additional months. Therefore the acid produced by the SOB is neutralized by the magnesium hydroxide instead of the calcium hydroxide binder lit the concrete (Sydney et al., 1996).

Crown Spray Operations

Depending on sewer size and depth of flow, various methods must be employed to deliver the magnesium hydroxide to the unsubmerged portions of the pipe. The basic components for crown spraying include: a spray head that can be pulled through the sewer at a controlled rate; a chemical storage tank that provides a reservoir for the slurry to be applied; and a pump capable of pumping the slurry to the spray head from the chemical tank. The Districts has found a minimum application thickness of 125 mils (1/8 inch) is adequate for corrosion control. To determine the pull rate of the float, the flow rate of the chemical pump is divided by the calculated volume of magnesium hydroxide required per linear foot.

In 1989, the Districts developed an experimental caustic spray system to apply the caustic solution to the unsubmerged portions of the pipe. The experimental caustic spray system consisted of a vacuum tank truck, chemical pump, high pressure hose, float, and spray head assembly with a series of nozzles. The spray float and nozzles were pulled through the sewer from the upstream manhole to the downstream manhole by using a cable winch at the downstream manhole. The upstream hose reel was used to control the travel speed of the float. In 1992, the Districts received a trailer-mounted caustic spray delivery system, which was similar to the experimental system except that a chemical tank, a chemical pump, a gas engine driven hydraulic pump and a hose reel were mounted on a trailer. In 1994, the Districts received a self-contained crown spray delivery system specifically designed to apply sodium hydroxide. The chemical tank, chemical pump and hose reel were mounted on the truck chassis. In March 2000, the Districts received a modified combination jet-vac sewer cleaning truck to spray the unsubmerged portions of the pipe (Figure 4). This system is improved from its predecessors by using a floating jet nozzle to propel the hose downstream instead of floating a tag line with flow. In this system the float is pulled upstream to apply the chemical, which eliminated the need for a downstream winch. This system improved the efficiency because it does not require a tag line to be floated downstream and there is no longer a need for a crewmember to control the downstream winch during the treatment.

Since the addition of the modified combination jet-vac mink, the spray application of the magnesium hydroxide slurry by the Districts has normally been performed in the following manner. The high pressure chemical/water pump first delivers water from the water tank through the high pressure hose to a floating jet nozzle which is used to propel the hose downstream. At the downstream manhole, the floating jet nozzle and hose are retrieved from the manhole. The jet nozzle is removed and the crown spray float with a spray head is then attached to the end of the hose (Figure 5). The float is returned to the sewer. The high pressure chemical/water pump will then deliver magnesium hydroxide slurry from the chemical tank compartment through the high pressure hose to the spray float. When all the water is purged from the hose, and magnesium hydroxide begins to spray through the nozzle, the chemical pressure is adjusted so that the magnesium hydroxide is sprayed a couple of feet above the pipe crown in the manhole. The float and hose are then pulled upstream at a velocity controlled by the operator while the magnesium hydroxide is sprayed on the unsubmerged portions of the pipe. Multiple sewer reaches can be sprayed in a single setup, although it is recommended that the float position be checked in the intermediate manholes to insure the float has not flipped over and the spray pattern is acceptable. A chemical/water spray gun is used to apply chemical on inverted siphon uplegs, manholes and other structures.

At the end of each day's operation, the operator runs the high pressure chemical/water pump with water to flush the chemical remaining hl the system. At the end of each week, the compartments of the dual chemical/water tank are thoroughly cleaned to remove any remaining chemical. Past experience has shown that magnesium hydroxide slurry adheres tenaciously to the tank surfaces if not removed at the end of each week. A wash-down system is used so that the operator can clean the compartments of the dual chemical/water tank and wash any unsightly magnesium hydroxide spillage and wastewater on the street into a manhole.

Equipment

Descriptions of the crown spray delivery system components are as follows:

- *Crown Spray Truck*—The Districts use a modified combination jet vac from Vactor Manufacturing. The truck has a two-compartment 304 stainless steel tank with a total capacity of 3,000 gallons with 1,500 gallons for magnesium hydroxide slurry and 1,500 gallons for water. The two compartments are interconnected with a valved exterior bypass piping system to allow manual switching from chemical to water. In addition, a fill connection is installed between the pump and the tank to allow gravity feed of chemical from a support vehicle.
- *Pump*—The high pressure chemical/water pump is a single piston, double acting positive displacement Vactor pump (80 gpm at 2,500 psi). The pump is operated and controlled hydraulically and is capable of varying flow of chemical/water throughout its operating range. The chemical/water pump is driven by a hydraulic oil pump, driven by the truck engine via a heavy-duty power take-off (PTO). An in-line Y-strainer with a 1/8 inch mesh screen is located on the inlet side of the pump to protect the pump from debris entering from the water or chemical tank.
- *Hydraulic Power System*—Two variable displacement piston type hydraulic pumps are powered by a heavy duty, air shift, standard 8-bolt type PTO, driven by the truck's transmission. One of the hydraulic pumps has the capacity to drive the high pressure chemical/water pump and the water washdown pump at maximum operating speed at the same time. The other hydraulic pump drives the hose reel assembly.

- *Operator Control Station*—The operator control station is located in the rear of the truck and has all the gauges, switches, controls, displays, etc., necessary to operate the unit. The control station has the following controls and displays: hose reel controls (pull rate speed and direction), hydraulic oil pressure, chemical/water pump On/Off switch, chemical/water pump discharge pressure, chemical/ water pump flow rate control, flow meter display (flow rate and totalized flow), footage meter display (footage and pull rate), utility work light switches, and an emergency hydraulic disengage switch.

Scheduling

The Districts has found that surface pH is the most reliable method of determining the level of corrosive activity. Therefore, concrete sewer pipes are selected for the crown spray program based on surface pH and general condition. If the pH of the sewer crown is measured at 5 or below, the sewer is placed on the crown spray program. If visible corrosion is found during closed circuit TV inspection, the sewer will also be placed on the crown spray program. Generally a sewer which has suffered visible corrosion will also have a pH lower than 6.

Sewers on the crown spray program are generally sprayed annually. Before the sewer is sprayed with magnesium hydroxide, the crown pH is measured and recorded. The frequency that a sewer needs to be sprayed will depend of the level of activity of the acid producing bacteria. This activity is affected by the sewer characteristics (air temp, humidity, air draft, etc.), rate of sulfide generation, mass transfer of hydrogen sulfide from the sewage to the head space, competition with other heterotrophic bacteria, residual effects of magnesium hydroxide on the SOB, axed the level of adherence of the magnesium hydroxide slurry on the sewer crown (Badia et al., 1991).

All concrete sewer reaches that have been recommended for repair, that have exposed reinforcing steel, or are located in geologically sensitive areas are crown sprayed annually or more frequently as required to control corrosion. For all other sewers, the frequency assigned to a particular sewer depends on historical crown pH data measured in that sewer. If, one year after the sewer is sprayed, the pH of the crown is 8 or above, the sewer's crown spray frequency will be reduced to every two years. If, one year after the sewer is sprayed, the pH of the crown is 4 or below, the sewer's crown spray frequency will be increased to every 6 months. A sewer, therefore, needs to be monitored after its first spraying before its frequency can be established.

Scheduling of the sewers to be crown sprayed should be coordinated with other routine maintenance programs. In general, cleaning will remove the magnesium hydroxide coating from the sewer. Therefore sewers should be scheduled for crown spray after they are cleaned. Also the magnesium hydroxide coating will cover some damage in the pipe. Therefore, if a sewer needs to be closed circuit TV inspected, then this should be done before crown spraying the sewer. Sewers with heavy root infiltration should not be crown sprayed because the float could become caught in the roots and cause an overflow. The treatment plant that the treated sewer is tributary to should be given prior notification that the influent could have an elevated pH. The treatment plant will then be aware that the elevated pH is from the crown spray process and not an illegal industrial source. Also, the Districts' small wastewater treatment plants have experienced decreased methane gas production from primary sludge anaerobic digestion due to an increased alkalinity caused by crown spray.

9

Corrosion in Cement Plants

INTRODUCTION

The type of corrosion mechanism and its rate of attack depend on the nature of the atmosphere in which the corrosion takes place. The first step in preventing corrosion is understanding its specific mechanism. The second and often more difficult step is designing an effective type of protection mechanism. Successful enterprises cannot tolerate major corrosion failures, especially those involving personal injuries, fatalities, unscheduled shutdowns and environmental contamination Typically, once a plant or any piece of equipment is put into service, maintenance is required to keep it operating safely and efficiently. This is particularly true for aging systems and structures, many of which may operate beyond the original design life.

Corrosion issues in cement plants are frequent and costly. Some have experienced more corrosion than others and some process areas can be more susceptible to its effects than others. In general, corrosion in cement plants occurs when process gases containing moisture, SOx, HCl, and NOx, operate at dew point temperatures. This paper will bring forth practical methods used to prevent corrosion and technology recently developed to control corrosion in a cement plant.

CEMENT

A cement is a binder, a substance that sets and hardens independently, and can bind other materials together. The word "cement" traces to the Romans, who used the term opus caementicium to describe masonry resembling modern concrete that was made from crushed rock with burnt lime as binder. The volcanic ash and pulverized brick additives that were added to the burnt lime to obtain a hydraulic binder were later referred to as cementum, cimentum, cäment, and cement. Cements used in construction can be characterized as being either hydraulic or non-hydraulic. Hydraulic cements (e.g., Portland cement) harden because of hydration, a chemical reaction between the anhydrous cement powder and water. Thus, they can harden underwater or when constantly exposed to wet weather. The chemical reaction results in hydrates that are not very water-soluble and so are quite durable in water. Non-hydraulic cements do not harden underwater; for example, slaked limes harden by reaction with atmospheric carbon dioxide. The most important uses of cement are as an ingredient in the production of mortar in masonry, and of concrete, a combination of cement and an aggregate to form a strong building material.

TYPES OF MODERN CEMENT

Portland Cement

Portland cement is by far the most common type of cement in general use around the world. This cement is made by heating limestone (calcium carbonate) with small quantities of other materials (such as clay) to 1450 °C in a kiln, in a process known as calcination, whereby a molecule of carbon dioxide is liberated from the calcium carbonate to form calcium oxide, or quicklime, which is then blended with the other materials that have been included in the mix. The resulting hard substance, called 'clinker', is then ground with a small amount of gypsum into a powder to make 'Ordinary Portland Cement', the most commonly used type of cement (often referred to as OPC). Portland cement is a basic ingredient of concrete, mortar and most non-specialty grout. The most common use for Portland cement is in the production of concrete. Concrete is a composite material consisting of aggregate (gravel and sand), cement, and water. As a construction material, concrete can be cast in almost any shape desired, and once hardened, can become a structural (load bearing) element. Portland cement may be grey or white.

Energetically Modified Cement

The grinding process to produce Energetically modified cement (EMC) yields materials made from pozzolanic minerals that have been treated using a patented milling process ("EMC Activation"). This yields a high-level replacement of Portland cement in concrete with lower costs, performance and durability improvements, with significant energy and carbon dioxide savings. The resultant concretes can have the same, if not improved, physical characteristics as "normal" concretes, at a fraction of the cost of using Portland cement.

Portland Cement Blends

Portland cement blends are often available as inter-ground mixtures from cement producers, but similar formulations are often also mixed from the ground components at the concrete mixing plant. Portland blastfurnace cement contains up to 70 per cent ground granulated blast furnace slag, with the rest Portland clinker and a little gypsum. All compositions produce high ultimate strength, but as slag content is increased, early strength is reduced, while sulfate resistance increases and heat evolution diminishes. Used as an economic alternative to Portland sulfate-resisting and low-heat cements. Portland flyash cement contains up to 35 per cent fly ash. The fly ash is pozzolanic, so that ultimate strength is maintained. Because fly ash addition allows a lower concrete water content, early strength can also be maintained. Where good quality cheap fly ash is available, this can be an economic alternative to ordinary Portland cement. Portland pozzolan cement includes fly ash cement, since fly ash is a pozzolan, but also includes cements made from other natural or artificial pozzolans. In countries where volcanic ashes are available (e.g. Italy, Chile, Mexico, the Philippines) these cements are often the most common form in use. Portland silica fume cement. Addition of silica fume can yield exceptionally high strengths, and cements containing 5–20 per cent silica fume are occasionally produced. However, silica fume is more usually added to Portland cement at the concrete mixer. Masonry cements are used for preparing bricklayingmortars and stuccos, and must not be used in concrete. They are usually complex proprietary formulations containing Portland clinker and a number of other ingredients that may include limestone, hydrated lime, air entrainers, retarders, waterproofers and coloring agents. They are formulated to yield workable mortars that allow rapid and consistent masonry work. Subtle variations of Masonry cement in the US are Plastic Cements and Stucco Cements.

These are designed to produce controlled bond with masonry blocks. Expansive cements contain, in addition to Portland clinker, expansive clinkers (usually sulfoaluminate clinkers), and are designed to offset the effects of drying shrinkage that is normally encountered with hydraulic cements. This allows large floor slabs (up to 60 m square) to be prepared without contraction joints. White blended cements may be made using white clinker and white supplementary materials such as high-purity metakaolin. Colored cements are used for decorative purposes. In some standards, the addition of pigments to produce "colored Portland cement" is allowed. In other standards (e.g. ASTM), pigments are not allowed constituents of Portland cement, and colored cements are sold as "blended hydraulic cements". Very finely ground cements are made from mixtures of cement with sand or with slag or other pozzolan type minerals that are extremely finely ground together. Such cements can have the same physical characteristics as normal cement but with 50 per cent less cement particularly due to their increased surface area for the chemical reaction. Even with intensive grinding they can use up to 50 per cent less energy to fabricate than ordinary Portland cements.

Non-Portland Hydraulic Cements

Pozzolan-lime cements. Mixtures of ground pozzolan and lime are the cements used by the Romans, and can be found in Roman structures still standing (e.g. the Pantheonin Rome). They develop strength slowly, but their ultimate strength can be very high. The hydration products that produce strength are essentially the same as those produced by Portland cement. Slag-lime cements. Ground granulated blast furnace slag is not hydraulic on its own, but is "activated" by addition of alkalis, most economically using lime. They are similar to pozzolan lime cements in their properties. Only granulated slag (i.e. water-quenched, glassy slag) is effective as a cement component. Supersulfated cements.

These contain about 80 per cent ground granulated blast furnace slag, 15 per cent gypsum or anhydrite and a little Portland clinker or lime as an activator. They produce strength by formation of ettringite, with strength growth similar to a slow Portland cement. They exhibit good resistance to aggressive agents, including sulfate. Calcium aluminate cements are hydraulic cements made primarily from limestone and bauxite. The active ingredients are monocalcium aluminate $CaAl_2O_4$($CaO \cdot Al_2O_3$ or CA in Cement chemist notation, CCN) and mayenite $Ca_{12}Al_{14}O_{33}$ (12 $CaO \cdot 7\ Al_2O_3$, or $C_{12}A_7$ in CCN). Strength forms by hydration to calcium aluminate hydrates. They are well-adapted for use in refractory (high-temperature resistant) concretes, e.g. for furnace linings. Calcium sulfoaluminate cements are made from clinkers that include ye'elimite ($(Ca_4(AlO_2)_6SO_4$ or C_4A_3S in Cement chemist's notation) as a primary phase.

They are used in expansive cements, in ultra-high early strength cements, and in "low-energy" cements. Hydration produces ettringite, and specialized physical properties (such as expansion or rapid reaction) are obtained by adjustment of the availability of calcium and sulfate ions. Their use as a low-energy alternative to Portland cement has been pioneered in China, where several million tonnes per year are produced. Energy requirements are lower because of the lower kiln temperatures required for reaction, and the lower amount of limestone (which must be endothermically decarbonated) in the mix. In addition, the lower limestone content and lower fuel consumption leads to a CO_2 emission around half that associated with Portland clinker. However, SO_2 emissions are usually significantly higher. "Natural" cements correspond to certain cements of the pre-Portland era, produced by burningargillaceous limestones at moderate temperatures. The level of clay components in the limestone (around 30–35 per cent) is such that large amounts of belite (the low-early strength, high-late strength mineral in Portland cement) are formed without the formation of excessive amounts of free lime. As with any natural material, such cements have highly variable properties. Geopolymer cements are made from mixtures of water-soluble alkali metal silicates and aluminosilicate mineral powders such as fly ash andmetakaolin.

CORROSION PROCESS

Carbon steel and even stainless steel corrodes in flue gas service. Equipment like electrostatic precipitators, bag houses, cooling ducts, conditioning towers and stacks frequently fail due to corrosion. Corrosion is worse when there is a presence of acidic compounds in the flue gas. The source can be from sulfur content in the feed or the fuel, chloride content in the feed or in air and CO2 and NOx from combustion. The moisture content in the gas produces hot acid condensation on the steel walls, in most cases intermittent for short periods of time, the cumulative impact can be up to 1.0 mm/yr, meaning less than 5 years life. Air pollution control devices, the fans and the stack are all candidates for corrosion. Water spray cooling towers used to control temperatures, amplify the problem. Some processes have acid gas scrubbers, which are also problematic if they are not protected. In these systems, the stack would also be a problem area. In general, equipment operating in the cooler end of the process is where most of the corrosion develops. These areas are sensitive to cold air in leakage, low external temperatures, and startups and shutdowns.

Corrosion Control

There are different ways to reduce the corrosion impact. Adjusting the operating conditions, for example higher temperature to avoid condensation. This is not always possible and is a substantial waste of energy. Different materials of construction, like the use of 316L or higher alloys leads to 3 to 5 times capital investment and involves the risk of stress corrosion cracking. Traditional coatings require a surface preparation that is expensive and critical and usually have a limited useful life due to blistering and delamination. In many cases operators attempt to maintain process gas temperatures above the dew point and or remove corrosive gas constituents. Maintaining sufficient gas temperature to avoid condensation on the equipment walls can be expensive and limited by environmental regulations. This approach does not eliminate condensation during start up and shutdown, when temperatures rise and fall through the dew point. Removing corrosive gas constituents is costly and sometimes not feasible.

Insulation

Fig. Severe corrosion damage in a reverse air bag house filter.

Proper insulation and its maintenance can sometimes solve corrosion problems under the right conditions. However, insulated equipment with operating gas temperatures around the dew point can still have significant corrosion. Figure 1 shows the inside of an insulated baghouse that has experienced severe corrosion. This baghouse operates in a high sulfur environment near the dew point. The walls are corroding and scale is falling on the tube sheet along the walls.

Corrosion Resistant Metals

Stainless steels are alloys of iron that generally have a minimum of 12 per cent chromium. More chromium can be added to increase the corrosion resistance. Alloys containing Molybdenum have improved resistance to pitting and crevice corrosion. The addition of nickel provides resistance to reducing environments. When nickel comprises more than 25 per cent of the metal, it improves the stress corrosion cracking resistance. If process gas temperatures operate above 300ºC (570°F), then these materials are frequently a solution to corrosion. Equipment, ductwork, and stacks fabricated of these alloys, however, are very expensive. Proper understanding of process variables, raw materials, and fuels is key to determining what type of materials to use to prevent corrosion. Problems may arise when raw material, fuels, and temperatures are not as expected. Examples are alternative fuel utilization, raw material substitution, or changing to high sulfur coal or petroleum coke.

Fig. Catastrophic failure of a silicone high temperature coating.

Conventional Protective Coatings

Many coatings have been developed in the past. Epoxy and silicone coating materials can resist the effects of acid condensation to some degree. Acrylics, alkyds, or polyesters will not withstand high operating temperatures. The failure modes for these types of coatings are oxidative degradation and delamination. Oxidation damage occurs when the process equipment operates above 150°C (300°F). Undercut corrosion, dis-bonding and delamination happen when there is any surface damage or imperfection in the surface preparation. Figure 2 illustrates a typical failure mode of a high temperature silicone coating on a baghouse after being exposed to high temperature condensing acid.

New Material Technologies

There are new coating technologies available, one is a hybrid organic-inorganic polymeric alloy material, suitable for continuous operation up to 225°C (437°F) that can handle peaks for several hours up to 260°C (500°F). This material, known as FlueGard-225SQC, has a tenacious bonding to steel and very good resistance to hot acids and abrasion.

Fig. Inspection of FlueGard-225SQC after 2 years in a bag house filter. No corrosion.

The first successful applications were made more than 4 years ago with numerous installations to date in baghouses, precipitators, fans, stacks and ducts. There are currently many ongoing projects in different industries like cement, oil refining, power generation, steel mills, metal smelting, lime, waste incineration, carbon black and battery recycling. Figure 3 shows the effectiveness of this corrosion protection system in a cement plant bag house filter after more than 2 years in service. This hybrid polymeric alloy coating technology is a revolutionary solution to corrosion protection in high temperature equipment. Research and development is further increasing the limits on operating temperatures. The coating can be applied during original fabrication as well as after the equipment is in operation, when actual conditions indicate excessive corrosion. A successful coating application requires proper surface preparation, correct application technique, and complete high temperature curing, all performed by a qualified contractor. Figure 3 shows a baghouse coated with the new polymer alloy coating. After two years in service, there is no corrosion.

Recent Material Development

There is a very recent material developed to address the corrosion problems at very high temperatures. This new system is a combination of an inorganic polymer binder and two reactive inorganic fillers that have particle sizes in the nanometer range. The available surface of these fillers is about one million times larger that conventional materials and the end result is a corrosion protection coating that works well at 425°C and resists exposures up to 600°C.

10

Environmental Impacts of CSP Desalination

Impacts of seawater desalination to the environment, which will be explained in this section, are caused by feed water intake, material and energy demand, and by brine discharge. The selection of the seawater intake system depends on the raw water source, local conditions, and plant capacity. The best seawater quality can be reached by beach wells, but in these cases the amount of water that can be extracted from each beach well is limited by the earth formation, and therefore the amount of water available by beach wells is very often far below the demand of the desalination plant. For small and medium reverse osmosis plants, a beach well is often used. For seawater with a depth of less than 3 m, short seawater pipes or an open intake are used for large capacities. Long seawater pipes are used for seawater with depths of more than 30 m.

The seawater intake may cause losses of aquatic organisms by impingement and entrainment. The effects of the construction of the intake piping result from the disturbance of the seabed which causes re-suspension of sediments, nutrients or pollutants into the water column. The extent of damage during operation depends on the location of the intake piping, the intake rate and the overall volume of intake water. The second impact category is linked to the demand of energy and materials inducing air pollution and contributing to climate change. The extent of impact through energy demand is evaluated by life cycle assessment, LCA. The impacts of this category can be mitigated effectively by replacing fossil energy supply by renewable energy and using waste heat from power generation for the thermal processes.

The third impact category comprises effects caused by the release of brine to the natural water body. On one hand the release of brine stresses the aquatic environment due to the brine's increased salinity and temperature. On the other hand the brine contains residuals of chemicals added during seawater pre-treatment and by-products formed during the treatment. These additives and their by-products can be toxic to marine organisms, persistent and/or can accumulate in sediments and organisms. Apart from the chemical and physical properties the impact of the brine depends on the hydrographical situation which influences brine dilution and on the biological features of the discharge site. For instance, shallow sites are less appropriate for dilution than open-sea sites and sites with abundant marine life are more sensitive than hardly populated sites. But dilution can only be a medium-term mitigation measure. In the long run the pre-treatment of the feed water must be performed in an environmentally friendly manner. Therefore alternatives to conventional chemical pre-treatment must be identified.

The environmental impacts of seawater desalination will be discussed separately for each technology because of differences in nature and magnitude of impacts. The technologies regarded here are MSF, MED and RO as they are, at least at the moment, the predominant ones of all desalination technologies and therefore these plants are responsible for almost all impacts on the environment caused by desalination. An excellent and highly recommendable compendium of environmental impact of MSF and RO desalination technologies is /Lattemann and Höpner 2003/. Much of the data used here has been taken from that source.

MULTI-STAGE FLASH DESALINATION (MSF)

SEAWATER INTAKE

Due to their high demand of cooling water, MSF desalination plants are characterized by a low product water conversion rate of 10 to 20 per cent. Therefore the required volume of seawater input per unit of product water is large, i.e. in the case of a conversion rate of 10 per cent, 10 m³ of seawater are required for 1 m³ of produced freshwater (see Figure 6 1). Combining the high demand of seawater input in relative terms with the high demand of seawater input in absolute terms due to the large average MSF plant size the risks of impingement and entrainment at the seawater intake site must be regarded as high. Therefore, the seawater intake must be designed in a way that the environmental impact is low.

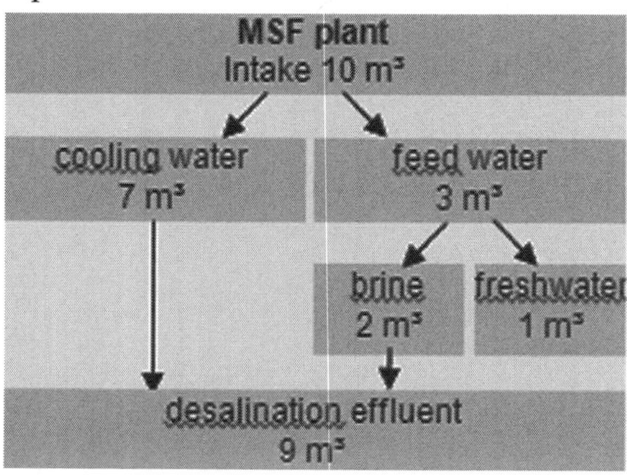

Fig. Flow chart of reference MSF process

Discharge of Brine Containing Additives

The discharge of brine represents a strong impact to the environment due to its changed physical properties, i.e. salinity, temperature and density, and to the residues of chemical additives or corrosion products. In MSF plants common chemical additives are biocides, anti-scalants, anti-foaming agents, and corrosion inhibitors. The conditioning of permeate to gain palatable, stable drinking water requires the addition of chlorine for disinfection, calcium, e.g. in form of calcium hydroxide, for remineralisation and pH adjustment /Raluy 2003/, /Delion et al. 2004/. In case of acidification as pre-treatment removal of boron might be necessary/Delion et al. 2004. Figure 6-2 Shows where the chemicals are added, and indicates the concentrations as well as the characteristics of the brine and its chemical load.

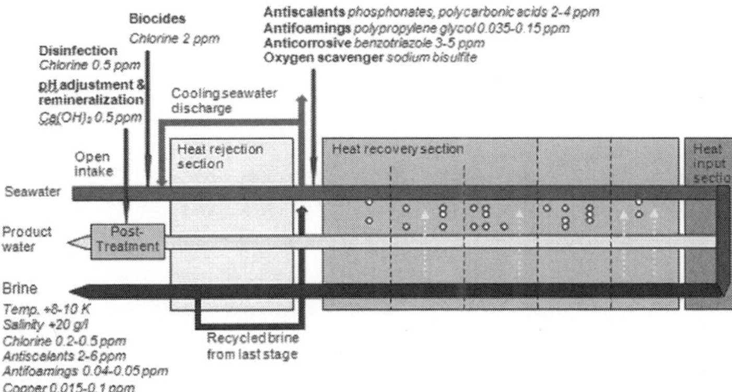

Fig. MSF process scheme with input and output concentrations of additives and brine characteristics, /Lattemann and Höpner 2003/, modified

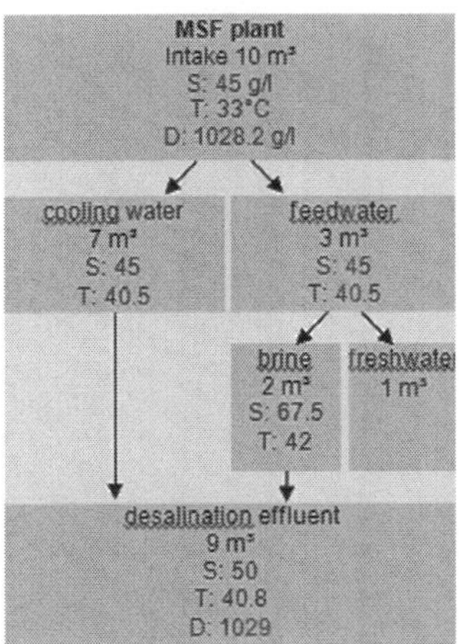

Fig. Flow chart of reference MSF process with salinity (S, in g/l), temperature (T, in °C) and density (D, in g/l), /Lattemann and Höpner 2003/, modified

Physical Properties of Brine

The physical parameters of the brine are different compared to the intake seawater. During the distillation process the temperature rises and salt accumulates in the brine. Taking the reference process (Figure 6 1) with a conversion rate of approx. 10 per cent (related to the seawater flow) as example the salinity of the brine rises from 45 g/l to 67.5 g/l (Figure 6 3). Brine and cooling water temperature rises by 9 and 7.5 K, respectively. Salinity of the brine is reduced by blending with cooling water, but still reaches a value of 5.4 g/l above ambient level. The resulting increase of density is small what can be attributed to balancing effects of temperature and salinity rise. In general, the increase of the seawater salinity in the sea caused by solar evaporation is normally much higher than by desalination processes. However, the brine discharge system must be designed in a way that the brine is well distributed and locally high temperature and salinity values are avoided.

Biocides

Surface water contains organic matter, which comprises living or dead particulate material and dissolved molecules, leads to biological growth and causes formation of biofilm within the plant. Therefore the seawater intake flow is disinfected with the help of biocides. The most common biocide in MSF plants is chlorine. A concentration of up to 2000 μg/l in the seawater intake flow is sustained by a continuous dosage. Chlorine reacts to hypochlorite and, in the case of seawater, especially to hypobromite. Residual chlorine is released to the environment with the effluents from cooling and distillation where it reaches values of 200-500 μg/l, representing 10–25 per cent of the dosing concentration. Assuming a product-effluent-ratio of 1:9 the specific discharge load of residual chlorine per m^3 of product water is 1.8–4.5 g/m^3. For a plant with a desalination capacity of 24,000 m^3/day, for instance, this means a release of 43.2–108 kg of residual chlorine per day.

Further degradation of available chlorine after the release to the water body will lead to concentrations of 20-50 μg/l at the discharge site. Chlorine has effects on the aquatic environment because of its high toxicity, which is expressed by the very low value of long-term water quality criterion in seawater of 7.5 μg/l recommended by the U.S. Environmental Protection Agency (EPA 2006, cited in /Lattemann and Höpner 2007a/) and the predicted no-effect concentration (PNEC) for saltwater species of 0.04 μg/l determined by the EU environmental risk assessment (ECB 2005, cited in /Lattemann and Höpner 2007a/). In Figure 6 4 the occurring concentrations near the outlet and at a distance of 1 km are compared to ecotoxicity values determined through tests with different aquatic species and to the EPA short-term and long-term water quality criteria. It is striking that most of the concentrations at which half of the tested populations or the whole population is decimated at different exposure times or show other effects are exceeded by the concentrations measured near the outlet and even at the distance of 1 km. The values are quoted in /Lattemann and Höpner 2003/ who took them from Hazardous Substance Databank.

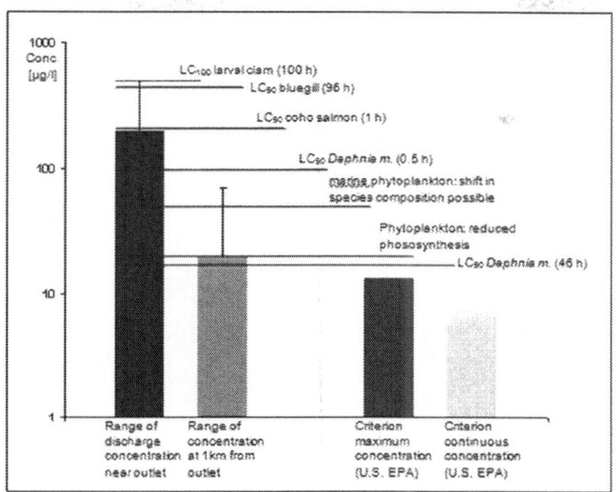

Fig. Chlorine: Ecotoxicity (LC50 = mean lethal concentration) compared to ranges of brine concentration and water quality criteria, /Lattemann and Höpner 2003/, modified

Another aspect of chlorination is the formation of halogenated volatile liquid hydrocarbons. An important species is bromoform, a trihalomethane volatile liquid hydrocarbon. Concentrations of up to 10 µg/l of bromoform have been measured near the outlet of the Kuwaiti MSF plant Doha West / Saeed et al. 1999/. The toxicity of bromoform has been proven by an experiment with oysters which have been exposed to a bromoform concentration of 25 µg/l and showed an increased respiration rate and a reduced feeding rate and size of gonads (Scott et al. 1982, cited in /Saeed et al. 1999/). Larval oysters are even more sensitive to bromoform, as significant mortality is caused by a concentration of 0.05–10 µg/l and acute, 48 h exposures.

Antiscalants

A major problem of MSF plants is the scale formation on the heat exchanger surfaces which impairs heat transfer. The most common scale is formed by precipitating calcium carbonates due to increased temperatures and brine concentration. Other scale forming species are magnesium hydroxide calcium sulphate, the latter being very difficult to remove as it forms hard scales. Therefore sulphate scaling is avoided in the first place by regulating the operation parameters temperature and concentration in such a way that the saturation point of calcium sulphate is not reached. Calcium carbonates and magnesium hydroxides, again, are chemically controlled by adding acids and/or antiscalants. In the past, acid treatment was commonly employed. With the help of acids the pH (acidity value) of the feed water is lowered to 2 or 3 and hereby the bicarbonate and carbonate ions chemically react to carbon dioxide which is released in a decarbonator.. Thus, the $CaCO_3$ scale forming ions are removed from the feed water. After acid treatment the pH of the seawater is re-adjusted. Commonly used acids are sulphuric acid and hydrochloric acid, though the first is preferred because of economic reasons. High concentrations and therefore large amounts of acids are necessary for the stoichiometric reaction of the acid.. Apart from a high consumption of acids further negative effects of using acids are the increased corrosion of the construction materials and thus reduced lifetimes of the distillers as well as handling and storage problems. The negative effects mentioned above have led to the development of alternatives: Nowadays antiscalants are replacing acids during operation. But before talking about antiscalants, the use of acids as cleaning agents needs to be mentioned because that's when significantly acidic effluents occur. During this periodic cleaning procedure the pH is lowered to 2–3 by adding citric, sulfamic or sulphuric acid, for instance, to remove carbonate and metal oxide scales.

In this context Mabrook (1994, in /Lattemann and Höpner, 2003/) explained an observed change in density and diversity of marine organisms by a decreased pH of 5.8 compared to 8.3 in coastal waters. Eco-toxic pH values range from 2-2.5 for starfish (LC50, HCl, 48 h) to 3-3.3 for salt water prawn (LC50, H_2SO_4, 48 h) and show the sensitivity of marine organisms to low pH values. Little mobile organisms, like starfish, are especially affected by an acid plume as they cannot avoid this zone. To mitigate these possible effects the cleaning solution should be neutralized before discharge or at least blended with the brine during normal operation. An antiscalant can suppress scale formation with very low dosages, typically below 10 ppm. Such low dosages are far from the stoichiometric concentration of the scaling species. Hence inhibition phenomena do not entail chemical reactions and stem from complex physical processes involving adsorption, nucleation and crystal growth processes. Scale suppression in the presence of minute concentrations of antiscalants is believed to involve several effects:

- Threshold effect: An antiscalant can slow down the nucleation process occurring in a supersaturated solution. Thereby, the induction period, which precedes crystal growth, is increased. The inhibition effect of anti-scalants is based on their ability to adsorb onto the surfaces of sub-microscopic crystal nuclei, which prevents them from growing any further or, at least, substantially slows down the growth process. Since anti-scalant molecules with a low molecular weight are more mobile, the extension of the induction period is more pronounced with molecules of comparatively low molecular weight.
- Crystal distortion effect: Adsorbed antiscalant molecules act to distort the otherwise orderly crystal growth process. A different degree of adsorption and retardation of the growth process on

different crystal faces results in alteration of the crystal structure. The scale structure can be considerably distorted and weakened. The distorted crystals are less prone to adhere to each other and to metal surfaces. When crystallisation has started either further growth is inhibited or the precipitates form a soft sludge that can be easily removed rather than hard scales /Al-Shammiri et al. 2000.

- Dispersive effect: Antiscalants with negatively charged groups can adsorb onto the surfaces of crystals and particles in suspension and impart a like charge, hence repelling neighbouring particles, thereby preventing agglomeration and keeping the particles suspended in solution.
- Sequestering effect: Antiscalants may act as chelating agents and suppress the particle formation by binding free Ca^{2+} or Mg^{2+} ions in solution. Anti-scalants with strong chelating characteristics cannot work at the sub-stoichiometric level, as the anti-scalant is consumed by the scale-forming ions. Sequestration is affected by chemicals that require relatively high concentrations and is not a physical inhibition effect.

Polyphosphates represent the first generation of antiscalant agents with sodium hexametaphosphate as most commonly used species. A procedural disadvantage is the risk of calcium phosphate scale formation. Of major concern to the aquatic environment is their hydrolytic decomposition at 60°C to orthophosphate which acts as a nutrient and causes eutrophication. The development of algae mats on the water body receiving the discharge could be ascribed to the use of phosphates /Abdel-Jawad and Al-Tabtabaei 1999/, in /Lattemann and Höpner 2003/). Because of these reasons they have partly been substituted by thermally stable phosphonates and polycarbonic acids, the second generation of antiscalants. Where phosphates have been replaced by these substances the problem of algae growth could be solved completely. Main representatives of polycarbonic acids are polyacrylic and polymaleic acids. Especially polyacrylic acid has to be dosed carefully if precipitation is to be avoided.

The reason for this is that, at lower concentrations, it enhances agglomeration and therefore also serves as a coagulant in RO plants (see below). Discharge levels of phosphonates and polycarbonic acids are classified as non-hazardous, as they are far below concentrations with toxic or chronic effects. They resemble naturally occurring humic substances when dispersed in the aquatic environment which is expressed by their tendency to complexation and their half-life of about one month, both properties similar to humic substances. Though they are generally assumed to be of little environmental concern, there is a critical point related to these properties. As they are rather persistent they will continue to complex metal ions in the water body. Consequently, the influence on the dissolved metal concentrations and therefore metal mobility naturally exerted by humic substances is increased by polymer antiscalants. The long-term effect induced hereby requires further research.

Experimental data on the bioaccumulation potential of polycarboxylates are not available. However, polymers with a molecular weight > 700 are not readily taken up into cells because of the steric hindrance at the cell membrane passage. Therefore a bioaccumulation is unlikely. Copolymers have a favourable ecotoxicological profile. Based upon the available short-term and long-term ecotoxocity data of all three aquatic trophic levels (fish, daphnia, algae) for a variety of polycarboxylates, it is considered that exposure does not indicate an environmental risk for the compartments water, sediment and sewage treatment plants. A MSF plant with a daily capacity of 24,000 m^3 releases about 144 kg of antiscalants per day if a dosage concentration of 2 mg per litre feedwater is assumed. This represents a release of 6 g per cubic meter of product water.

Antifoaming Agents

Seawater contains dissolved organics that accumulate in the surface layer and are responsible for foaming. The use of antifoaming agents is necessary in MSF plants, because a surface film and foam -increase the risk of salt carry-over and contamination of the distillate. A surface film derogates the thermal desalination process by increasing the surface viscosity. An elevated surface viscosity hampers deaeration. Furthermore, if the surface tension is too high, brine droplets will burst into the vapour phase during flashing. Deaeration is essential for thermal plants as it reduces corrosion; salt carry-over with brine droplets must be avoided for a clean distillation..

As the antifoaming agents are organic substances, too, they must carefully be chosen and dosed. Blends of polyglycol are utilized, either containing polyethylene glycol or polypropylene glycol. These substances are generally considered as non-hazardous and low discharge concentrations of 40-50 μg per litre of effluent further reduce the risk of environmental damage. However, highly polymerized polyethylene glycol with a high molecular mass is rather resistant to biodegradation. On this account it has been replaced in some industrial applications by substances, such as dialkyl ethers, which show a better biodegradability. Addition of usually less than 0.1 ppm of an antifoaming agent is usually effective. Concentrations in the discharge were found to be half this level, which is mainly due to mixing of brine with cooling water /Lattemann and Höpner 2003/. While the brine contains residual antifoaming agents, the cooling water is not treated and thereby reduces the overall discharge concentration. Under the assumption of a product-feedwater-ratio of 1:3 and 0.035-0.15 ppm dosing 0.1-0.45 g per cubic meter of product water are released.

Corrosion Inhibitors and Corrosion Products

An important issue for MSF plants is the inhibition of corrosion of the metals the heat exchangers are made of. The corrosive seawater, high process temperatures, residual chlorine concentrations and corrosive gases are the reason for this problem. Corrosion is controlled by the use of corrosion resistant materials, by deaeration of the feed water, and sometimes by addition of corrosion inhibitors. Especially during acidic cleaning corrosion control by use of corrosion inhibitors is essential for copper-based tubing. In a first step oxygen levels are reduced by physical deaeration. The addition of chemicals like the oxygen scavenger sodium bisulfite can further reduce the oxygen content. Sodium bisulfite should be dosed carefully as oxygen depletion harms marine organisms.

Corrosion inhibitors generally interact with the surfaces of the tubes. Ferrous sulphate, for example, adheres to the surface after having hydrolized and oxidized and hereby protects the alloy. Benzotriazole and its derivates are special corrosion inhibitors required during acid cleaning. They possess elements like selenium, nitrogen, sulphur and oxygen with electron pairs which interact with metallic surfaces building a stable protective film. However, it is assumed that in the end the major amount is discharged with the brine. Due to the slow degradation of benzotriazole, it is persistent and might accumulate in sediments if the pH is low enough to allow adsorption to suspended material. Acutely toxic effects are improbable because the expected brine concentrations are well below the LC50 values of trout and Daphnia magna. Still the substance is classified as harmful for marine organisms. The release of benzotriazole per cubic metre product water, corresponding to a continuous dosage of 3-5 ppm to the feed water, amounts to 9-15 g.

The most important representative of heavy metals dissolved from the tubing material is copper, because copper-nickel heat exchangers are widely used. In brines from MSF plants it represents a major contaminant. Assuming a copper level of 15 ppb in the brine and a product-brine-ratio of 1:2 /Höpner and Lattemann 2002/, the resulting output from the reference MSF plant with a capacity of 24,000 m^3/d is 720 g copper per day. Generally, the hazard to the ecosystem emanates from the toxicity of copper at high levels. Here, levels are low enough not to harm the marine biota, but accumulation of copper in sediments represents a latent risk as it can be remobilised when conditions change from aerobic to anaerobic due to a decreasing oxygen concentrations.

To illustrate the latent risk posed by discharge of untreated brine Figure 6 5 compares reported discharge levels to eco-toxicity values and the EPA water quality criteria. The eco-toxicity values have been derived from values which have been determined during tests with copper sulphate under the assumption that copper sulphate is of less concern for saltwater organisms / Lattemann and Höpner 2003/. Diluting discharge water with cooling water does not produce relief as reported levels are still above water quality criteria and total loads stay the same.

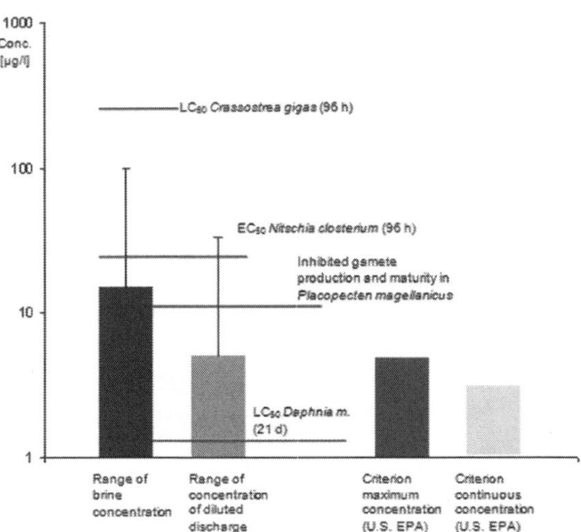

Fig. Copper: Eco-toxicity (LC50 = mean lethal concentration, EC50 = mean effective concentration) compared to ranges of brine concentration and water quality criteria, /Lattemann and Höpner 2003/, modified

MULTI-EFFECT DISTILLATION DESALINATION (MED)

Seawater Intake

The flow rate of the cooling water which is discharged at the outlet of the final condenser depends on the design of the MED distiller and the operating conditions. In the case of a conversion rate of 11 per cent (related to the seawater intake flow), 9 m^3 of seawater are required for 1 m^3 of fresh water (Figure 6 6). Due to the smaller unit sizes the seawater intake capacity for a single MED unit would be lower than for a single MSF unit, but in the majority of cases the required distillate production is reached by installing several units in parallel. Thus, the seawater intake capacity for MED plants and MSF plants would be similar. Nevertheless, the potential damage caused by impingement and entrainment at the seawater intake must be regarded as high.

Discharge of Brine Containing Additives

The discharge of brine represents a strong impact to the environment due to its changed physical properties and to the residues of chemical additives or corrosion products. In MED plants common chemical additives are biocides, antiscalants, antifoaming agents at some plants, and corrosion inhibitors at some plants. The conditioning of permeate to gain palatable, stable drinking water requires the addition of chlorine for disinfection, calcium, e.g. in form of calcium hydroxide, for remineralization and pH adjustment /Raluy 2003/, /Delion et al. 2004/. Figure 6 7 shows where the chemicals are added and at which concentrations as well as the characteristics of the brine and its chemical load.

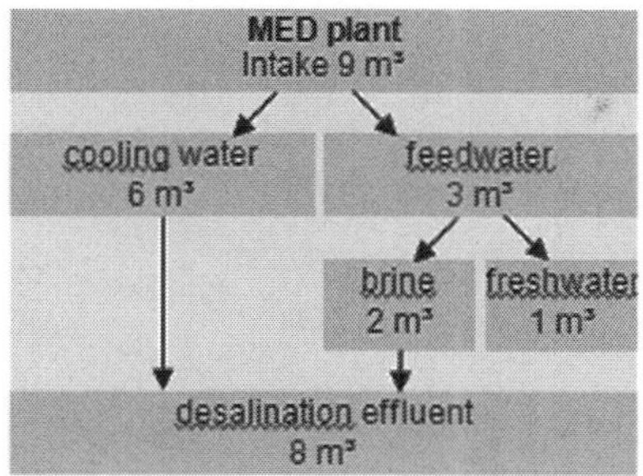

Fig. Flow chart of reference MED process

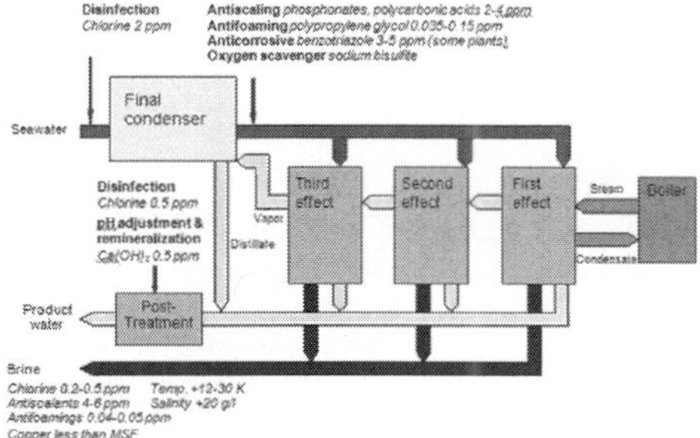

Fig. MED process scheme with input and output concentrations of additives and brine characteristics

Physical Properties of Brine

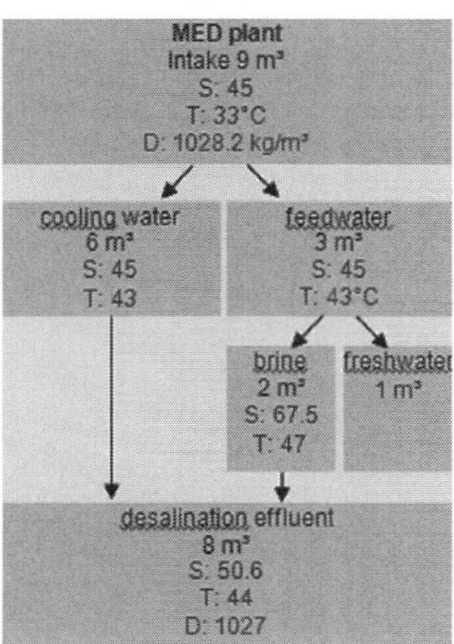

Fig. Flow chart of reference MED process with salinity (S, in g/l), temperature (T, in °C) and density (D, in g/l), /Lattemann and Höpner 2003/, modified

The physical parameters of the brine are different compared to the intake seawater. During the distillation process the temperature rises and salt accumulates in the brine. Taking the reference process (Figure 6 6) with a conversion rate of approx. 11.2 per cent as example the salinity rises from 45 g/l to 66 g/l (Figure 6 8). Brine and cooling water temperature rises by about 14 and 10 K, respectively. Salinity of the brine is reduced by blending with cooling water, but still reaches a value of 5.6 g/l above ambient level. The resulting decrease of density is very small what can be attributed to balancing effects of temperature and salinity rise.

Biocides

Surface water contains organic matter, which comprises living or dead particulate material and dissolved molecules, leads to biological growth and causes formation of biofilm within the plant. Therefore both the feed water and the cooling water are disinfected with the help of biocides. The most common biocide in MED plants is chlorine. A concentration of up to 2000 μg/l is sustained by a continuous dosage. Chloride reacts to hypochlorite and, in the case of seawater, especially to hypobromite. Residual chloride is released to the environment with the brine where it reaches values of 200-500 μg/l, representing 10–25 per cent of the dosing concentration. Assuming a product-effluent-ratio of 1:8 the specific discharge load of residual chlorine per m^3 of product water is 1.6–4.0 g/m^3. For a plant with a daily desalination capacity of 24,000 m^3, for instance, this means a release of 38.4-96.0 kg of residual chlorine per day.

Antiscalants

A major problem of MED plants is the scale formation on the heat exchanger surfaces which impairs the heat transfer. The most common scale is formed by precipitating calcium carbonates due to increased temperatures and brine concentration. Other scale forming species are magnesium hydroxide, and calcium sulphate, the latter being very difficult to remove as it forms hard scales. Therefore sulphate scaling is avoided in the first place by regulating the operation parameters temperature and concentration in such a way that the saturation point of calcium sulphate is not reached. Calcium carbonates and magnesium hydroxides, again, are chemically controlled by adding acids and/or antiscalants.

In the past, acid treatment was commonly employed. With the help of acids the pH (acidity value) of the feed water is lowered to 2 or 3 and hereby the bicarbonate and carbonate ions chemically react to carbon dioxide which is released in a decarbonator. Thus, the $CaCO_3$ scale forming ions are removed from the feed water. After acid treartment the pH of the feed water is re-adjusted. Commonly used acids are sulphuric acid and hydrochloric acid, though the first is preferred because of economic reasons. High concentrations and therefore large amounts of acids are necessary for the stoichiometric reaction of the acid. Apart from a high consumption of acids further negative effects of using acids are the increased corrosion of the construction materials and thus reduced lifetimes of the distillers as well as handling and storage problems. The negative effects mentioned above have led to the development of alternatives: Nowadays antiscalants are replacing acids during operation. But before talking about antiscalants, the use of acids as cleaning agents needs to be mentioned because that's when significantly acidic effluents occur. During this periodic cleaning procedure the pH is lowered to 2-3 by adding citric or sulfamic acid, for instance, to remove carbonate and metal oxide scales. In this context Mabrook (1994, in /Lattemann and Höpner, 2003/) explained an observed change in density and diversity of marine organisms by a decreased pH of 5.8 compared to 8.3 in coastal waters. Ecotoxic pH values range from 2-2.5 for starfish (LC50, HCl, 48 h) to 3-3.3 for salt water prawn (LC50, H2SO4, 48 h) and show the sensitivity of marine organisms to low pH values. Little mobile organisms, like starfish, are especially affected by an acid plume as they cannot avoid this zone. To mitigate these possible effects the cleaning solution should be neutralized before discharge or at least blended with the brine during normal operation.

The mode of action of antiscalants is described in Chapter 6.1.2. They react substoichio-metrically which is the reason why they are effective at very low concentrations. Polyphosphates represent the first generation of antiscalant agents with sodium hexametaphosphate as most commonly used species. A procedural disadvantage is the risk of calcium phosphate scale formation. Of major concern to the aquatic environment is their hydrolytic decomposition at 60°C to orthophosphate which acts as a nutrient and causes eutrophication. The development of algae mats on the water body receiving the discharge could be ascribed to the use of phosphates (Abdel-Jawad and Al-Tabtabaei 1999, in /Lattemann and Höpner 2003/). Because of these reasons they have partly been substituted by thermally stable phosphonates and polycarbonic acids, the second generation of antiscalants. Where phosphates have been replaced by these substances the problem of algae growth could be solved completely. Main representatives of polycarbonic acids are polyacrylic and polymaleic acids. Especially polyacrylic acid has to be dosed carefully if precipitation is to be avoided. The reason for this is that, at lower concentrations, it enhances agglomeration and therefore also serves as a coagulant in RO plants.

Discharge levels of phosphonates and polycarbonic acids are classified as non-hazardous, as they are far below concentrations with toxic or chronic effects. They resemble naturally occurring humic substances when dispersed in the aquatic environment which is expressed by their tendency to complexation and their half-life of about one month, both properties similar to humic substances. Though they are generally assumed to be of little environmental concern, there is a critical point related to these properties. As they are rather persistent they will continue to complex metal ions in the water body. Consequently, the influence on the dissolved metal concentrations and therefore metal mobility naturally exerted by humic substances is increased by polymer antiscalants. The long-term effect induced hereby requires further research. A MED plant with a daily capacity of 24,000 m^3 releases about 144-288 kg of antiscalants per day if a dosage concentration of 2-4 mg per litre feedwater is assumed. This represents a release of 6 g per cubic meter of product water.

Antifoaming Agents

MED plants also use antifoaming agents, but compared to MSF plants, it's less usual. The use of antifoaming agents can be necessary if foam forms in the presence of organic substances concentrated on the water surface which derogates the thermal desalination process by hampering the falling film flow onto the horizontal evaporator tubes and thus the wetting of the tubes. As the agents are organic substances, too, they must carefully be chosen and dosed. Blends of polyglycol are utilized, either containing polyethylene glycol or polypropylene glycol. These substances are generally considered as non-hazardous and low discharge concentrations of 40-50 μg/l per litre brine further reduce the risk of environmental damage. However, highly polymerized polyethylene glycol with a high molecular mass is rather resistant to biodegradation. On this account it has been replaced in some industrial applications by substances, such as dialkyl ethers, which show a better biodegradability. Under the assumption of a product-feedwater-ratio of 1:3 and 0.035-0.15 ppm dosing 0.1-0.45 g per cubic meter of product water are released.

Corrosion Inhibitors and Corrosion Products

The corrosion inhibitors that are used in MSF plants are also necessary in MED plants. However, it is assumed that the copper load is smaller compared to MSF plants as operation temperatures are lower and piping material with lower copper contents are used, such as titanium and aluminium-brass.

REVERSE OSMOSIS (RO)

Seawater Intake

The conversion rate of RO processes ranges between 20 and 50 per cent /Goebel 2007/, signifying an intake volume of less than 5 m^3 of seawater per cubic meter of freshwater.

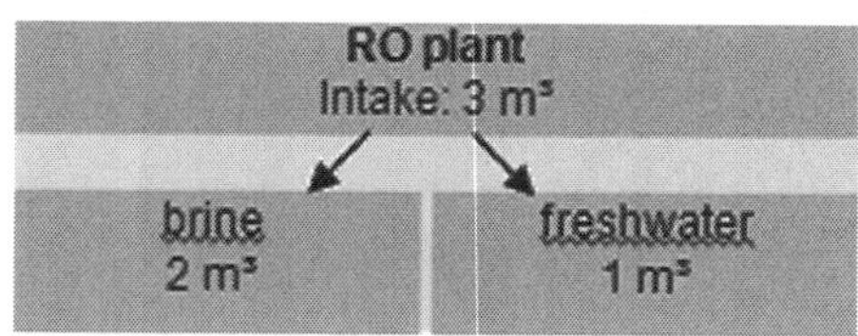

Fig. Flow chart of reference RO process

Therefore, compared to the thermal processes the mechanical process of RO requires significantly less intake water for the same amount of product water. Consequently the loss of organisms through impingement and entrainment is lower. The flows, shown in Figure 6 9, result from a conversion rate of 33 per cent.

Discharge of Brine Containing Additives

The discharge of brine represents a strong impact to the environment due to its changed physical properties and to the residues of chemical additives or corrosion products. In RO plants common chemical additives are biocides, eventually acids if not yet substituted by antiscalants, coagulants, and, in the case of polyamide membranes, chlorine deactivators. The conditioning of permeate to gain palatable, stable drinking water requires the addition of chlorine for disinfection, calcium, e.g. in form of calcium hydroxide, for remineralization and pH adjustment /Raluy 2003/, /Delion et al. 2004/. Figure 6 10 shows where the chemicals are added and at which concentrations as well as the characteristics of the brine and its chemical load.

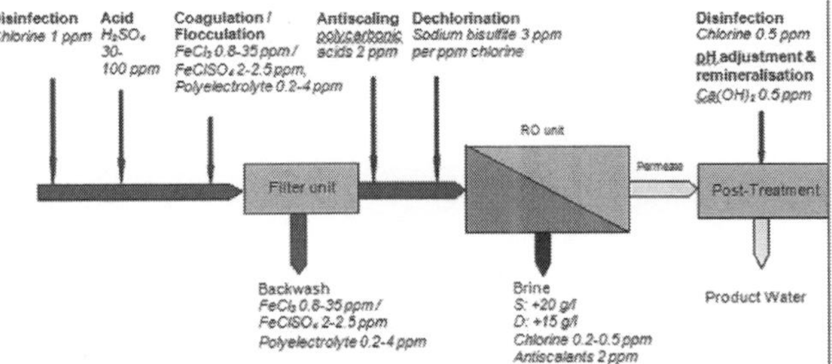

Fig. RO process scheme with input and output concentrations of additives and brine characteristics, /Lattemann and Höpner 2003/, modified

Physical Properties of Brine

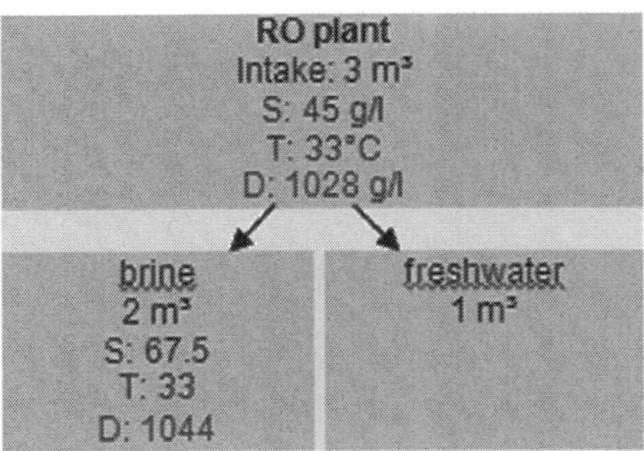

Fig. Flow chart of reference RO process with salinity (S, in g/l), temperature (T, in °C) and density (D, in g/l), /Lattemann and Höpner 2003/, modified

The salinity of the brine is increased significantly due to high conversion rates of 30 to 45 per cent. The conversion rate of 32 per cent of the process presented in Figure 6 9 leads to a brine salinity of 66.2 g/l (Figure 6 11). As the temperature stays the same during the whole process, also density increases significantly from 1028 g/l to 1044 g/l. If the RO process is coupled with electricity generation and the effluent streams are blended, the warmed cooling water from the power plant reduces the overall density slightly compared to the ambient value and the overall salinity is almost reduced to the ambient level.

Biocides

Surface water contains organic matter, which comprises living or dead particulate material and dissolved molecules, leads to biological growth and causes formation of biofilm within the plant. Therefore the RO feed water is disinfected with the help of biocides. The most common biocide in RO plants is chlorine. A concentration of up to 1000 μg/l is sustained by a continuous dosage. Chloride reacts to hypochlorite and, in the case of seawater, especially to hypobromite. In RO desalination plants operating with polyamide membranes dechlorination is necessary to prevent membrane oxidation. Therefore the issue of chlorine discharge is restricted to the smaller portion of plants which use cellulose acetate membranes. Regarding these plants residual chlorine is released to the environment with the effluents where it reaches values of 100-250 μg/l, representing 10–25 per cent of the dosing concentration. Assuming a product-effluent-ratio of 1:2 the specific discharge load of residual chlorine per m^3 of product water is 0.2–0.5 g/m^3. For a plant with a daily desalination capacity of 24,000 m^3, for instance, this means a release of 4.8–12 kg of residual chlorine per day. Again, the problem of chlorine discharge is restricted to plants with cellulose acetate membranes. In contrast, the release of chlorination by-products is an issue at all RO plants regardless of the material of their membranes, as by-products form up to the point of dechlorination.

Coagulants

The removal of suspended material, especially colloids, beforehand is essential for a good membrane performance. For this purpose coagulants and polyelectrolytes are added for coagulation-flocculation and the resulting flocs are hold back by dual media sand-anthracite filters. Coagulant substances are ferric chloride, ferrous sulphate, and ferric chloride sulphate or aluminium chloride. To sustain the efficiency of the filters, they are backwashed regularly. Common practice is to discharge the backwash brines to the sea. This may affect marine life as the brines are colored by the coagulants and carry the flocs. On the one hand the decreased light penetration might impair photosynthesis. On the other hand increased sedimentation could bury sessile organisms, especially corals. The dosage is proportional to the natural water turbidity and can be high as 30 mg/l. This extreme dosage results in a specific load of 90 g per m^3 of product water and a daily load of a 24,000 m³/d plant of 2200 kg which adds to the natural turbidity.

Polyelectrolytes support the flocculation process by connecting the colloids. Possible substances are polyphosphates or polyacrylic acids and polyacrylamides respectively, which are also used as antiscalants. The concentration decides whether they have a dispersive or coagulative effect. Compared to their use as antiscalants the dosage of polyelectrolytes is about a tenth of the concentration required for dispersion. These substances are not toxic; the impact they cause is connected to the increased turbidity. A dosage of 500 μg/l implies a discharge of 1.5 g per m³ of product water and a daily load of a 24,000 m³/d plant of 36 kg which adds to the natural turbidity.

Fig. Red brines containing ferric sulphate from filter backwash at Ashkelon RO desalination plant; backwash with 6,500 m³ in 10-15 minutes every hour.

Antiscalants

The main scale forming species in RO plants are calcium carbonate, calcium sulphate and barium sulphate. Acid treatment and antiscalant dosage are used for scale control. Here, sulphuric acid is most commonly used and dosed with a range of 30–100 mg/l. During normal operation the alternative use of antiscalants, such as polyphosphates, phosphonates or polycarbonic acids, has become very common in RO plants due to the negative effects of inorganic acid treatment explained in Chapter 6.1.2. As it is explained there, these antiscalants react substoichiometrically and therefore low concentrations of about 2 mg/l are sufficient. A RO plant with a daily capacity of 24,000 m³ releases about 144 kg of antiscalants per day if dosage concentration of 2 mg per litre feedwater and product-feedwater-ratio of 1:3 are assumed /Höpner and Lattemann 20002/. This represents a release of 6 g per cubic meter of product water.

Membrane Cleaning Agents

Apart from acid cleaning, which is carried out with citric acid or hydrochloric acid, membranes are additionally treated with sodium hydroxide, detergents and complex-forming species to remove biofilms and silt deposits. By adding sodium hydroxide the pH is raised to about 12 where the removal of biofilms and silt deposits is achieved. Alkaline cleaning solutions should be neutralized before discharge, e.g. by blending with the brine. Detergents, such as organo-sulfates and –sulfonates, also support the removal of dirt particles with the help of both their lipophilic and hydrophilic residues. Regarding their behaviour in the marine environment, organo-sulfates, e.g. sodium dodecylsulfate (SDS), and organo-sulfates, e.g. sodium dodecylbenzene sulfonate (Na-DBS), are quickly biodegraded. Apart from the general classification of detergents as toxic no further information is available on toxicity of Na-DBS, but it's assumed to be relatively low once the decomposition has started with cutting off the hydrophilic group. In contrast, LC50 for fish, Daphnia magna and algae are available in the case of SDS confirming the categorization as toxic substance. But, again, fast degradation reduces the risk for marine life. This risk could be further reduced by microbial waste treatment which destroys the surface active properties and degrades the alkyl-chain.

Complex-forming species, such as EDTA (Ethylendiamine tetraacetic acid) are employed for the removal of inorganic colloids and biofouling. From comparing the calculated maximum estimate of discharge concentration (46 mg/l) and an LC50 for bluegill (159 mg/l, 96 h) it can be deduced that in the case of EDTA direct toxicity is of minor concern. In contrast, persistent residual EDTA in the marine environment might provoke long-term effects in connection with its chelating and dispersing properties. Consequences of increased metal solubility and mobility and thereby reduced bioavailability still need further investigation. Generally, total amounts are of bigger interest than concentrations.

During the periodic membrane cleaning process also further disinfectants such as formaldehyde, glutaraldehyde, isothiazole, and sodium perborate, are used. These substances are toxic to highly toxic and reach toxic concentrations if discharged all at once. Therefore deactivation should be compulsory. Several deactivation substances are available: formaldehyde can be deactivated with hydrogen peroxide and calcium hydroxide or sodium hydroxide and isothiazole is neutralized with sodium bisulfite. Sodium perborate has to be handled carefully as it breaks down to sodium borate and hydrogen peroxide. The latter is the actual biocide and therefore may not be overdosed, also for reasons of membrane protection as it has an oxidizing effect.

Corrosion Products

In RO plants corrosion is a minor problem because stainless steels and non-metal equipment predominate. There are traces of iron, nickel, chromium and molybdenum being released to the water body, but they do not reach critical levels /Lattemann and Höpner 2007a/. Nevertheless, an environmentally sound process should not discharge heavy metals at all; therefore alternatives to commonly used material need to be found.

Dechlorination

The removal of chlorine is performed with sodium bisulfite, which is continuously added to reach a concentration three to four times higher than the chlorine concentration, the former amounting to 1500-4000 μg/l. The corresponding amount per cubic metre of product water is 4.5-12 g/m³. As this substance is a biocide itself and harms marine life through depletion of oxygen, overdosing should be prevented. Alternatively sodium metabisulfite is used.

LIFE-CYCLE ASSESSMENT OF MATERIALS AND EMISSIONS

Methodology of LCA and Material Flow Networks

Generally accepted guidelines for carrying out a Life Cycle Assessment (LCA) can be found in ISO 14040 ff. /Guinée 2002/. In an LCA the production, the operation, and the dismantling of the considered products are modelled. Included are the upstream processes of the most important fuels and materials. In Figure 6 13 this is demonstrated by the example of a solar thermal power plant's life cycle. Starting with the production of the solar thermal power plant the upstream processes of both the used materials and the used electricity are modelled up to the mining processes of the crude materials. To operate the plant, some more materials are used (for example reimbursement of broken mirrors, make-up heat transfer fluid, water for cleaning the mirrors). For the time being the plant's end of life is only considered partly because there does not exist adequate data and concepts up to now.

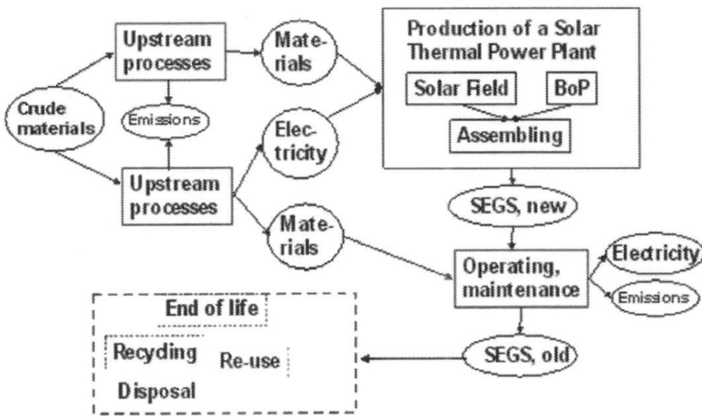

Fig. Life cycle of a solar thermal power plant (type SEGS)

After modelling the relevant material and energy flows in a material flow net the life cycle inventory is created. The input-output balance of the whole system is calculated using upstream processes taken from commercial LCA databases. Finally, the environmental impacts are calculated by allocating the resulting emissions to different impact categories (global warming potential, acidification, or resource consumption, for example). By scaling the results to a functional unit (1 kWh electricity or 1 m3 water) different production processes can be compared and the best technology with regard to an impact category can be selected.

Frame Conditions and Data Sources

In a broad sense the question should be answered if environmental impacts associated with the provision of desalted water using fossil primary energy carriers can be reduced by a system using concentrated solar power. Furthermore, it is also interesting to what extent a changed electricity mix for the production of the facilities could have an effect on the balance, like e.g. the electricity mix in MENA used for RO or heat and power provided by a gas-fired combined generation (CHP) plant for MED and MSF. The LCA considers exploration, mining, processing and transportation of the fuels, especially for the electricity mix as well as materials for the required infrastructure. Furthermore, the production of single components is considered. This comprises the solar field, steam generator, mechanical and electrical engineering, constructional engineering, thermal energy storage, steam turbine and the desalination plant. Modelling of the facility operation includes maintenance, i.e., cleaning and material exchange. The disposal of the facility is composed of the demolition, depository and recycling.

The function studied in the LCA is that of cleaning seawater with a salinity of 45 g/l to produce freshwater with 200 ppm salt included. The functional unit has been defined as 1 m3 of freshwater delivered from the plant. In the following paragraphs it is differentiated between materials, modules, and components. While a material means stainless steel or molten salt an (LCA) module means the process of manufacturing these materials (and representing this process as LCA data). A component consists of different materials and modules, for example the solar field of a power plant. The reference period for this study is the year 2007 that means the most actual available LCA modules are used. Reference area is the MENA region. Since LCA modules are not available for this region modules representing the European situation are used. This means that the results are based on production processes with a better performance and efficiency than usually available in the MENA region.

The MENA electricity mix is modelled to be able to compare RO using MENA electricity with RO using solar electricity. Notwithstanding the former assumptions it is modelled regarding a possible situation in 2010. This means that the results of RO using MENA electricity become better than today's situation because the 2010 electricity mix considers more renewable energies than used today. As sources for the LCA modules in general the Swiss LCA database ECOINVENT® /ecoinvent 2007/ is used. For some modules not available in /ecoinvent 2007/ the LCA tool UMBERTO® /IFEU/IFU 2007/ is used. The material and energy flow network as well as the life cycle impact assessment is modelled with Umberto. The study uses the most recent inventory data available both for solar thermal power plants and the desalting processes:

Solar Thermal Power Plant

As CSP plant a direct steam based trough is chosen. As reference the pre-commercial 5 MW INDITEP power plant planned to be built in Spain is taken and scaled up to 20 MW. The data is taken from /NEEDS 2007/. Instead of using parabolic troughs as designed for INDITEP the solar field is exchanged by a linear Fresnel collector field. Data for one m^2 solar field provided by the company Novatec-Biosol who developed Fresnel mirrors with a very light design was implemented /Novatec 2007/. The solar field is linearly scaled up to the necessary extent. Since for direct steam technology a latent heat storage medium is needed for evaporation, DLR provided data for a 6 hours 50 MWel storage system using phase change materials (PCM) based on PCM developments in laboratory scale. The storage system is linearly scaled up to the necessary extent. It operates in three steps /Michels and Pitz-Paal 2007/: During the preheating step a conventional concrete storage is used which is heated up (sensible heat storage). This step is followed by the evaporation phase served by a (cascaded) latent heat storage. The increasing heat causes (several) phase changes (e.g. from solid to liquid) but does not increase the storage temperature by itself. In the last step, the superheating phase, a concrete storage is used again. For the applied storage system $NaNO_3$ is used, but in general different mixtures of $NaNO_3$, KNO_3 and KCL are possible. To increase the thermal conductivity aluminium plates are placed into the salt. To refer the resulting emissions to one kWh the yearly expected output has to be multiplied with the expected life time of the power plant. The following lifetimes are assumed: solar field and power block: 30 years, storage system 25 years, building 60 years.

Desalination Plants

The same desalination plants as described in the former chapters are modelled within the LCA. The inventory data is taken from /Raluy et al. 2006. Table 6 1 shows the relevant energy consumptions of the desalting plants based on 46 000 m3/d capacity. The lifetime of the desalination plants is assumed to be 25 years, that of the building to be 50 years.

Table. Energy consumption of seawater desalination plants. MSF Multi-Stage Flash, MED Multi-Effect Desalination, RO Reverse Osmosis

Energy source	Unit	MED	MSF	RO
Electricity	kWh / m^3 desalted water	2	4	4
Heat	MJ / m^3 desalted water	237	300	

MENA Electricity Mix

The MENA electricity mix was built using electricity production modules available in the ecoinvent database and suitable to the MENA situation assumed for 2010. For example, the electricity generation from oil was modelled using the Greek module because of its low energy efficiency. Table 6 2 presents details on the assumed MENA mix.

Table. Composition of the Modelled MENA Electricity Mix

Energy source	Share		LCA module (ecoinvent name)	Efficiency
	%	TWh/a		%
Renewables	6	50	electricity, hydropower, at power plant [GR]	
Oil	63	500	electricity, oil, at power plant [GR]	37.9
Natural Gas	25	200	electricity, natural gas, at power plant [IT]	37
Hard Coal	6	40	electricity, hard coal, at power plant [ES]	35.8

Natural Gas Fired Power Plants

Both natural gas fired power plants (the combined cycle power station as well as the combined heat and electricity power station) are taken from ecoinvent representing the best available technology within this group. Figure 6 14 shows the evaluated seawater desalination technologies and their possible combination with energy from solar thermal power plants and fossil fuels. The Reverse Osmosis (RO) Membrane Technology is combined with electricity from the solar thermal power plant and compared with the same technology using electricity from the MENA mix and – as best available technology – electricity from a gas-fired combined cycle power station. Multi-Effect-Distillation (MED) and Multi-Stage Flash Desalination both need power and steam. MED is combined with electricity and steam delivered by the CSP plant. This combination is compared with MED and MSF both using electricity and steam from a natural gas fired CHP plant using the best available technology.

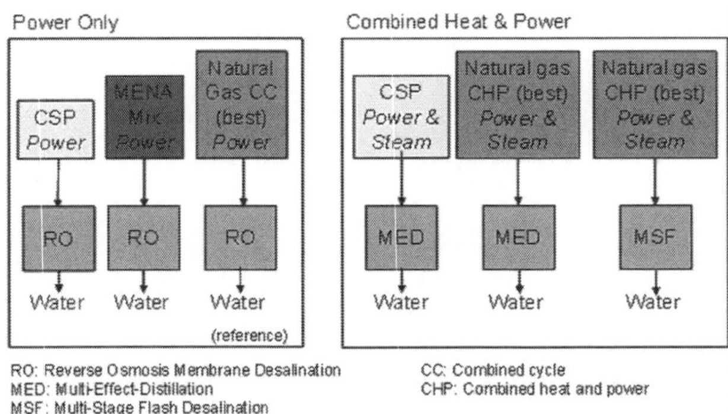

Fig. Considered seawater desalination technologies based on solar energy or fossil fuels

The results are compared to desalted water stemming from a reverse osmosis plant that receives electricity from a gas-fired combined cycle power station (third version in the figure below) because in terms of environmental impact it represents the best possible conventional solution for desalination based on fossil fuel available today. According to ISO 14 042 requirements impact categories have to be chosen to assess the results of the inventory analysis (so called life cycle impact assessment). The impact categories applied in this study are taken from the method "UBA-Verfahren" provided by the German Federal Environmental Agency (UBA) /UBA 1995, UBA 1999/. The parameters result from the impact categories and are shown in Table 6 3.

Table. Impact Categories and Inventory Parameters Applied in this Study.

Impact Category	Inventory Parameter	Aggregated Impact Parameter	Ratio
Resource consumption	Cumulated Energy Demand (CED)	MJ (inventory parameter)	
Global warming a	CO_2	g CO_2–Equivalents	1
	CH_4		21
	N_2O		310
Acidification	SO_2	mg SO_2–Equivalents	1
	NOX		0.7
	NH_3		1.88
	HCl		0.88
Eutrophication	NOX	mg $PO4^{3}$--Equivalents	0.13
	NH_3		0.33
Summer Smog (Photochemial oxidant)	NMHC	mg Ethen-Equivalents	0.416
	CH_4		0.007
Cancerogenic potential, human-toxicity (inventory parameter)	Particles and dust	mg (inventory parameter)	

a Time horizon 100 years

Results

They are scaled to the best possible conventional solution (reverse osmosis plant combined with a gas-fired combined cycle power station, 100 per cent line). The figure clearly shows that the environmental impact of MSF, even if operated by steam stemming from combined heat and power (CHP), would have a five-fold impact with respect to energy and global warming, and even an eleven-fold impact with respect to eutrophication when compared to the best conventional case. The next strongest impact is caused by MED operated with steam from fossil fuel fired CHP which is still three- to seven-fold with respect to the best case. Reverse osmosis powered by the electricity mix available in MENA has also considerably higher emissions than the best case and even represents the worst case in the category acidification due to high consumption of electricity, rather low efficiencies of power generation, and the intensive use of fuel oil in the MENA electricity mix.

The figure shows clearly that for all categories conventional reverse osmosis has lower impacts than conventional MSF and MED, and that MED is also considerably better than MSF. It also shows that, depending on the category, in case of operating RO and MED using concentrating solar power as energy source, between 90 per cent and 99 per cent of the overall emissions can be eliminated. Therefore, CSP eliminates one of the major causes of environmental impact of seawater desalination: the emissions related to its large energy demand.

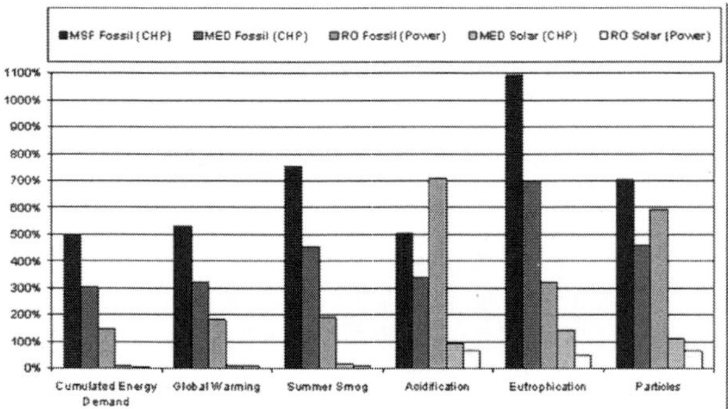

Fig. Life-cycle emissions of seawater desalination technologies in the MENA region based on fossil fuel and concentrating solar power compared to the best possible conventional solution based on a gas-fired combined cycle power plant providing electricity for reverse osmosis (100 per cent).

Table. Life-cycle emissions of seawater desalination plants in the MENA region based on fossil fuel vs. plants based on concentrating solar power.

Impact Category Fossil (CHP)	Unit	MED (CHP)	Ro Solar Solar	RO Fossil (Power)	MED Fossil (Power)	MSF (CHP)
Cumulated Energy Demand	kJ/m^3	3,579	2,298	63,790	131,767	218,417
Global Warming	kg CO2/m3	0.27	0.21	4.41	7.75	12.83
Summer Smog	kg Ethen/m3	5.89E-05	3.30E-05	6.63E-04	1.54E-03	2.53E-03
Acidification	kg SO2/m3	4.48E-03	3.16E03	3.35E-02	1.61E-02	2.27E-02
Eutrophication	kg PO4/m3	4.50E04	1.49E-04	9.90E-04	2.16E-03	3.38E-03
Particles	kg PM10/m3	1.01E-03	5.97E-04	9.97E-03	4.13E-03	6.31E-03

In case of the CSP/RO plant, the remaining emissions related to the construction of the solar field, the thermal energy storage and the power block are comparable to those related to the construction of the RO plant itself (Figure 6 16). The same is true for CSP/MED, in fact in this case the emissions related to the construction of the solar field, the thermal energy storage and the power block are clearly smaller than those related to the MED plant itself, due to its large material demand (Figure 6 18).

Compared to the presently used standard solution for seawater desalination in the MENA region, a multi-stage flash plant connected to a combined heat and power station, CSP/RO and CSP/MED reduce the cumulated energy consumption and the emission of greenhouse gases to about

1 per cent. Thus, CSP desalination offers a cost-effective and environmental-friendly solution for the MENA water crisis and can solve the problem of water scarcity in a sustainable way, taking also into account all necessary measures for water efficiency and re-use. If the electricity mix used for the production of the plants can be changed to more renewable energy in the future, the overall emissions will be reduced even further. However, there are also considerable environmental impacts related to the concentrated brine and to the chemicals contained in the effluent of both RO and MED seawater desalination plants. In the following we will investigate a series of possible solutions to mitigate those emissions to a compatible level.

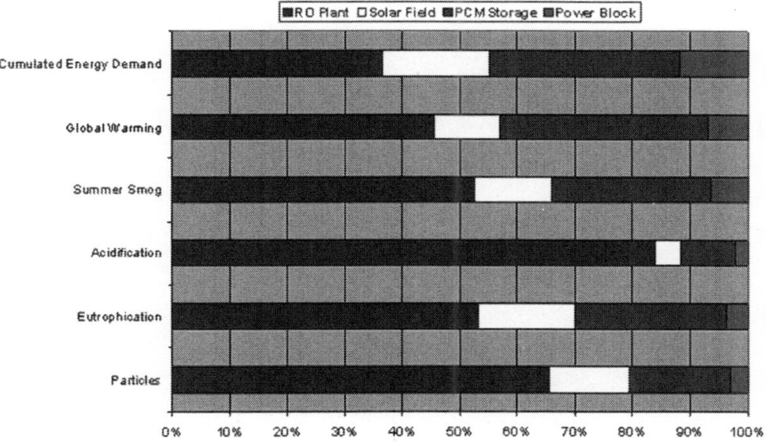

Fig. Contribution of the different components of a solar CSP/RO plant to life-cycle emissions.

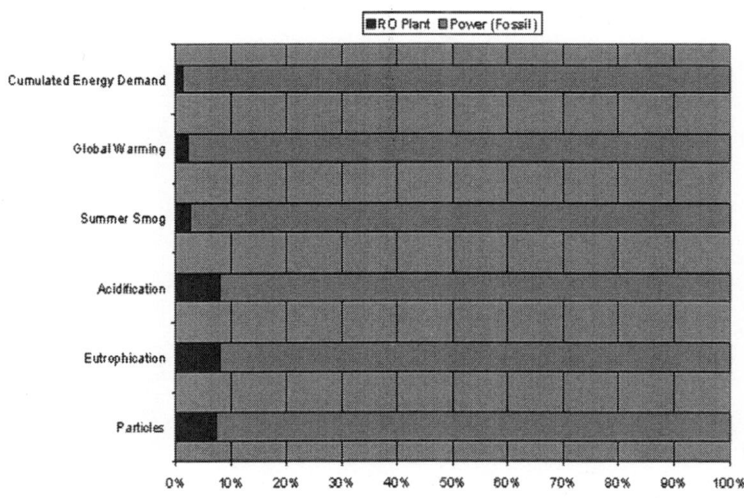

Fig. Contributions to the life-cycle emissions of a conventional RO plant receiving power from the MENA electricity grid.

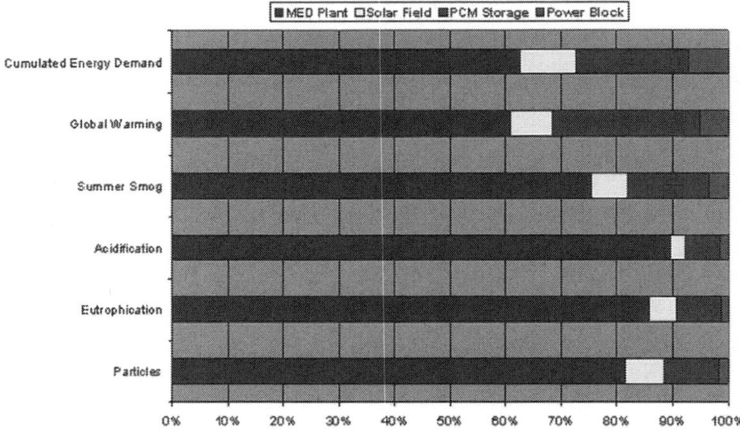

Fig. Contribution of the different components of a solar CSP/MED plant to the total life-cycle emissions.

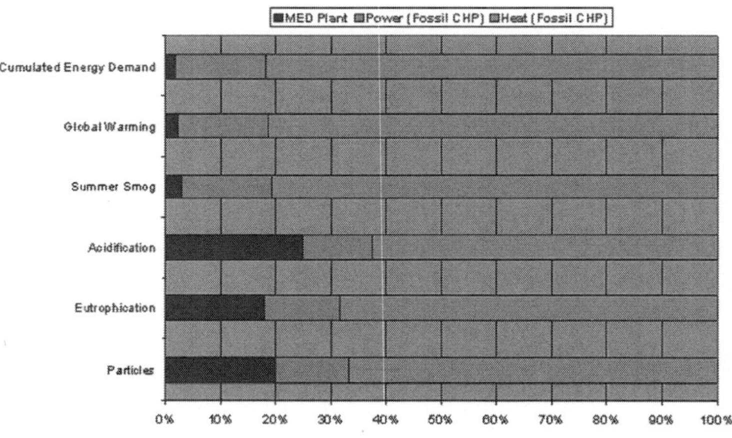

Fig. Contributions to the life-cycle emissions of a conventional MED plant receiving energy from a natural gas fired combined heat and power station (CHP).

MITIGATION MEASURES

Capacities of seawater desalination are expected to rise significantly in the short- and medium term. The growing impacts caused by increasing numbers of desalination plants cannot be accepted. Therefore mitigation measures have to be taken to reduce the impacts drastically. In this chapter possible mitigation measures are identified and finally an outlook for an environmentally sound desalination plant will be presented. The first step is to stop generating desalted water with fossil energy and to switch to renewable energy. As explained earlier in this report concentrated solar power is the ideal alternative to fossil fuels, especially in the context of desalination. By using concentrated solar power for desalination the impact categories energy demand and air pollution are mitigated strongly. Analogically, the impacts caused by seawater intake and brine discharge need to be mitigated. Apart from impact-specific measures there are general measures such as environmental impact assessment and site selection that need to be taken into account in the course of planning desalination plants.

General Measures

During the planning process, all impacts the desalination project could have on the environment should be evaluated and mitigation measures should be taken into account. By carrying out an Environmental Impact Assessment (EIA) all potential impacts can be identified and evaluated and adequate mitigation measures and process alternatives can be developed in a systematic manner /Lattemann and Höpner 2007a,c/. An EIA is a project- and location specific instrument. In order to regard and evaluate the cumulative impacts of all plants in a region a regional water management is necessary. Strategic Environmental Assessment (SEA) is the instrument for such a purpose, because it helps to achieve sustainable development in public planning and policy making. An important mitigation measure is the careful selection of the plant site. There are environmental, technical and economic aspects that should be taken into account. Regarding the environmental aspects, the WHO recommends to avoid ecosystems or habitats that are unique within a region or globally worth protecting, that are inhabited by protected, endangered or rare species, that are important feeding or reproduction areas or that are highly productive or biodiverse (WHO in review, cited in /Lattemann and Höpner 2007a/).

Technical requirements are sufficient capacities for dilution and dispersion of the discharged brine. Here, apart from the discharge practice, the main factors of influence are the oceanographic features of the site, such as currents, tides, surf, water depth, and shoreline morphology (WHO in review, cited in/Lattemann and Höpner 2007a/). An important economic aspect is the distance of the site to the sea, to infrastructure, such as water distribution networks, power grid, road and communication network, and to the consumers. A co-use of existing infrastructure is both economically and environmentally desirable. Another aspect is the potential of conflicts with other uses and activities /Lattemann and Höpner 2007a/.

Seawater Intake

The practice of water intake influences both the direct impacts on marine organisms and the quality of intake water, which defines the pre-treatment steps. A modification of open source water intake consists in the reduction of the intake velocity and a combination of differently meshed screens, but also in locating the intake in deeper waters or offshore /Lattemann and Höpner 2007a/. Desirable alternatives to open source water intake represent beach well intake and seabed filters with directed drilled horizontal drains, the latter being applicable in aquifers, i.e. permeable, porous and fractured geological formations.

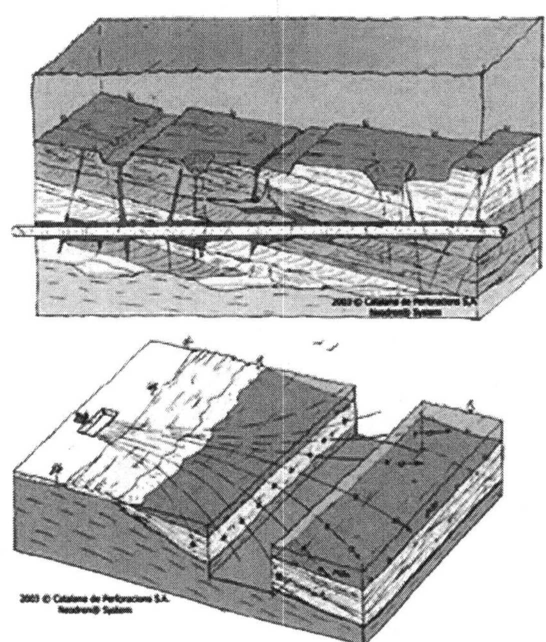

On the one hand, these measures decrease the loss of organisms through impingement and entrainment of both larger organisms and smaller plankton organisms. On the other hand multimedia and cartridge filters are not necessary and the amount of pre-treatment chemicals can be reduced or chemical pre-treatment becomes dispensable at all as the seabed acts as a natural pre-filter. This accounts especially to the technique of horizontal drain seabed intake, offered for example by Catalana de Perforacions, Fonollosa, Spain, under the trade name Neodren, which is equipped with high efficient filtering devices. The filtration pipes run in separate boreholes executed from the back of the coastline into the subsoil under the sea /Catalana de Perforacions 2007/. However, these alternative locations of source water intake mean a higher impact during construction due to unavoidable soil disturbance if drilling or excavation is necessary.

Co-location of desalination and power plant reduces overall intake water volume as cooling water from the power plant can be used as feed water to the desalination plant. Therefore impacts from entrainment and impingement, as well as from construction and land use are minimized. Reduced volumes of intake water also mean reduced chemicals in case they are still necessary. Less pre-treatment chemicals in turn represent less negative effects. The concept still has to be proven for large scale applications.

Pre-treatment

As shown in Chapter 6.1 conventional pre-treatment of input seawater including media filtration requires a variety of chemicals partially at concentrations harming the environment. Thus, alternative pre-treatment methods need to be identified allowing to reduce or to avoid the use of hazardous chemicals. A possibility is to substitute the chemical additives by electricity from renewable energies needed for additional filtration steps. Another way to avoid environmental impacts represents the substitution of hazardous chemicals by environmentally sound, so-called green additives.

Filtration Technologies

Common filtration technologies in desalination plants are media filters, e.g. dual-media or single-medium filters retaining sand particles and macrobacteria. They rely on gravity removal mechanism and require the addition of coagulants for maximum efficiency. However, scaling and fouling of tubes and RO membranes is caused by particles mostly of smaller size, such as microbacteria, viruses, colloids, dissolved salts and dissolved organics. Adequate filter technologies for these small fractions are membrane filtration systems. These systems are further divided into microfiltration, ultrafiltration and nanofiltration respectively. Figure 6 21 shows which filters apply for which particle classes.

As a first step of filtration pre-treatment microfiltration (MF) can be applied to remove colloids and suspended matter larger than 0.1 μm. For metal membrane MF system ozone backwashing is applicable having proven to be more effective than permeate or air backwashing /Kim et al. 2007. Ultrafiltration (UF) can be an effective pre-treatment against fouling of RO membranes as it retains colloids and dissolved organics. Depending on the operation scheme, a benefit for the environment is the reduction of RO membrane cleaning frequency and therefore the consumption of chemical / Vedavyasan 2007/. Another benefit can be the elimination of chlorine, sodium bisulfite for dechlorination and coagulants /Wilf and Klinko 1998/. No usage of chlorine means any formation of hazardous trihalomethanes, thus in this context, the impacts on marine organisms are significantly reduced. If the UF membrane is backwashed regularly and thoroughly with permeate, the use of chemicals both in the UF and the RO step can be eliminated completely / Xu et al. 2007/. In the pilot plant tested here the UF shows excellent performance with a backwash executed every 40 minutes lasting for 30 seconds and a backwash flow rate of 1800 l/h. Apart from a sand filter upstream of the UF no further pre-treatment step was required.

Pre-treatment with MF/UF is recommendable if the intake is designed as open water intake, as in this case the water contains bacteria and colloids. However if the feed water intake is designed as beach well intake, a pre-treatment with MF/UF is not necessarily required /Pearce 2007/. In combination with horizontal drain seabed intake UF pre-treatment including an upstream micro-bubble flotation is recommended /Peters and Pintó 2007/. Micro-bubble flotation uses a nozzle-based system for micro-bubbles with a narrowly distributed diameter. The UF unit is suggested to operate in dead-end modus.

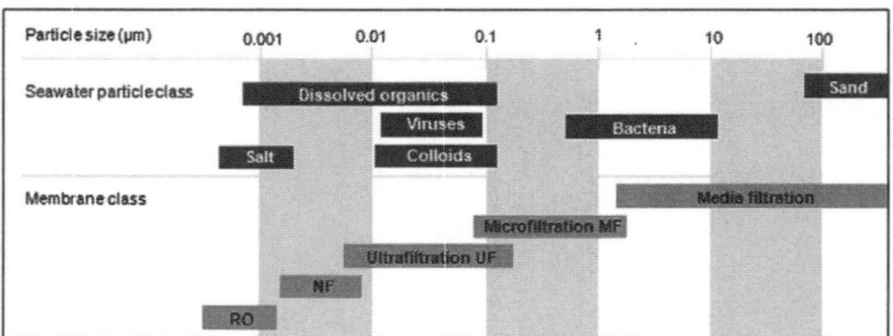

Fig. Typical seawater particles and filtration technologies compared in size (RO = reverse osmosis, NF = nanofiltration), /Goebel 2007/, modified.

Nanofiltration (NF) removes very fine suspended matter and residual bacteria, but above all it is a water softening treatment as it retains divalent ions. As these ions, e.g. Ca^{2+} and SO^{2-}, contribute significantly to scaling, nanofiltration prevents the formation of scales and replaces the conventional softening treatment /Hassan et al. 1998/. According to /Al-Shammiri et al. 2004/ nanofiltration can be considered as revolution in scale inhibition, as it prevents scale formation like no other treatment method. But NF does not only reduce hardness ions by up to 98 per cent, it also lowers the values of total dissolved solids by more than 50 per cent /Hassan et al. 1998./. Consequently, NF substitutes antiscalants, no matter which desalination process, and antifoamings in the case of thermal processes. Additionally it raises the performance of RO membranes as the RO feed water from nanofiltration contains less total dissolved solids.

Even though chemicals can be reduced or even avoided in the actual desalination process, pre-treatment filters need to be cleaned periodically as fouling occurs on the filtration membranes themselves. However, the use of chemicals at this point of the process would, again, mean impact to the environment. Therefore the development of filtration pre-treatment needs to lead into the direction that has been shown by /Xu et al. 2007/ through proving that filter cleaning can be carried out effectively by backwashing without chemicals. A higher overall consumption of electricity, e.g. due to additional pumping capacities required for membrane filtration or due to loss of permeate for membrane backwashing, can easily be accepted if it is generated from renewable energies such as concentrating solar power.

In the literature, membrane filtration systems are mentioned mainly in the context of RO desalination systems. However, they should definitely be considered for thermal processes, too, as they contribute to a reduction of chemical usage and therefore to the mitigation of impacts to the marine environment. Nano-filtration would add 300-350 $/m³/d to the investment of a desalination plant and 1 ct/m³ to the operating cost for labour, 2 ct/m³ for the replacement of membranes and 1.5 ct/m³ for chemicals, adding a total of 10-15 ct/m³ to the cost of water. To this the cost of 1.2 kWh/m³ for additional power consumption would add /MEDRC 2001.

Green Additives

In single cases, where chemicals cannot be avoided through additional filtration steps, they need to be substituted by so-called "green" chemicals. Criteria for the classification of chemicals are set by the Oslo and Paris Commission (OSPAR, cited in /Ketsetzi et al. 2007/):

- Biodegradability: > 60 per cent in 28 days Chemicals with a biodegradability of < 20 per cent in 28 days should be substituted.
- Toxicity: LC_{50} or EC_{50} > 1 mg/l for inorganic species, LC_{50} or EC_{50} > 10 mg/l for organic species
- Bioaccumulation: Logpow < 3, pow = partition in octanol/water

A chemical, that fulfils two out of three requirements and whose biodegradability is higher than 20 per cent in 28 days, is qualified for the PLONOR list (Pose little or no risk). In the future, only PLONOR listed additives should be allowed. A step into that direction is made by /Li et al. 2006/ by developing the non-toxic, rapidly biodegradable antiscalant PAP-1, synthesized from polyaspartic acid and further polycarboxylic acids. It showed very good results in the efficiency of magnesium and calcium scale inhibition as well as a fast biodegradability with 38.25 per cent reached at day 8 and 58.3 per cent at day 20 (Figure 6 22). Furthermore its impact on organisms has been tested with an algae growth inhibition test. Figure 6 23 shows the results of the measurements of chlorophyll-a concentration as an indicator of algae growth. As no limitation of growth by PAP-1 can be observed the authors classified it as environmentally friendly.

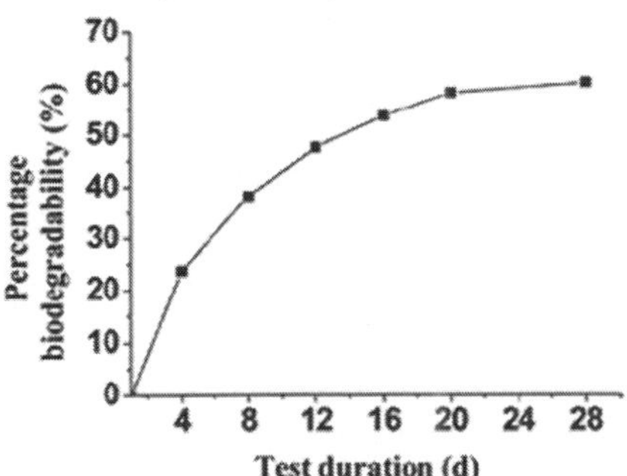

Fig. Biodegradability of PAP-1 as a function of time /Li et al. 2006/

Another approach to green antiscalants is presented by /Ketsetzi et al. 2007/. They tested the efficiency of silica scale inhibition of cationic macromolecules, i.e. inulin-based polymers modified with ammonium. The tested inhibitors showed the highest efficiency at a relatively high dosage of 40 ppm and were able to keep silica soluble at a concentration of about 300 ppm depending on the design of the polymeric inhibitor, i.e. on the average number of cationic groups per monomeric unit. There is no statement on the environmental compatibility of the inhibitor. However, inulin is of vegetable origin; therefore negative impacts on the environment are not expected.

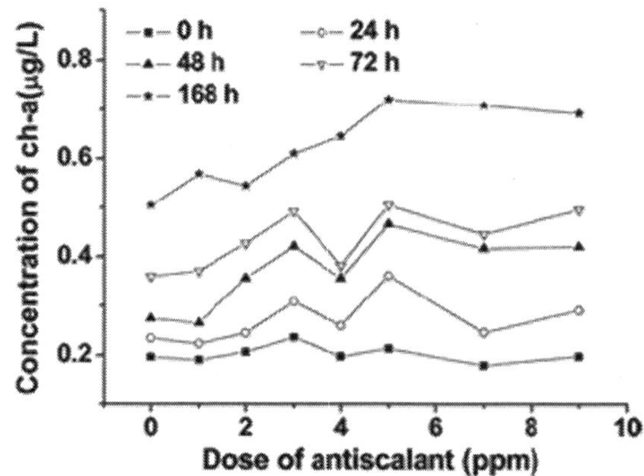

Fig. Concentration of chlorophyll-a (μg/l) as a function of antiscalant dosing concentration (ppm) with a varying test duration /Li et al. 2006/

Tubing Material

The heavy metal discharge of thermal desalination plants needs also to be eliminated. This can be achieved by substituting less corrosion-resistant tubing materials, such as copper alloys, aluminium brass and low alloyed stainless steel by more resistant materials. Among stainless steel grades the high alloyed austenitic grades and austenitic-ferritic grades possess the highest corrosion resistance /Olsson and Snis 2007/. The latter is a new generation of stainless steel, which is called duplex stainless steel due to its austenitic-ferritic microstructure. Titanium is already a commonly used tube material with a high corrosion resistance, but the prices and the lead times have significantly increased in the last years. A third alternative are polymeric materials, provided that their thermal conductivity can be increased by innovative solutions or that polymer films with a very low wall thickness can be used in order to reduce the heat transfer resistance. At present polymers are sometimes used for pipes, nozzles and droplet separators /El-Dessouky and Ettouney 1999/. By contrast, duplex steel and titanium are already used in various plants. A mid- to long-term solution could be the development of protective coatings.

A high corrosion resistance of stainless steel is achieved by either a high grade of alloying, such as the 254SMO grade with 6 per cent of molybdenum, or by the austenitic-ferritic microstructure of duplex stainless steel (DSS). Due to its structure DSS possesses higher strength, at least twice as high as austenitic steel enabling gauge, weight and cost reductions /Olsson and Snis 200/). With rising prices of alloying elements DSS is less costly than highly alloyed austenitic grades. Therefore DSS represents a real alternative to 254SMO for the replacement of corroding low alloyed stainless steel. In Table 6 5 some DSS grades and their possible locations in desalination plants are listed. In SWRO plants the high pressure parts require the most resistant grade

S32750 with the highest grade of alloying. Where pressure and salinity is lower, e.g. in the second pass parts, the lower alloyed grades S32205, S32101, and S32304 are sufficient. In MSF evaporator shells a dual duplex design has been implemented consisting of the more resistant grade S32305 for hostile conditions and of less resistant grades (S32101, S32304) for less hostile conditions.

Table. Duplex Stainless Steel Grades and their Possible Applications/Olsson and Snis 2007.

		SWRO		MSF		MSF & MED	
steel	High pressure		second pass		condensers	brine	Evaporator
shells							
designation	parts, energy	TDS: >500	TDS: 300-	(heat	heater	Dual duplex design	
(ASTM)	recovery	ppm	500 ppm	recovery)		more	less hostile
	system					hostile	
S32750	x						
S32205		x		x		x	
S32101			x		x		x
S32304			x		x		x
examples	Singapore			Aruba	Aruba	Taweelah B, Jebel Ali, Ras Abu Fontas	

Under conditions occurring in SWRO plants, i.e. at temperatures of 25°C and 45°C and salinities of 35,000 and 55,000 mg/l, titanium shows the highest corrosion resistance compared to austenitic stainless steels and nickel alloys/ Al-Malahy and Hodgkiess 2003/. Consequently the use of titanium for the high pressure parts of a SWRO plant is recommended.

Polymeric materials, e.g. PTFE, show many advantages compared to steel and to copper alloys, such as easy construction, lower construction and installation costs and the ability to operate at higher top brine temperatures without the risk of effects of scale formation /El-Dessouky and Ettouney 1999. The major advantage, of course, is the corrosion resistance making corrosion inhibitors dispensable, thus reducing the environmental impact twofold. However, there are drawbacks on the engineering side due to certain properties of polymeric, such as a thermal expansion ten times higher than metals requiring special design considerations and material aging especially at high operation temperatures that has to be taken into account. At present, the use of polymeric heat exchangers is limited by lack of practice codes, fouling concerns, limited choice and also the conservative nature of users. However, experience is made with polymeric material in a single-effect mechanical vapour compression desalination plant and described in /El-Dessouky and Ettouney 1999/. In contrast to thermal processes, polymeric materials have already entered RO plants. Here their use represents a reliable and cost effective strategy, but the high pressure parts are in the focus of the durability issue /Al-Malahy and Hodgkiess 2003/. Table 6 6 summarises which alternative materials can be used for the critical components of the different desalination processes.

Table. Overview of Suitability of Alternative Materials for the Processes MSF, MED, and RO

Material	MSF	MED	RO
high alloyed stainless steel	x	x	x
DSS	x	x	x
Titanium	x	x	x
Polymers	prospective	prospective	x

TREATMENT OF EFFLUENT BEFORE DISCHARGE

Dechlorination

If chlorine cannot be substituted as biocide right from the start, it is indispensable for environmentally friendly desalting to dechlorinate the brine before discharge. This can be carried out with the help of the chemicals described in Chapter 6.3.2. Further chemicals in discussion for dechlorination are sulphur dioxide and hydrogen peroxide. The former yields hydrochloric and sulphuric acid, which will be neutralized by seawater alkalinity, and should be of no concern if dosage is low. The latter yields water, oxygen, and chloride and is of concern if overdosed as it is an oxidant like chlorine. Additionally residual chlorine can be depleted by activated carbon filters.

Removal of Metal Cations

Releasing heavy metal to the sea represents a risk to the environment that needs to be avoided. Therefore heavy metal cations should be removed from the effluent before discharge, if corrosion cannot be stopped by the substitution of conventional piping material by material resistant to corrosion. From various industries producing wastewater polluted by heavy metals different techniques for metal ion recovery are known. Possible techniques are precipitation, complexation, adsorption, biosorption, and ion exchange. Apart from the latter these techniques require an integrated filtration step to separate the bound metal from the brine, which can be carried out by either micro- or ultrafiltration. Precipitation of certain heavy metals can be achieved by adding lime /Masarwa et al. 1997/. In their test iron and manganese is coprecipitated in the course of removing silica.

Another method of heavy metal removal is complexation-ultrafiltration. A possible complexing agent is carboxyl methyl cellulose (CMC), a water-soluble metal-binding polymer /Petrov and Nenov 2004/. CMC possesses good complexation ability especially towards Cu2+ but also towards Ni2+, the quantitatively most important elements in the context of corrosion in desalination plants. A metal:CMC mole ratio of 1:6 is suggested for low metal concentrations characterising thermal desalination brines. A high retention rate of complexed, ultrafiltrated copper of up to 99 per cent is reached. For metal recovery decomplexation and subsequent ultrafiltration proves to be highly effective at pH 2. A method of combined complexation and filtration is the filtration with chelating membranes made of polyvinyl alcohol (PVA) as polymer matrix and polyethyleneimine and polyacrylic acid as chelating poly-electrolytes /Lebrun et al. 2007/. The process of adsorption is a commonly applied technique in the field of wastewater and exhaust air treatment. A well-known example is activated carbon which might be an adequate adsorbent for the treatment of desalination discharge.

Removal of heavy metals by adsorption using a powdered synthetic zeolite as bonding agent is described by /Mavrov et al. 2003/. The advantages of this new bonding agent are its high bonding capacity and its selectiveness even in the presence of other metal ions, such as Ca^{2+}, Mg^{2+}, and Na^{+}. Depending on the contamination grade of the discharge the bonding agent separation is carried out either by cross-flow microfiltration (for concentrations of up to 60 ppm) or by membrane microfiltration followed by flotation (for concentrations of up to 500 ppm). The membrane material is polypropylene and aluminium oxide respectively.

The possibility of biosorption has been tested with Streptomyces rimosus biomass /Chergui et al. 2007/ obtained as waste from an antibiotic production plant. The biomass samples were prepared by washing with distilled water, drying at 50°C for 24 h, grinding, and sieving to receive the fraction between 50 and 160 μm of particle diameter. The biosorption of Cu^{2+} has been most efficient in the sodium form, i.e. after NaOH treatment of the biomass. Desorption and regeneration showed to be most efficient with sulphuric acid reducing the biosorption capacity by 17 per cent in the case of copper. This method needs further research concerning the influence of surface-active and complexing agents or other metal ions, which might be present in desalination effluents. Furthermore, metal ions can be removed with the help of ion exchangers. A laboratory-scale treatment system with a strongly acidic cation resin showed high removal efficiencies for chromium and zinc /Sapari et al. 1996/.

Natural Evaporation and Disposal as Solid Waste

An interesting approach to avoiding the impacts through brine discharge is natural evaporation and disposal as solid waste. A laboratory-scale test has been conducted by /Arnal et al. 2005/ showing the possibility of natural evaporation enhanced by capillary adsorbents, especially for plants where discharge to the sea is impossible or difficult, e.g. brackish water desalination plants. The comparison with the reference sample without adsorbent showed that the capillary adsorbents lead to significantly higher evaporation rates. However, a general drawback is the low evaporation rate, thus requiring large areas.

ENHANCED PRACTICE OF DISCHARGE TO THE WATER BODY

To eliminate the negative effects of strongly elevated temperature (thermal processes) and salinity (mechanical processes) in the mixing zone of the desalination discharges the increase of temperature and salinity should be limited to 10 per cent /Lattemann and Höpner 2007c/. There are several ways to achieve this goal. First of all, maximum heat dissipation before entering the mixing zone and effective dilution in the mixing zone are essential. The oceanographic properties of the site influence the dilution capacities of a site. Dilution requires good natural mixing conditions and transport. Therefore ideal discharge sites are on high energy coasts or offshore. In the case of horizontal drain seabed intake, an elegant solution to the problem of discharge is the combination of intake pipes and discharge pipes, the latter designed with a smaller diameter running inside the intake drain /Peters and Pintó 2007/ . Due to the possible length of horizontal drains the point of discharge can be situated in sufficient distance to the coast line where mixing conditions are good.

Dilution can be enhanced further by the installation of diffuser systems. The type of diffuser system needed depends on the characteristics of the discharge jet, i.e. a buoyant effluent diffuser for a buoyant jet and a dense effluent diffuser for a dense jet respectively /Cipollina et al. 2004/. Another effective measure is to dilute the desalination effluents by blending with the waste streams of other industrial activities. An example is the common practice of blending with cooling water from power plants. The temperature of the effluents can be reduced with the help of evaporation cooling towers.

Changing Operation Parameters

The best solution to a problem is to avoid it right from the start. In the context of seawater desalination this means to change the operation parameters is such a way, that scaling, fouling and corrosion do not occur or at least that they can be reduced. Fossil fuelled plants are optimized in respect to their energy efficiency; therefore efforts were made to increase the water recovery rate. But high recovery rates lead to the problems mentioned above which need then to be solved by chemicals. By using renewable energies, however, the water recovery rate can be reduced to facilitate the desalination process.An example following this approach is the concept of RO desalination using wind energy with a reduced water recovery rate that does not require any chemicals at all /Enercon 2007/. Disinfection of source water happens merely with the help of UV rays. With the help of an integrated energy recovery system energy efficiency is increased.

OPTIONS FOR ENVIRONMENTALLY ENHANCED SEAWATER DESALINATION

By using heat and electricity from concentrating solar power plants the major impacts from energy consumption and air pollution are avoided. Enhancing the practice of seawater intake and hereby achieving higher quality input seawater leads to less chemical-intensive or even chemical-free pre-treatment and consequently less potential waste products in the effluents. The pre-treatment process itself can be advanced to further reduce the use of chemicals. Finally the practice of discharge needs to be improved in such a way that optimum dilution is guaranteed. Among the market-dominating desalination technologies, MSF performs worst regarding efficiency, costs and overall impact, which is why it falls out of consideration. Therefore future concepts will only be illustrated for MED and RO.

ENHANCED CSP/MED PLANT

The future advanced MED plant would run completely with heat and electricity from concentrating solar power (CSP/MED). The impacts from energy consumption are reduced to a minimum originating from the upstream processes of the CSP plant, i.e. production and installation of collector field, heat storage and conventional steam power station. The related emission can only be reduced by increasing the renewable share of power generation of the total energy economy. During operation of the plant there is no use of fossil energy carriers and there are no emissions to the atmosphere. The features characterizing the future MED plant are summarized schematically in Figure 6 24 and are presented in the following.

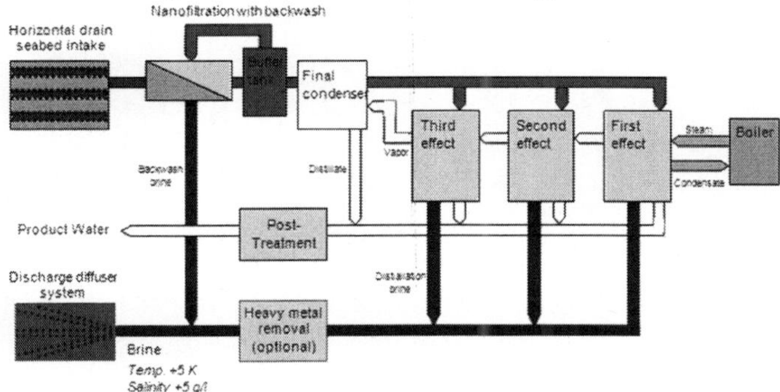

Fig. Scheme of A-MED process including horizontal drain seabed intake, nano-filtration unit, buffer tank for backwash of nano-filtration membranes and discharge diffuser system

The seawater intake is designed as a seabed filter intake through directed drilled horizontal drains. This system is environmentally compliant, because it does not affect aquatic organisms neither through impingement nor through entrainment. Where this system cannot be realised beach wells are the suggested alternative. Open source water intake is considered only on sites where neither horizontal seabed filters nor beach wells are possible. Due to the filtrating effect of seabed intake the source water is largely free from suspended inorganic and organic matter.

Optimally, the pre-filtered seawater does not require chlorination due to the long passage through the subsoil. In that case the pre-treatment consists of a nano-filtration system to eliminate colloids, viruses and hardness, i.e. divalent ions. As these ions are largely removed no antiscalants are necessary. Furthermore anti-foaming is dispensable as hardly any organic matter passes the nano-filtration membranes. The nano-filtration system comprises a permeate buffer tank where the NF permeate is stored for membrane backwashing. Backwashing is the essential measure to retain the performance of the NF membrane and has to be done regularly with a sufficient backwash flow rate. The backwash brine is blended with the distillation brine. In case of sub-optimally pre-filtered source water and unfiltered open source water, further pre-treatment steps consisting of micro-filtration and ultra-filtration become necessary each with a backwashing facility.

The tubing is made of corrosion-resistant material, such as titanium, or of conventional material coated with a durable protection film respectively. Anyway, the risk of corroding tubes is reduced by the enhanced pre-treatment that does not require acid cleaning anymore. However, to guarantee effluents free from heavy metals a post-treatment step can be inserted optionally where the heavy metals are removed applying one of the techniques described in Chapter 6.1. The practice of effluent discharge is enhanced with a diffuser system providing optimal and rapid dilution. In the future advanced CSP/ MED plant, the use of chemicals and the concentration of brine will be avoided to a great extent by increased filtering and diffusion. Additional energy for this process will be obtained from solar energy. For a first estimate, we will assume that the chemicals required per cubic metre of desalted water will be reduced to about 1 per cent of present amounts and that on the other hand an additional 40 per cent of electricity will be required for pumping.

ENHANCED CSP/RO PLANT

A future advanced RO plant would run completely with electricity from concentrating solar power plants. The impacts from energy consumption are reduced to a minimum originating from the upstream processes of the CSP plant, i.e. production and installation of collector field, heat storage and conventional steam power station. During operation there is no use of fossil energy carriers and there are no emissions to the atmosphere. The features characterizing the future RO plant are summarized schematically in Figure 6 25 and are presented in the following.

The seawater intake is designed as a seabed filter intake through directed drilled horizontal drains. Where this system cannot be realised beach wells are the suggested alternative. Open source water intake is considered only on sites where neither horizontal seabed filters nor beach wells are possible. Optimally, the pre-filtered seawater does not necessitate chlorination due to the long passage through the subsoil. In that case the pre-treatment consists of a nano-filtration system to eliminate colloids, viruses and hardness, i.e. divalent ions. As these ions are largely removed no antiscalants are necessary. The nano-filtration system comprises a permeate buffer tank where the NF permeate is stored for membrane backwashing. Backwashing is the essential measure to retain the performance of the NF membrane and has to be done regularly with a sufficient backwash flow rate. The backwash brine is blended with the RO brine.

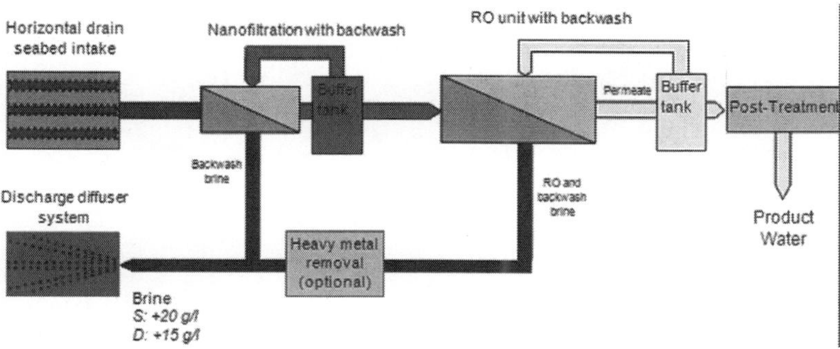

Fig. Scheme of A-RO process including horizontal drain seabed intake, nano-filtration unit, buffer tank for backwash of nano-filtration membranes and discharge diffuser system

In case of sub-optimally pre-filtered source water and unfiltered open source water further pre-treatment steps consisting of micro-filtration and ultra-filtration become necessary each with a backwashing facility. Thanks to the high quality of NF permeate, i.e. the feed to the RO membranes, the number of RO stages can potentially be decreased /Hassan et al. 1998/ thus reducing the investment costs and energy consumption of the RO. In analogy to the NF system, the RO unit requires a backwashing facility including a RO permeate buffer tank. The piping is made of corrosion-resistant material, such as stainless steel and PVC for high and low pressure piping respectively, or of conventional material coated with a durable protection film respectively. Anyway, the risk of corroding tubes is reduced by the enhanced pre-treatment that does not require acid cleaning anymore. However, to guarantee effluents free from heavy metals a post-treatment step can be inserted optionally where the heavy metals are removed applying one of the techniques described in chapter 6.1. The practice of effluent discharge is enhanced with a diffuser system providing optimal and rapid dilution.

In the future advanced CSP/RO plant, the use of chemicals and the concentration of brines will be avoided to a great extent by increased filtering and diffusion, and energy input will be delivered by solar energy. For a first estimate, we will assume that the chemicals required per cubic metre of desalted water will be reduced to about 1 per cent of present amounts and that on the other hand an additional 20 per cent of electricity will be required for pumping.

IMPACTS OF LARGE-SCALE DESALINATION IN THE MENA REGION

In this chapter we will assess to total absolute emissions and impacts of seawater desalination in the Middle East and North Africa as for today and for the AQUA-CSP scenario until 2050. The world wide desalination capacity rising rapidly reached 24.5 million m^3/d by the end of 2005 (IDA 2006, cited in /Lattemann and Höpner 2007a/). With 87 per cent of all plants the EUMENA region is by far the most important region in the context of desalination. "The largest number of desalination plants can be found in the Arabian Gulf with a total seawater desalination capacity of approximately 11 million m3/day which means a little less than half (45 per cent) of the worldwide daily production.

The main producers in the Gulf region are the United Arab Emirates (26 per cent of the worldwide seawater desalination capacity), Saudi Arabia (23 per cent, of which 9 per cent can be attributed to the Gulf region and 13 per cent to the Red Sea) and Kuwait (< 7 per cent)" (cited from /Lattemann and Höpner 2007a/). Regarding the emissions through the brine discharge, from all MSF plants the Arabian Gulf receives a daily load of copper of 292 kg, amounting to more than 100 t/y. The chlorine load emitted daily by MSF and MED plants reaches up to 23 t/d and more than 8000 t/y. The Red Sea region shows the third highest concentration of desalination plants worldwide with an overall capacity of 3.4 million m3/day (Figure 6 28, / Lattemann and Höpner 2007a/). With a capacity share of 23 per cent RO plays a significant role compared to the Arabian Gulf, where this technology reaches only 5 per cent. Still enormous amounts of copper and chlorine are released yearly: 28 *t* of copper and 2100 t of chlorine from both MSF and MED plants.

In the Mediterranean, the total production from seawater is about 4.2 million m3/day, representing 17 per cent of the worldwide capacity / Lattemann and Höpner 2007a/. The largest producer of this region is Spain with 30 per cent of the capacity not including its RO plants on the Canary Islands with an additional capacity of 411,000 m^3/d (Figure 6 29). While in the Gulf region thermal processes account for 90 per cent of the production, the predominant process in the Mediterranean is RO with almost 80 per cent of the capacity. The only exception to this trend is Libya where the dominating process is MSF. Consequently the release of copper and chlorine is less of concern compared to the Arabian Gulf.

According to /IDA 2006/ the MENA region had in 2005 a total desalination capacity of about 16.3 Mm^3/day. If we consider the specific air pollutants from Table 6 4 for conventional MSF, MED and RO taking as reference background the MENA electricity mix according to Table 6 2, and the chemicals typically contained in the effluents of each desalination system as shown before, we obtain the daily emissions of pollutants from desalination in the MENA region in the year 2005. For simplicity, MSF and MED plants have been calculated as if always coupled to power generation (Table 6 7). Therefore, estimates for 2005 are rather optimistic.

The AQUA-CSP scenario foresees an increase of desalination capacity in the MENA region from today 7 billion m³ per year to 145 billion cubic metres per year by 2050. This means a twenty-fold increase of desalted water within a time span of about 40 years. In 2050 almost all desalination plants will be of the type of advanced plants powered by CSP (and to a lesser extent by other renewable sources) with only 1 per cent of energy related emissions and only 1 per cent of the chemicals contained in the effluents compared to present standards. Roughly, this means that the overall load to the environment from power consumption and from chemicals can be reduced to about 20 per cent of the present load in spite of dramatically increasing desalination volumes. However, with the growth perspectives for desalination until 2015, all pollutants will approximately double by that time (Table 6 7). It will take until 2025 to achieve a majority of 55 per cent of solar powered, advanced desalination plants, and pollution by that time will increase by 3-4 times compared to 2005. This would only be acceptable if it would be a transitional effect. Luckily, this is the case in our scenario, and by 2050, when advanced, solar powered desalination will provide the core of desalted water, pollutants like carbon dioxide can be brought back below present levels (Figure 6 26).

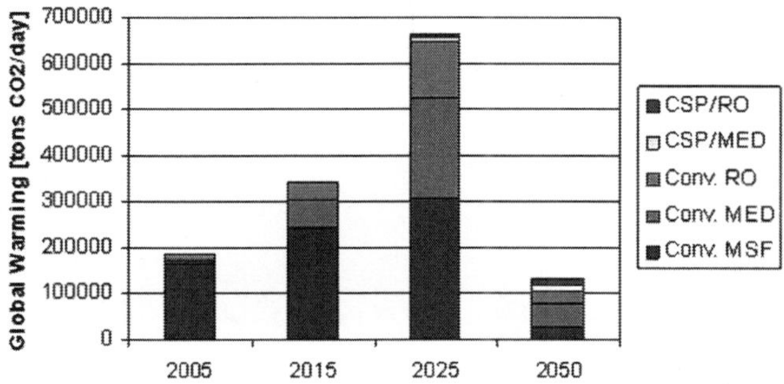

Fig. Greenhouse gas emissions from desalination in the AQUA-CSP scenario taking as basis the electricity mix of the MENA countries according to /MED-CSP 2005/.

Pollutants from CSP/RO and CSP/MED remaining in 2050 would be mainly caused by the construction of the plants. However, if the composition of the electricity mix in MENA would change to a mainly renewable supply according to the scenario dveloped in /MED-CSP 2050/, most of these pollutants would also be removed to a large extent, leading to an almost clean desalination system by that time (Table 6 7). The remaining conventional desalination plants using fossil fuels, which will cause most environmental impacts by that time, will subsequently be replaced by advanced systems. The only chemical pollutants that would increase by 2050 with respect to 2005 would be antiscalants and coagulants, that are however not considered as toxic substances. Nevertheless, their environmental impacts by causing turbidity and sediments could become critical (Chapter 6.1) and should be totally removed by further research and development. Also it must be considered that the advanced CSP/MED and CSP/RO concepts described here – and their low environmental impacts – are not yet state of the art today and their development and commercialization should be a primary target of R&D for desalination.

11

Corrosion Protection And Modern Ship Design

MINIMUM THICKNESSES AT SPECIAL SURVEY

Minimum thicknesses at the special survey are accepted (given) by the class societies. Years back these minimum thicknesses were calculated as a percentage of the newbuilding scantling, both for local and global strength members. Today for ships based on 1st principal calculation methods, such as Post-Marpol tankers designed with a low length/depth ratio to achieve the necessary volume in an efficient way, the scantling are often limited by the buckling criteria. For ships based on 1st principal calculation methods, 20 year North-Atlantic trade is used as design criteria for newbuildings, while this is reduced to 5 years for ships in operation due to survey intervals. From a safety point of view when a design/structure starts to buckle the safety factor for total collapse is reduced, while the similar safety factor is not affected to the same level if a crack occurs. Consequently, the buckling criteria will have a great influence on the result when minimum scantlings for ships in operation are calculated. The design of newbuildings should also take this into account. This is shown by an example below.

EXAMPLE

When DNV is calculating minimum thicknesses based on buckling criteria this is done by direct calculation methods. However, a simple evaluation illustrates the points mentioned above fully. The deck plating for a tanker is used in the example. The simple buckling criteria used are expressed as follows:

Deck Plate

$$t_{\min} = \frac{S}{K} + 0,5\ t_k$$

Where

t_{min} = min. acceptable thickness due to buckling

S = Stiffener/spacing

K = 56 mild steel

= 52 HT–32 steel

= 49 HT–36 steel

t_k = Corrosion addition, newbuildings

0,5 t_k= min. 0,5 mm

	Cargo tank 0,5 t_k = 0,5 mm			Ballast tank 0,5 t_k = 1,0 mm		
S	**Mild Steel**	**HT-32**	**H-36**	**Mild Steel**	**HT-32**	**HT-36**
	t_{min}	t_{min}	t_{min}	t_{min}	t_{min}	t_{min}
600	11,2	12,0	12,7	11,7	12,5	13,2
700	13,0	14,0	14,8	13,5	14,5	15,3
800	14,8	15,9	16,8	15,3	16,4	17,3
850	15,7	16,8	17,8	16,2	17,3	18,3
900	16,6	17,8	18,9	17,1	18,3	19,4
950	17,5	18,8	19,9	18,0	19,3	20,4
1000	18,4	19,7	20,9	18,9	20,2	21,4

Table 1

A typical spacing for a VLCC will be 850 – 900 mm with HT-32 steel, giving a minimum thickness of 18,3 mm in way of cargo tanks and 18,8 in way of ballast tank. A typical thickness from newbuilding of a VLCC will be approximately 20 mm in cargo tanks and 21 mm in the ballast tanks. As a result, the acceptable reduction will be 2-3 mm or 10-15 per cent due to buckling. As you can see from the table, it is big ships carried out in HT-steel where the buckling criteria is most critical.

Longitudinals

$$t_{\min web} = \frac{h}{c} + 0,5\, t_k$$

Where

$t_{min\ web}$ = minimum acceptable thickness due to buckling

h = height of web

c = 56 mild steel T, L and bulb profile

= 52 HT-32 steel T, L and bulb profile

= 49 HT-36 steel T, L and bulb profile

c = 17 mild steel flatbar

= 16,5 HT-32 steel flatbar

= 16 HT-36 steel flatbar

Flatbar

Height of web – S	Cargo tank 0,5 t_k = 1,0 mm Mild Steel	HT-32	HT-36	Ballast tank 0,5 t_k = 1,5 mm Mild Steel	HT-32	HT-36
	$t_{min\ web}$	$t_{min\ web}$	$t_{min\ web}$	$t_{min\ web}$	$t_{min\ web}$	$t_{min\ web}$
300	18,6	19,2	19,8	19,1	19,7	20,3
350	21,6	22,2	22,9	22,1	22,7	23,4
400	24,5	25,2	26,0	25,0	25,7	26,4
450	27,5	18,3	29,1	28,0	28,8	29,6
500	30,4	31,3	32,3	30,9	31,8	32,8
550	33,4	34,3	35,4	33,9	34,8	35,9
600	36,3	37,4	38,5	36,8	37,9	39,0
700	42,2	43,4	44,8	42,7	43,9	45,3

Table 2

L, T, bulb profile

Height of web - S	Cargo tank 0,5 t_k = 1,0 mm Mild Steel	HT-32	H-36	Ballast tank 0,5 t_k = 1,5 mm Mild Steel	HT-32	HT-36
	$t_{min\ web}$	$t_{min\ web}$	$t_{min\ web}$	$t_{min\ web}$	$t_{min\ web}$	$t_{min\ web}$
300	6,4	6,8	7,1	6,9	7,3	7,6
350	7,3	7,7	8,1	7,8	8,2	,6
400	8,1	8,7	9,2	8,6	9,2	9,7
450	9,0	9,7	10,2	9,5	10,2	10,7
500	9,9	10,6	11,2	10,4	11,1	11,7
550	10,8	11,6	12,2	11,3	12,1	12,7
600	11,7	12,5	13,2	12,2	13,0	13,7
700	13,5	14,5	15,3	14,0	15,0	15,8

Table 3

A typical deck long in a bulk carrier may be in the range of a height approximately 350 – 400 mm resulting in a minimum thickness of web in a ballast tank assuming HT-32 of 8-9 mm. For older ships the limitation due to torsional buckling may be decisive.

CORROSION ADDITION, TK

DNV has introduced long time ago a corrosion addition factor, tk. In other words, the strength evaluation is based on net scantling and then a corrosion addition, tk is added. This corrosion addition you will find in the Rules Part 3, Ch. 1, Sec. 2 and is based on 10 years average corrosion rate. In 1993 DNV carried out some research to verify the corrosion addition in ballast tanks given in the Rules. The result is shown in table 4 below:

Corrosion Addition – Ballast Tanks	Project D mm	DNV Rule mm
Location		
Deck Plating	2	2

Deck Longit. Web	1,8	3
Deck Longit. Flange		3
Side Plating, upper 2 m	3,5	2
Side Plating, remaining areas	2	1
Side Longit. Web	2,4	1,5
Side Longit. Flange		1,5
Bottom Plating	1,5	1
Bottom Longit. Flange	0,8	1,5
Bottom Longit. Web	1,4	1,5
Longit. Bkh. Plating, upper 2 m	2,3	2,5
Longit. Bkh. Plating, remaining areas	2	1
Longit. Bkh. Longit. Web, upper 2 m	3	3
Longit. Bkh. Longit. Web, remaining areas	2,3	1,5
Longit. Bkh. Longit. Flange	1,6	1,5
Deck and Side Transv. Web Plating, upper 2 m	3,5	3
Transv. Web Plating, other categories	2,1	1,5
Transv. Bkh. Plating	1,9	1
Transv. Bkh. Vertical Stiffener Web	2,1	1,5
Transv. Bkh. Horiz. Stringer Web Plating	1,9	1,5
Swash Bkh. Web Plating		1,5
Swash Bkh. Horiz. Stringer Web Plating		1,5
Swash Bkh. Vertical Girder Web Plating		1,5

The result from the research corresponds very well to what is given in the DNV Rules. However, for some horizontal areas and the upper part of the side and inner shell for double hull tankers, the tk value has been considered for a possible increase of 0,5 – 1,0 mm. Up till now it has not been done, especially since the research indicated that the standard deviation of the mean value evaluated was close to 80 – 100 per cent giving tk for deck plating in ballast tank.

For cargo oil tanks some research has been done to verify if there is a connection between corrosion rates and the specific oil type. To the knowledge of the undersigned no such connection has been found. However, Mr. Bjarne Thygesen of Osprey Maritime Limited in London has pointed out that the from time, to time reports of a relatively high corrosion rate in the upper part of the cargo tank may be due to

- Sulfur from the fuel gas (Inert gas)
- CO_2 combines with water and forms H_2CO_3, carbonic acid
- Hydrogen sulfide, H_2S, released from the crude oil
- High temperature in double hull tankers
- Insufficient cowing under deck to remove acidic deposits and corrosion products and at the same time building up a fatty oil film

A light deck color may improve the situation by lowering the temperature in the tanks. So far we have talked about general corrosion (electrochemical corrosion) in seawater and cargo tanks. This general corrosion multiplies with increasing:

- Salinity
- Temperature
- Oxygen content
- Water velocity
- Content of contaminants promoting corrosion

Later on we will touch upon local corrosion such as:

- Pitting
- Bacterial corrosion
- Galvanic corrosion

LIFE CYCLE COST CALCULATION

During the 1990's DNV has carried out a significant number of so called "Life Cycle Cost" calculations a of probabilistic nature. The figure below indicates the steel replacement for various locations given as a percentage of total steel replacement for a double hull VLCC of HT-36/32 design and minimum scantlings acceptable for a newbuilding. The ship has never been built, but used here as an illustration for what has been discussed previously. The target life time has been 30 years, and coating of ballast tanks is assumed to be a 15 years coating system with no anodes. The total steel replacement is calculated to be 1655 tons over a 30 year period. This corresponds to approximately 5 per cent of the steelweight.

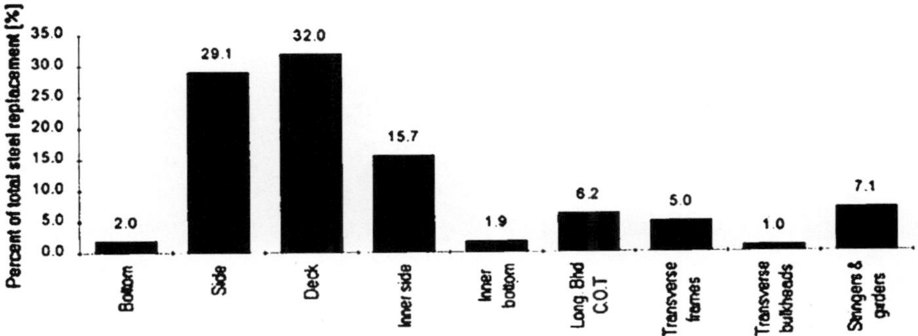

Fig. Steel replacement for various location given as a percentage of total steel replacement for NV36/32-design. Paint system 3, (P3).

As can be seen from the figure, the share of the deck area is 32 per cent, side 29 per cent and innerside 16 per cent. These high shares are due to:

- Limiting buckling capacity in the areas
- High annual corrosion rates in the areas

Both of these points have been discussed previously.

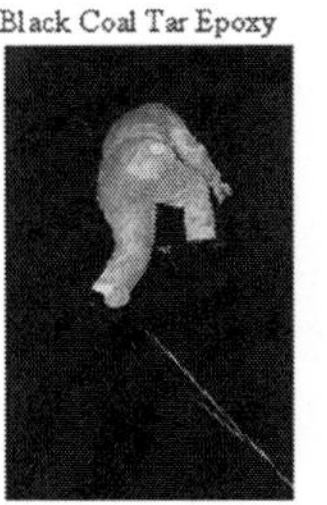

Fig. Coating in Water Ballast Tanks

ACCELLERATED CORROSION RATES

According to OCIMF, their members have indicated problems in new single and double hull tonnage from excessive pitting corrosion of up to 2,0 mm per year in the uncoated bottom plating in cargo tanks due, inter alia, to microbial induced corrosion processes. In addition, they claimed the same as indicated by Mr. Thygesen general corrosion up to 0,24 mm per year has been found in vapor spaces. The last one has been discussed previously in this paper and the value is within the corrosion addition given by DNV taking into account the standard deviation. For bulkers accelerated corrosion has occurred at the main frames especially at lower end towards the hopper tank where a mean value of 0,3 mm/year has been observed for ships carrying mainly coal and iron-ore (Capesize bulkers). This corrosion rate has now been taken care of by the IACS Rules.

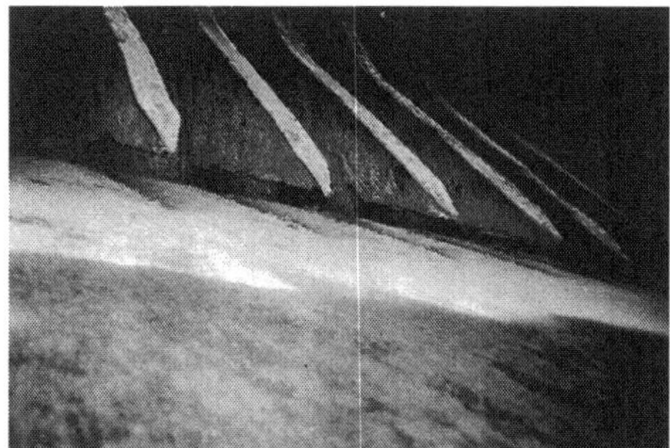

Fig. Heavy corrosion of main frames and hopper tank – bulk carrier

Pitting

Pitting is a form of local corrosion in which limited areas are attacked. Pitting is particularly likely to occur in the bottom of cargo and ballast tanks or other horizontal surfaces. Severe pitting corrosion creates a tendency for the pits to merge to form long grooves or wide scabby patches with an appearance resembling that of general corrosion. The mean corrosion rate in pits and grooves can be very high. When grooves occur for example close to a weld between longitudinal and deck (HAZ-zone) this may create a higher deformation (flexibility) of the stiffener resulting in an accelerated corrosion rate of the groove. This effect is from time to time introduced as "necking" effect. Due to the complexity and effects of groove corrosion, special repair methods are normally required. The figure below shows typical grooving corrosion in a stiffener connection.

Fig. Grooving corrosion in a stiffener connection

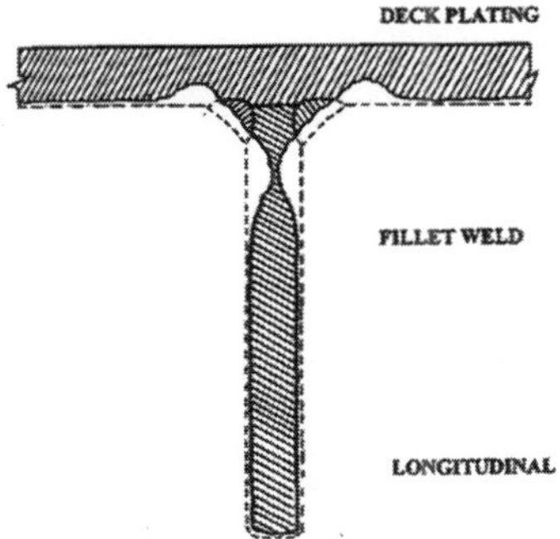

Fig. Typical grooving corrosion in stiffener connection

Measurements have also shown that annual general/local corrosion rates will increase when the ship gets older. This due to the "necking" effect described above.

In the following other potential causes of accelerated corrosion are listed:

- Coating not applied
- Excessive crude oil cargoes can cause a waxy layer to form the cargo tank steel structure and this layer helps to inhibit corrosion. Crude oil washing will give the same effect if the frequency is limited. Comments are previously given to the deck structure. However, washing mediums such as hot and cold sea water may remove this protective layer with the result that the corrosion process will start.
- High sulfur content of cargo oil
- Inert gas (dry atmosphere) has been looked upon as a positive effect for reduced corrosion. However, to have this effect the oxygen content should be below 1 per cent. Sulphurous components and soot in the gas may lead to acid in the tanks causing accelerated corrosion, which may occur quite frequently.

- Inadequate earthing of electrical equipment
- Local breakdown of coating
- Sludge and scale accumulation
- High humidity
- High temperature
- Bacterial corrosion
- Cargoes such as Nafta, Bensene, etc. which destroy the scale of the steel structure

BACTERIAL CORROSION

Small animals eating steel is a myth. However, it is a fact that high local corrosion rates are the result of chemical processes initiated by bacterial activity. This phenomena is often described as "bacterial corrosion" or "microbial corrosion" or perhaps more correctly" microbial influenced corrosion". This type of corrosion most frequently occurs on cargo oil tank bottom plating. It may also occur on horizontal surfaces in water ballast tanks.

It seems to be a fact that double hull ships are more exposed due to the "thermos bottle effect", keeping the temperature at a level where the bacteria are productive. The bacteria most frequently associated with corrosion of steel are those that generate sulfides (H2S) and these are commonly called sulfate reducing bacteria (SRB). Acid is the result of the chemical process and consequently pittings with an extreme corrosion rate may occur.

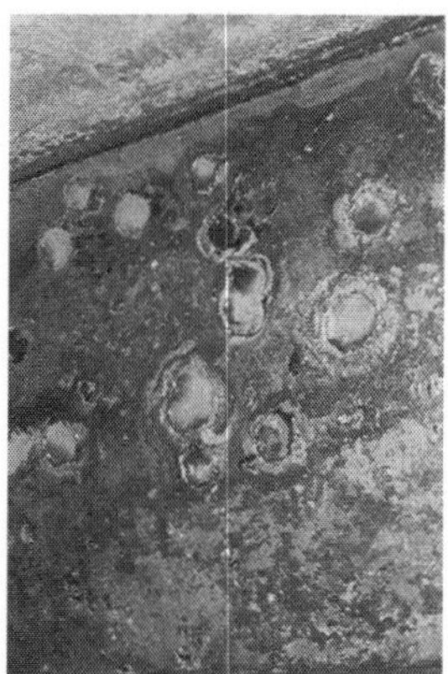

Fig. Bacterial corrosion

Experience feedback indicates that the process described above takes time and must have a given temperature normally between 20oC and 40oCo/50oCo. This is most probably the explanation that VLCC (long trade) are much more exposed to this type of corrosion than a shuttle tanker, where this type of corrosion does not seem to occur. Since this type of corrosion has occurred frequently on double hull tankers built in the Far East, where Thermal Mechanical Control Process (TMCP) is used for manufacturing the steel, TMCP steel has been listed by some as a potential cause of accelerated corrosion.

The investigation DNV has done as well as the experience feedback we have with the steel does not indicate anything wrong with the TMCP steel. The corrosion reported is "bacteria" corrosion only. In some test pieces from ships we have discovered a high content of salt on the surface which in addition will have a negative effect on the corrosion rate.

HOW TO DESIGN A SHIP FOR GOOD OPERATION AND TO MINIMUM COSTS

To achieve a good, safe ship operable in the market over a long period the steel replacement or avoiding of same will be the main parameter. To avoid steel replacement the following factors must be taken into account:

- Coating and anodes
- Increased scantlings
- Design
- Combination of the above
- Access

Fig. With a staging like this do you think the surveyor is able to look for cracks and corrosion?

The items above will initially increase the newbuilding costs, and the life-span considered will have a great impact on what the owner wants to do. Before we go any further, it is worthwhile to note that:

- If pitting problems are to be avoided , coating is the remedy
- If general corrosion, both increased scantling and coating will improve the situation

The minimum standard ship according to the class Rules is as follows:

- Minimum scantling according to the Rules
- Coating of ballast tanks, no requirement to the life-span of the coating
- Coating of cargo holds (inner bottom not included) for bulk carriers

In order to give opportunities to the owner, DNV has developed 2 voluntary class notations.

ICM (Increased Corrosion Margins)

For this class notation a further corrosion addition is added in the ballast tanks, cargo tanks (cargo holds). The result is that ICM gives the double corrosion addition compared with the main class. The notation is made flexible and the following notations may be chosen:

- ICM (BT), ICM(Btu), ICM (BTs) for ballast tank
- ICM (CT), ICM (Ctu), ICM (CTs) for cargo oil tank
- ICM (CH), ICM (Chu), ICM (CHs) for cargo holds in bulk carrier

or a combination of these.

Where

BT $\Rightarrow$ All ballast tanks
CT $\Rightarrow$ All cargo oil tanks
CH $\Rightarrow$ All cargo holds in a bulk carrier
u $\Rightarrow$ Upper part of the ships (above D/S)
s $\Rightarrow$ Strength deck of the ship and 1,5 m below

The full ICM for the whole ship will increase the ship weight with approximately 7,5 per cent while f.ex. strength deck(s) will increase the weight with approximately 1 per cent. The notation covering the strength deck(s) is the notation used most frequently.

Fig. Cracks and corrosion. Is this the start for the results shown in Figure 9?

Fig. Side shell disapproved in ballast tank

COAT-1 and COAT-2

In order to take care of possible bacterial corrosion of horizontal surfaces in cargo tanks as well as act as an alternative to ICM (CTs) the coat notation is introduced, as described:

Ship	Location	Coating system	
		COAT-1	COAT-2
Crude Oil Tankers	Ballast water tanks	II	III
	Inner bottom in cargo tank	II	II
	Upper part of stringers and deckhead in cargo tank	I	II
Bulk Carriers	Ballast water tanks	II	III
	Cargo hold, upper part	I	II
Other	Ballast water tanks	II	III

Fig. COAT-1/2 Requirements

Coating system I. is a relatively simple specification. The steel plates are blast cleaned and primed before being used in production. Damaged primer on the built structure is powertooled before coating. The main coat is 200 ?m. Coating system I is not recommended in ballast tanks.

Coating system II. is a good specification well suited for ballast tanks. The steel plates are blast cleaned and primed before being used in production. Damaged primer on the built structure is blast cleaned before coating. Two coats of totally 300 μm.

Coating system III. is a very good specification well suited for ballast tanks. The steel plates are blast cleaned and primed before being used in production. The built structure is blast cleaned and primed before coating. The salt content on surfaces to be coated is to be blow a given limit. Two or three coats of totally 300-400 ?m are to be applied.

Design Aspects

In addition to design evaluation for coating, looking at the accept criteria for ships in operation now given by DNV may be used. As emphasized before, the buckling criteria for large ships will be of great importance, since this will be the limiting criteria at least in the deck area.

Why not include in the specification from owner:

$$\text{Deck plate } t_{\min} = \frac{S}{K} + t_{corr}$$

Where K = 56, 52, 49 for mild steel, HT-32, HT-36 respectively t_{corr} = to be chosen by the owner, but to be in the range of 3-5mm. Similar can be done for longitudinals using the formula

$$t_{\min} = \frac{h}{c} + t_{corr}$$

where h and c is defined earlier.

This way of thinking starting with the accept criteria for ships in operation can of course be utilized for other parts of the ship.

IMPORTANCE OF ICM AND COATINGS FOR LIFE CYCLE EVALUATION

Previously in this paper a life cycle cost calculation was referred to based on a double hull VLCC minimum newbuilding scantlings carried out in HT-36/HT-32 steel. This probabilistic calculation has been evaluated further in order to find the design giving minimum net present value (NPV) of costs additional to basic costs.

Fig. Coating is the best solution when pitting occurs

Table 5 below shows the result.

Mean NPV of Costs Additional to Basis Costs (USD)

Lifetime (Years)	20	25	30
NV36/NV32, P3, A 0%	9.9	10.2	12.7
NV36/NV32, P2, A 60%	10.4	12.0	16.0
NV36/NV32, P3-P2, Bottom, A 0%	8.8	9.4	14.2
NV32/MS, P2, A 100%	8.7	10.2	17.2
NV36/NV32, ICM (CT s/BTs) P2, A 0%	8.4	10.5	17.6
NV36/NV32, ICM (CTs/BTs) P3, A 0%	10.2	10.3	11.6

Table 5

Notes:

- Double Hull VLCC
- Internal Rate of Return = 10 per cent, inflation 4 per cent
- P1, P2, P3 = 5, 10, 15 years paint systems
- A 0 per cent, 60 per cent, 100 per cent = Anode Life
- ICM = Increased Corrosion Margin, CTs = Cargo TJ DKHEAD (-1.5 m), WBTK DKHD (-1.5 m)
- NV36/NV32 = NV36 DK & BTM, NV22 Side and Longl. BHDS and Trans Structure

As could be expected both a good coating and increased scantling in deck area in cargo and ballast tank have a favorable effect.

12

Deterministic Prediction of Corrosion Damage in Low Pressure Steam Turbines

INTRODUCTION

The accumulation of damage due to localized corrosion [pitting, stress corrosion cracking (SCC) and corrosion fatigue (CF)] in low-pressure steam turbines components, such as blades, discs, and rotors, has been consistently identified as being among the main causes of turbine unavailability [1, 2]. Accordingly, the development of effective localized corrosion damage prediction technologies is essential for the successful avoidance of unscheduled downtime in steam turbines (and other complex industrial and infrastructural systems) and for the successful implementation of life extension strategies.

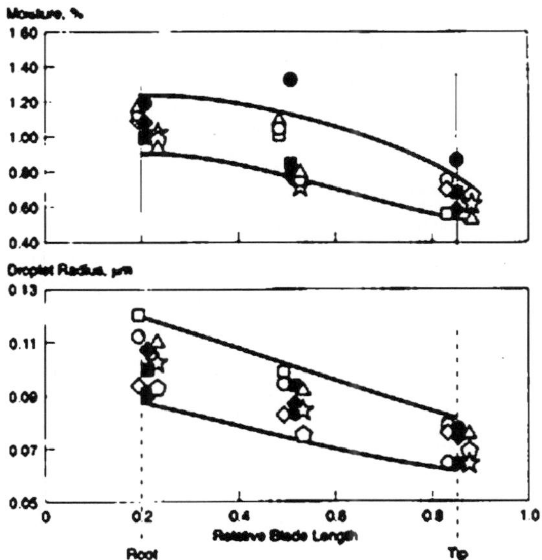

Fig. Moisture and droplet radius measurements along the length of the blade in a model turbine as a function of different inlet steam conditions. The symbols represent different steam chemistries.

Currently, corrosion damage is extrapolated to future times by using various empirical corrosion or fracture mechanics models coupled with damage tolerance analysis (DTA). In this strategy, known damage is surveyed during each outage, and the damage is extrapolated to the next inspection period allowing for a suitable safety margin. As previously noted [3], this strategy is inaccurate and inefficient, and in many instances it is too conservative. Instead, it was suggested that damage function analysis (DFA) is a more effective method for predicting the progression of damage, particularly when combined with periodic inspection. DFA is based upon deterministic prediction of the rates of nucleation and growth of damage, with particular emphasis on compliance of the embedded models with the natural laws. Although corrosion is generally complicated mechanistically, a high level of determinism has been achieved in various treatments of both general and localized corrosion, which can be used to predict accumulated damage in the absence of large calibrating databases.

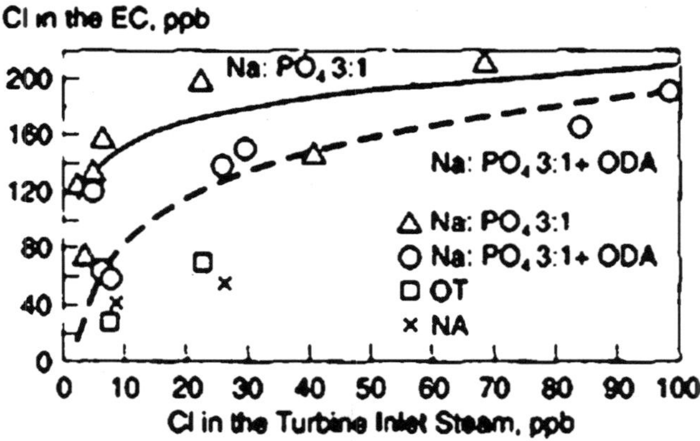

Fig. Chloride content in the early condensate (EC) versus level of chloride in the turbine inlet steam. Samples were taken at the mid-section of the blade height. Na = no addition to the cycle chemistry, OT = oxygen treatment.

At first glance it would seem that the deterministic description of corrosion damage reduces to the prediction of the size of the largest corrosion events (pit or crack), Lmax, as a function of time and operational conditions. However as shown below, such an approach appears to be inadequate for predicting damage in operating systems, because the localized corrosion in these systems is a progressive nucleation/growth/death phenomenon.

Extensive work in the laboratory and on operating test turbines [1, 2] has shown that the failure of turbine blades and disks is a strong function of the chemical conditions that exist in the steam. Of particular importance is the deposition of a thin liquid film on the turbine surfaces in the phase transition zone (PTZ), where the conditions of P and T are such that film deposition is thermodynamically possible. Deposition is preceded by the existence of significant moisture in the steam, as shown in Figure 1 (upper plots). As seen from this figure, the moisture content and the average drop radius decreases with distance from the blade root to the blade tip; these changes corresponding to a drying of the steam partly due to deposition of a liquid film on the blade surface and partly due to expansion of the steam as it moves outwards along the blade. In any event, thin liquid (electrolyte) films form on the metal surface and these films support a variety of corrosion processes that ultimately lead to premature failure. These electrolyte films concentrate steam impurities, most notably chloride ion (Figure 2), which is known to induce passivity breakdown. Passivity breakdown is the precursor to the nucleation of all localized corrosion damage, and it is the inhibition of passivity breakdown where great gains are to be made in the battle against corrosion-induced premature failure of LP steam turbine components.

Examination of the data summarized in Figure 2 suggests that the concentration factors are only modest, of the order of 2 to 8, depending upon the concentration of chloride ion in the inlet steam. This finding is important, because it is commonly assumed or postulated that concentrated solutions form on the surface, with the result that many laboratory studies are carrier out in concentrated solutions yielding results that suggest that failure should occur well before they are observed. To be sure, concentrated solutions are thought to form on the metal surfaces in restricted regions (e.g., blade root crevices and in keyways) or even on bold surfaces under specific steam conditions, but the norm is the formation of relatively dilute solutions as indicated in Figure 2. An important finding of the studies reported in Reference 2 is that oxygen does not concentrate in the early condensate, and hence, because the oxygen level in the inlet steam under normal deoxygenated operating conditions is normally very low the oxygen level in the condensate is also very low (sub-ppb levels). This factor has important consequences for the nucleation and growth of localized corrosion damage (see below).

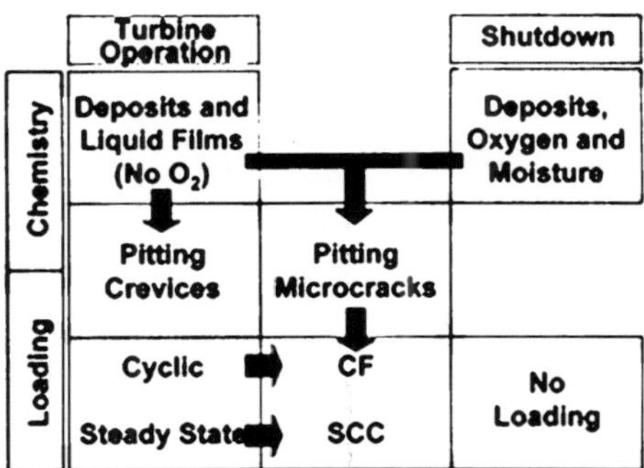

Fig. Outline of conditions that exist in Low Pressure Steam Turbines under operating and shutdown conditions and the relationship of these conditions to the development of corrosion damage.

The development of damage due to localized corrosion on steam turbine surfaces involves complex interactions between deposits and electrolyte films that form on the surfaces in the phase transition zone, passivity breakdown and the nucleation of stable pits, pit propagation, pit repassivation, the transition of pits into cracks, and the propagation of stress corrosion and corrosion fatigue cracks, as captured in Figure 3. The corrosion evolutionary path (CEP) includes periods of shutdown and operation, with various chemistries being possible during both periods. For example, during shutdown exposure to the air produces an aggressive environment in which pits may nucleate and grow on the disk and blade surfaces, with the result that during subsequent operation the pits will transition into SCC or CF cracks, ultimately resulting in failure. Indeed, an important conclusion from previous work [2], which is reinforced by the findings of the present study, is that control of the environment during shutdown is the key to trouble-free operation over long periods.

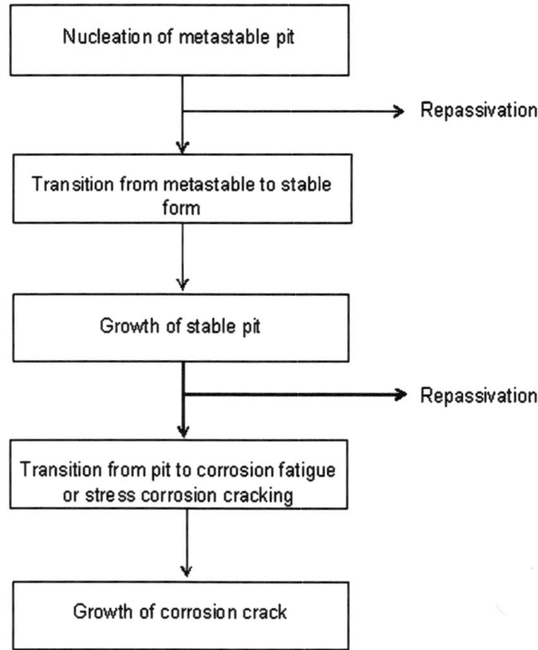

Fig. Schematic history of the nucleation and propagation of corrosion damage.

It is often reported that SCC and CF damage in low pressure steam turbines initiates in highly localized areas, most commonly at corrosion pits that act as stress raisers. After nucleation, the corrosion events develop and pass through distinct stages, as schematically indicated in Figure 4. It is evident, if we wish to describe the accumulation of damage quantitatively, that we must be able to describe each of the stages in mechanistic/deterministic form; namely, initiation of metastable pits, the survival of a (few) metastable pits to form stable pits, the growth of stable pits, the transition of pit into cracks, the growth of subcritical cracks, and finally unstable fracture. As outlined below, in spite of the obvious complexity of the processes involved, a high degree of determinism has been achieved in modeling the individual steps and practical algorithms are now available to predicting the accumulation of damage over realistic CEPs (as described later in this paper). In this regard, it is important that the reader be conversant with the term "deterministic model" as the word "determinism" is used rather loosely in engineering disciplines. In the present context, a deterministic model represents a physically viable model whose output (predictions) are constrained by the natural laws (conservation of mass, charge, etc). The viability of such a model depends heavily upon being able to accurately describe the physico-electrochemical processes involved in the system that impact the nucleation and accumulation of damage. The successful prediction of damage also requires accurate definition of the corrosion evolutionary path, which describes in parameter space (T, [O2], pH, [Cl-], etc) how the system evolves to the future state. We are somewhat fortunate in this regard that as the result of significant investment over many decades, a great deal is known about LP steam turbine environments, particularly with respect to the parameters that are of prime importance in the present work.

It is evident that some stages of corrosion propagation must have a statistical nature for a large ensemble of corrosion events. For example, it is assumed that pit repassivation obeys a first order decay law, i.e. that pit repassivate accidentally, which is fundamentally statistical in form. The statistical nature was also previously noted for the stages of pit nucleation [5] and for the transition of a pit into a crack, but that does not mean that these processes are not fundamentally deterministic in nature (c.f., quantum mechanics, which is deterministic, yields a probabilistic result). Furthermore, we cannot assume that the rate of pit propagation is unequivocally determined by its depth and by the external conditions, i.e. there is no distribution in pit propagation rate for pits with equivalent depth. It is well known that the morphology of pits on any given surface can vary significantly, with some shapes favoring more rapid mass and charge transfer, and hence greater propagation rates. In addition, some pits will initiate at metallurgical features that may favor more rapid propagation, e.g. MnS inclusion. Again, this distribution may have deterministic underpinnings. Thus, the distribution in pit propagation rate might arise, for example, from the spatial distribution in electrochemical activity of the anodic and cathodic sites on the corroding surface. Experimentally, this statistical nature of corrosion damage manifest itself as distribution in the depth of the deepest pit (crack), when measured on different samples (or when measured on the same sample but in different areas) under identical external conditions.

Accordingly, the task of deterministically describing corrosion damage under real conditions does not reduce to the prediction of the maximum depth, Lmax(t), but to the prediction of the probability, P(L,t), that the depth of the largest corrosion events has a value between L and L + dL for given observation time, *t*, or prediction of the probability of failure, Pf(L,t), i.e. the probability that the depth of the of the largest corrosion events is greater than the critical dimension L for given time, t. Thus, we can see a crude analogy with the "deterministic" description of the position of a quantum particle, when only the probability of particle having coordinates between x and x +dx (but not the exact value of x) can be predicted. The principal objective of this work is to develop, on the basis of DFA, "deterministic" models and associated computer codes for predicting the evolution of corrosion damage (i.e., "integrated" damage) in steam turbine blades and discs.

THEORETICAL FOUNDATION OF DAMAGE FUNCTION ANALYSIS (DFA)

The cases of relatively concentrated solutions, when the potential and concentration drops in corrosion cavities are relatively low, and hence when the rate of propagation of an individual cavity can be considered constant. The concentration of impurities in the electrolytes film that covers the metal surface in the phase transition zone in a LP steam turbine can be very low, in some cases. Under these conditions the potential and concentration drops in corrosion cavities can attain substantial values and the resultant decrease in the corrosion propagation rate of individual cavities with time must be taken into account. Below we briefly describe DFA for the case of very low electrolyte concentration (constant propagation rate).

Localized corrosion damage in an arbitrary system is completely defined if we know how many pits or other corrosion events (per cm^2) have depths between x_1 and x_2 ($x_1 < x_2$) for a given observation time, t. Let us denote this number by $\Delta Nk(x_1,x_2,t)$. Here, index k denotes different types of corrosion event (e.g., pits or cracks). Instead of employing a function with three variables, ΔNk, it is more convenient to use a function of two variables - the integral damage function (IDF), $F_k(x, t)$. This function is defined as the number (per cm2) of corrosion events with depths larger than x for a given observation time, t. It is evident that

$$\Delta N_k(x_1, x_2, t) = F_k(x_1, t) - F_k(x_2, t)$$

In turn, it is convenient to express the integral DF, Fk(x,t), via the differential DF, fk(x,t), using the relation

$$F_k(x,t) = \int_x f_k(x',t)dx'$$

or

$$f_k(x,t) = -\frac{\partial F_k(x,t)}{\partial x}$$

The differential DF for defects of type k, f_k, is defined from the condition that $f_k(x, t)dx$ is the number of defects k (per cm^2) having depths between x and $x + dx$ for a given observation time, t. It is evident that the set of functions fk for all types of defects yields complete information about damage in the system. The advantage of using the differential DF lies in the fact that it obeys a simple differential equation (see below) and accordingly can be calculated for any given set of conditions. Of course, all of the functions $\Delta N_k(x_1,x_2, t)$, $F_k(x, t)$, and $f_k(x, t)$ may depend, in the general case, on position on the metal surface, but in this analysis we will assume that all sites belong to a given ethnic population.

The function fk has dimension of #/(cm^2 cm) = #/cm^3 (analogous to the concentration of a particle). Accordingly, it is very convenient to regard each defect as a "particle" that moves in the x direction (perpendicular to the surface, with $x = 0$ being at the metal surface). The coordinate of this particle, x, coincides with the depth of penetration of the defect. Accordingly, f_k can be regarded as the concentration of particles and hence must obey the law of mass conservation,

$$\frac{\partial f_k}{\partial t} + \frac{\partial j_k}{\partial x} = R_k\,,\ k = 1, 2,\ldots,K$$

where jk and Rk are the flux density and the bulk source (sink) of the "particles" k, respectively. Thus, the subscript "k" enumerates the corrosion defect and "K" is the total number of different corrosion defects in the system. By definition, $R_k(x, t)$dxdt yields the number of defects k (per cm^2) with depths between x and $x + dx$ that arise (or disappear) during the period of time between t and t + dt, due to the transformation (repassivation, in the case of pits). The system of Equations (above) can be solved with the corresponding boundary and initial conditions.

$$j_k = n_k(t) \quad \text{at}\ x = 0,\ t > 0$$

and

$$f_k = f_{k0}(x) \quad \text{at}\ x > 0,\ t = 0$$

where fk_0(x) is the initial distribution of defect k [usually we can assume that fk_0(x) = 0, i.e., no damage exists at zero time] and $n_k(t)$ is the nucleation rate of the same defect [i.e, nk(t)dt is the number of stable defects (per cm^2) that nucleate in the induction time interval between t and t + dt].

According to the theory developed above, calculation of the damage functions requires the determination of three independent functions for each kind of corrosion defect, k: The rate of defect nucleation, nk, the flux density (growth rate) of the defect, jk, and the rate of transition of one kind of defect to another, R_k, (for example, the transition of an active pit into a passive pit or the transition of a pit into a crack). We will discuss briefly the feasibility of calculating each of these three functions..

Rate of pit nucleation

In many practical cases, it is possible to assume that all pits on a given surface nucleate during an initial period of time that is much less than the observation time, t, or the service life of the system, ts (instantaneous nucleation). For example, for the case of the pitting corrosion of aluminum in tap water, practically all of the pits were found to nucleate within the first two weeks. If pit nucleation is not "instantaneous", the simplest assumption concerning the pit nucleation rate, $n(t)$ = dN/dt, [N is the number of stable pits (per cm^2) that nucleate in the time interval between 0 and t.] is that $n(t)$ is proportional to the number of available sites, $N_0 - N(t)$, which yields.

$$N(t) = N_0[1 - \exp(-t/t_0)]$$

where N_0 is the maximum number of the stable pits (per cm^2) that can exist on the metal surface and t0 is some characteristic time that depends on the corrosion potential, temperature, and electrolyte composition. A more general and sophisticated prediction of the pit nucleation rate is given by the Point Defect Model (PDM) [13–17]. In accordance with this model, it can be shown that total number of nucleated pits can be described by the following expression.

$$N(t) = N_0 \ \mathrm{erfc}\left(\frac{a}{t} + b\right)/\mathrm{erfc}(\mathrm{b})$$

Parameters a and b specifically depend on temperature, pH, activity of halide ion, and the metal potential, but do not depend on the induction time. These dependencies can be found in the original work. However, it is also important to note that Equation (above) describes pit nucleation for the case when the external conditions do not depend on time. This restriction can be a serious impediment in predicting the pit nucleation rate under practical conditions, for example, when external conditions differ substantially for operating and shutdown periods. Generalization of PDM to the case of variable external conditions.

Rate of Pit (Crack) Propagation

The quantitative description of pit (or crack) growth remains as one of the key problems in predicting corrosion damage in many practical systems. This follows from the fact that the calculated corrosion damage that is based only on this (growth) stage can be compared with experiment, in many limiting cases. For example when all pits nucleate "instantaneously", or when the induction time for pit nucleation is much smaller than the observation time, it is possible to ignore the initial stage of pit nucleation when estimating the damage. In addition, if the probability of survival of a corrosion defect is sufficiently high, we must take into account the possibility that a stable corrosion defect (pit or crack) nucleates immediately after the start of operation and propagates without repassivation. In any case, calculations based only on the growth stage yield the most conservative estimate of the service life, ts,min, of the system. We can be sure that, if calculation of the service life is based on growth alone, the real service life, ts, will at least be not less than ts,min. In the simplest case, it is assumed that the pit propagation rate, v, depends only on the depth of the pit, and accordingly the flux density of the active pits can be presented in the form

$$j_a(x,t) = f_a(x,t)v(x)$$

where, f_a is the damage function of the active pits.

It is well known from both experiment and theory that the dependence of the characteristic pit size (for example, depth, L) on time, *t*, can be expressed by a simple equation of the following form

$$L = kt^m$$

where *k* and m are empirical constants, and, usually, m = 1. However, this dependence of L on t cannot be used directly in mathematical calculations for small times, because of the non-physical limit

$$v = \frac{dL}{dt} = kmt^{m-1} \to \infty, \qquad \text{at } t \to 0 \text{ and } m < 1$$

This is why, instead of Equation (above), we use the following interpolation equation for pit propagation rate

$$v = \frac{dL}{dt} = v_0(1 + t/t_0)^n$$

where n = m–1 and t_0 are constants, and v_0 is the initial, finite rate of pit propagation. Equation (above) yields $v = v_0$ at $t/t_0 << 0$ and $v = v_0(t/t_0)n$ at $t >> t_0$.

In many cases, the period of time over which the approximation

$$v(t) \approx v_0 = \text{constant}$$

is valid can be comparable with the observation time (or even with the service life of the system). The reason is that corrosion is, generally speaking, a slow process and under real, practical conditions, values of the critical pit depth of the system, xcr, and typical service life, ts, impose significant restrictions on the values of the initial and average corrosion current densities and, thus, on the potential and concentration drops that might be observed in a corrosion cavity. For constant external conditions, the dependence of cavity propagation rate as a function of the cavity depth, x, can be written in the form

$$v = \frac{dL}{dt} = v_0\varsigma(x)$$

where function $\zeta(x)$ satisfies the evident boundary condition $\zeta(x) \to 0$ at $x \to 0$. For the particular case, when Equation (above) holds, $\zeta(x)$ can be expressed in the form:

$$\varsigma(x) = \frac{1}{(1 + x/x_0)^{(1-m)/m}}$$

where $x_0 = v_0\, t_0/m$.

It is important to note that parameters m, t_0, (or x_0) can be estimated from "first principles" as a result of solving the relevant system of mass and charge conservation equations for the species in the solution. Thus, it can be shown that

$$m = \frac{\alpha_{eff} + 1}{2\alpha_{eff} + 1}$$

where aeff is the effective anodic transfer coefficient for the metal or alloy (for the case of diffusion limitations m = ½) and x_0 depends specifically on a_{eff}, electrolyte composition and the thickness of the electrolyte film on the metal surface. After transition of a pit into a crack or into a corrosion fatigue crevice, the total rate of corrosion crevice propagation, V, can be presented as the sun of two parts

$$V = V_{el} + V_{mec}$$

where V_{el} is the electrochemical (environmental) component, which is determined by Faraday' law, and V_{mec} is the mechanical component (crack advance associated with mechanical fatigue/creep). Generally speaking, for the case of corrosion fatigue, V_{mec} can be written in the form

$$V_{mec} = C(\Delta K)^n$$

where ΔK is the stress intensity range and C and n are empirical constants that depends on the alloy. For the case of stress corrosion cracking in disc materials, it is widely accepted that the following relation holds

$$\ln V_{mec} = -4.968 - 7302/T + 0.0278\sigma_y$$

where V_{mec} is measured in in/hr, T is the temperature in ^{0}R, and σ_y is the room temperature yield strength of the disk, in ksi [21]. In many practical cases, the condition $V_{mec} >> V_{el}$ exists and the crack propagation rate reduces to its mechanical component. For the cases when environmental component V_{el} is comparable with V_{mec} the value of V_{el} can be found as a result of solving the relevant system of mass and charge conservation equations for the species in the solution (by analogy with pit propagation rates). For this purpose, the various "coupled environment" models [Coupled Environment Pitting Model (CEPM), Coupled Environment Crevice Model (CECM), Coupled Environment Fracture Model (CEFM), and the Coupled Environment Corrosion Fatigue Model (CECFM)], which are based on the on the coupling of the internal and external environments by the need to conserve charge in the system, can be used [20, 25]. It has been tacitly assumed above that the rate of pit propagation is unequivocally determined by its depth and by the external conditions, i.e. there is no distribution in pit propagation rate for pits with equivalent depth. However, as noted above, a distribution in pit propagation rate might be observed in practical systems, because of underlying distributions in system parameters that affect the growth rate.

In generalizing DFA, for this case, we assume that the pits that propagate with initial rate v_0 are nucleated in accordance with the equation.

$$n(t) = \int_0^\infty \lambda(t, v_0) dv_0$$

Here, the function $\lambda(t, v)$ yields the number of pits (per cm^2) that have initial propagation rates between v_0 and $v_0 + dv_0$ that nucleate in the period of time between t and t + dt. It can be shown that, in this case, the expression for the flux of pits can be expressed in general form as

$$j_a(x, t) = \int_0^\infty \exp(-\gamma x/v_0)\lambda(t\text{-}x/v_0, v_0) dv_0$$

where

$$g(x) = \int_0^x \frac{dx'}{\varsigma(x')}$$

Let us assume that the distribution in pit propagation rate does not depend on time, i.e.

$$\lambda(t,v) = n(t)\ \psi(v_o)$$

To move further, we must assume a distribution function, ?(v0), for the pit growth rate, in order to account for those factors that result in a distribution in growth rate that are not captured by the (present) deterministic models. For our purposes, it is most convenient to approximate $\psi(v_0)$ by Laplace's distribution function

$$\psi(v_0) = \frac{\exp(-|v_0 - V_0|/\beta)}{2\beta}$$

where V_0 is the mean initial pit propagation rate and $\sigma^2 = 2\beta^2$ is the dispersion.

Rate of pit repassivation and transition of pits into cracks. As noted above, we assume that the repassivation process obeys a first-order decay law

$$R_a(x, t) = -\gamma f_a(x,t)$$

where γ is the delayed repassivation ("death") constant (i.e., the rate constant for repassivation of stable pits). The repassivation constant, γ, is, in general, expected to be a function of the external conditions, including the corrosion potential, temperature, and electrolyte composition. Generally speaking, g is also expected to be a function of the depth of the pit, x, because the local potential in the solution at the cavity surface depends on the IR potential drop in the cavity, i.e. γ might be a function of both the spatial coordinates and time. Of course, if the potential and concentration drops inside the corrosion cavity are insignificant during pit propagation, it is possible to neglect changes in parameter γ (see above). However, the value of this constant still depends on the external conditions, such as potential, pH, and concentration of aggressive species in the bulk electrolyte. Finally, active pits may no longer be viable if the potential, E, at the pit internal surface is less then some critical value, E_{cr}. Accordingly, if the value of E_{cr} is reached at some pit depth, xcr, active pits passivate and cannot penetrate further into the metal.

Regarding the transition of a pit into a crack, we assume that a pit immediately transforms into crack if its depth exceeds some critical value x_{cr}. According to Chen, et. al., two conditions must be satisfied for crack nucleation to take place; namely, $K_I > K_{ISCC}$ (for SCC) or $\Delta K_I > \Delta K_{I,th}$ (CF) and $V_{cr} > V_{pit}$ (V_{cr} and V_{pit} are crack and pit propagation rates, respectively). The first requirement defines the mechanical (fracture mechanics) condition that must be met for the prevailing stress and geometry, while the second simply says that the nucleating crack must be able to "out run" the pit. The analytical solution of Equation for the differential damage functions for active pits, f_a, passive pits, f_p, and cracks, f_{cr}, are found as (albeit in rather complicated form):

$$f_a = \int_0^\infty \left\{ \begin{array}{l} \dfrac{\exp[-\gamma g(x)/v_0]\, n\,(t-g(x)/v_0)}{v_0 \varsigma(x)} \times \\ \times U[(\max(x_{tr}, x_{v0}) - x] \end{array} \right\} \psi(v_0)\, dv_0$$

$$f_p = \gamma\, U(x_{tr} - x) \times$$
$$\times \int_0^\infty \frac{\gamma \exp[-\gamma g(x)/v_0] N(t - g(x)/v_0)}{v_0 \varsigma(x)} \psi(v_0) dv_0$$

$$f_{cr} = U(x - x_{tr}) \times$$
$$\times \int_0^\infty \left\{ \begin{array}{l} U(x_{tr} - x_{v0}) A(x, x_{tr}, t) + \\ + U(x_{tr} - x_{v0}) U(x_{v0} - x) B(x, t) + \\ + U(x_{tr} - x_{v0}) U(x - x_{v0}) \times \\ \times [B(x_{v0}, t - \theta\,(x, x_{v0})) + A(x, x_{v0}, t)] \end{array} \right\} \psi(v_0) dv_0$$

where

$$A(x, y, t) = \frac{\exp[-\gamma g(y)/v_0]\, n\,[t - g(y)/v_0 - \theta_{cr}(x, y)]}{V_{cr}(x)}$$

and

$$B(x, t) = \frac{\gamma}{V_{cr}(x)} \times \int_{x_{tr}}^{x} \frac{\exp[-\gamma g(x')/v_0]\, n\,[t - g(x')/v_0 - \theta\,(x, x')]}{v_0\, \varsigma(x')} dx'$$

U(x) is the unit function [U(x) = 0 at x=1 and U(x) = 1 at x>0], max(a,b) = a for a=b and max(a,b) = b for a<b. Here, we denote by xtr the depth of a pit at which the condition of $K_I = K_{ISCC}$ (for SCC) or $\Delta K_I = \Delta K_I$,th (corrosion fatigue, CF) is fulfilled and by x_{v0} the depth of the pit at which the condition $V_{cr} = V_{pit}$ for a pit that was born with an initial propagation rate, v_0, is fulfilled. In doing so, we assume that Condition holds. We also assume that the crack propagation rate, $V_{cr}(x)$, depends only on the crack depth, x. Accordingly, the function

$$\theta_{cr} = \int_{x'}^{x} \frac{dy}{V_{cr}(y)}$$

is the age of a crack with depth x, which was born at distance x' from the metal surface (i.e., in the bottom of a pit). It is important to note that the solution obtained recognizes the fact that, if the depth of the pit increases to x_{tr}, and after that pit is passivated, it must immediately transition into a crack, because the condition $V_{cr} > V_{pit}$ is fulfilled.

Using the solutions obtained, the percentage of pits with cracks at specific pit depths, ε, is readily calculated from the relation

$$\varepsilon = \frac{f_{cr}}{f} \times 100\%$$

where $f = f_a + f_p + f_{cr}$ is the total differential damage function for corrosion defects. The value of e has been measured on 3NiCrMoV steam turbine disk steel employing self loaded specimens to 90 per cent of s0.2 for aerated pure water and aerated water containing 1.5 ppm of chloride ion at 90 oC. It was noted that a remarkable similarity exists between the percentage of pits with cracks at specific pit depths, despite changes in exposure time and presence or absence of chloride. It was also mentioned that, whilst the crack depth usually exceeded the pit depth, there were examples where the pit depth extended significantly beyond the crack; i.e. the crack may not have initiated from the base of the pit. Accordingly, simple application of fracture mechanics principles based on the pit acting as the effective crack of the same depth would not be pertinent.

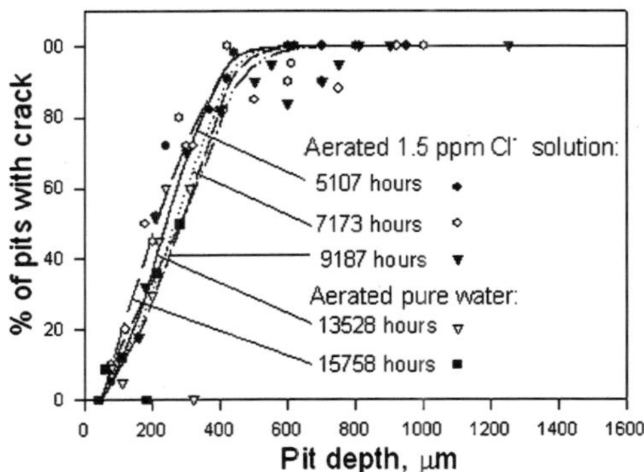

Fig. Likelihood of cracks developing from pits as a function of depth for 3NiCrMoV disk steel.

It can be shown that, for the same values of the kinetic parameters, Thus, in accordance with the experimental data, we note that xtr is of the order of 50 μm, and $V_{cr} \approx 10^{-10}$ m/s for corrosion in aerated, 1.5 ppm Cl^- solution (in the case of pure aerated water, the crack growth rate is lower by factor of 2). In addition, we assume that the distribution in the pit propagation rate is described by Equation (23) with $V_0 \approx 3V_{cr}$ and $\beta \approx 0.1V0$ (very sharp distribution), $x_0 \approx x_{tr}$ and $\gamma \approx 1/(1000\ \text{h})$. Finally, we assume that instantaneous nucleation of pits occurs in the system.

Figure 5 shows that satisfactory coincidence is observed between the calculated and measured values of ε. Of course, this procedure cannot be regarded as an independent experimental determination of the unknown kinetic parameters (for example, γ). These measurements will be reported in consequent papers. Figure 5 only shows that DFA allows us to explain satisfactorily the principal experimental data for the transition of pits into cracks during the stress corrosion cracking of steam turbine disk steel on the basis of the proposed mechanical conditions.

STATISTICAL PROPERTIES OF THE DAMAGE FUNCTION

There exists a close correspondence between Damage Function Analysis (DFA), which has been described at some length above, and Extreme Value Statistics (EVS). The latter technique has been used extensively to extrapolate damage (maximum pit or crack depth) from small samples in the laboratory to larger area samples in the field. Furthermore, DFA provides a means of calculating the central and scale parameters and their time-dependencies in EVS from first principles, and hence represents a unification of the two prediction philosophies.

From a statistical point of view, all distributed properties of the system are completely determined by the Cumulative Distribution Function (CDF), $\Phi(x)$. By definition, $\Phi(x)$ is the probability that the depth of a randomly selected pit (crevice) is $\leq x$. We postulate that the pit distribution on the metal surface is uniform. Accordingly, the total number of nucleated pits in the entire system is SN(t), where S is the area of the system and, from the definition of the integral damage function, *F*, the number of pits that have the depth $\leq x$, is $S[N(t)-F(x, t)]$. Accordingly, from the definition of probability we have

$$\Phi(x,t) = \frac{S[N(t) - F(x,t)]}{SN(t)} = 1 - \frac{F(x,t)}{N(t)}$$

We see that the CDF for a given observation time, $\Phi(x,t)$, can be predicted if we know (can calculate) the integral damage function of the system [note that the number of nucleated, stable pits, $N(t)$ simply equals $F(0, t)$]. This relationship can be regarded as being the bridge between the statistical and deterministic approaches for estimating the accumulation of localized corrosion damage on a surface.

From the practical point of view, the most important value for characterizing corrosion damage is the failure probability, P_f, of the system. By definition, P_f, is the probability that at least one corrosion event in any form (pit, crevice, stress corrosion crack, or fatigue crack) reaches a depth, x, at a given observation time, *t*, where *x*, in this case, is the critical dimension. It is evident that the probability of a corrosion event not achieving the depth

x at the tine, t, is $\Phi(x, t)$. Accordingly, the probability that none of the corrosion events will reach the depth of x is, $\Phi(x, t)^{SN(t)}$ where S is the surface area of the system and, finally, the probability of failure is

$$P_f(x, t)=1- \{1 - F(x, t)/N(t)\}^{SN(t)}$$

Using the well known limit relation, $\lim(1+a/x)^x \rightarrow \exp(a)$ as $x \rightarrow \infty$, by assuming that $F(x,t)/N(t) << 1$, and by noting that $SN(t)$ is a large number, we can rewrite Equation (above) in the form

$$P_f(x,t) = 1 - \exp\{-SF(x,t)\}$$

Equation allows us to calculate the probability of failure if the integral damage function, F, is known. The latter function can be found as a solution. In particular, for the case of pitting corrosion, the total integral damage function, F, is obtained by analytically solving Equations to yield

$$f_a = \int_0^\infty \frac{\exp[-\gamma g(x)/v_0]\lambda[t-g(x)/v_0, v_0]}{v_0\varsigma(x)} dv_0$$

$$f_p = \int_0^\infty \frac{\gamma \exp[-\gamma g(x)/v_0)\Lambda[t-g(x)/v_0, v_0]}{v_0\varsigma(x)} dv_0$$

and

$$F = \int_0^\infty \exp[-\gamma g(x)/v_0)\ \Lambda[t-g(x)/v_0, v_0]dv_0$$

where f** yields the number of pits (per cm2) that have initial propagation rates between v0 and v0 + dv0 and nucleate in the period of time between 0 and t. F is the sum of the integral damage functions for active and passive pits for the given case. If, in addition, Equation holds, we have for the integral and the cumulative damage functions the following expressions

$$F(x, t) = \int_0^\infty N(t-g(x)/v_0)\exp(-\gamma g(x)/v_0)\ \psi(v_0)dv_0$$

and

$$\Phi(x,t) = 1 - \int_0^\infty N[t-g(x)/v_0)]\exp[-\gamma g(x)/v_0]\ \psi(v_0)dv_0/N(t)$$

In particular, for the case of instantaneous nucleation, we have for the CDF

$$\Phi(x, t) = 1 - \int_{x/t}^\infty \exp[-\gamma g(x)/v_0]\ \psi(v_0)dv_0$$

As has been shown experimentally, in many practical cases, the asymptotic behavior (for large values of x) of the CDF can be described by the exponential relationship [7, 27]:

$$\Phi(x, t) = 1 - \exp[-(x - u)/\alpha]$$

where u is the central parameter (the most frequent value) and a is the scale parameter, which defines the width of the distribution. Accordingly, as follows from Equation, the extreme value distribution (EVD), Ψ (x, t) (the probability that the largest value of pit depth Ψ x), is described by a double exponent (Gumbel Type I extreme value distribution) in the form:

$$\Psi(x, t) = 1\text{-} P_f = \exp[-\exp(-y)]$$

where

$$y = (x - h)/\alpha$$

and $h = u + \alpha \ln(\mathrm{SN})$.

Numerical calculations shows that the dependencies of ln(1-CDF) on pit depth, x, can be approximated by straight lines, at least for sufficiently large values of x, and therefore provide a theoretical basis for applying the Type I extreme value distributions to real corrosion systems. However, all calculations from Ref. 9 were performed for the case where the pit propagation rate does not depend on the pit depth, i.e. $\zeta(x) = 1$ (concentrated or moderately dilute solutions). Our numerical calculations confirm this conclusion, at least for the case where the function $\zeta(x)$ is described by Equation. In some cases parameters a and h can be expressed in analytical form. Thus, for the case of instantaneous nucleation and constant pit propagation rate, these parameters can be presented in the form:

$$\alpha = \frac{\beta t}{1 + \gamma\beta t/V_0}, \quad \text{and} \quad \mathrm{h} = \frac{[\mathrm{V}_0 + \beta \ln(0.5SN_0)]t}{1 + \gamma\beta\, t/V_0}$$

APPLICATION OF DFA TO THE PREDICTION OF CORROSION DAMAGE IN STEAM TURBINES

First of all, we will demonstrate how DFA can help us to extrapolate corrosion damage into the future. i.e. by using the results of short term experiments we will predict the results of long term experiments on the same system. As opposed to damage tolerance analysis (DTA), prediction can be made over times that are much longer than the experimental calibrating time, because of the deterministic nature of the embedded models. Thus, our predictions compared with experimental data for the pitting of aluminum in tap water. Note that only data at the two shortest times (and zero time) were used to calibrate the model.

Practically, the calibration was done in the following way. The model was fitted to the short-term experimental data (those for 7 and 30 days) to determine values for the three parameters:

$$a_1 = \mathrm{b},\ a_2 = V_0 + \mathrm{b}\ \ln\ (0.5\mathrm{SN0}) \text{ and } a_3 = \mathrm{gb}/V_0$$

and values for parameters a and h were obtained. From these parameters, we calculated the mean depth of the deepest pit (the most probable value), $\overline{x}_{max}$, as a function of time by using the relationship

$$\overline{x}_{max} = h + \gamma_e \alpha$$

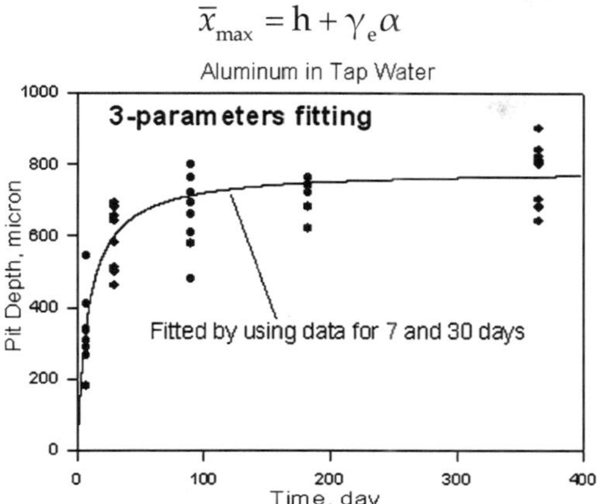

Fig. The mean depth of the deepest pit as a function of time.

where γ_e = 0.5772… is Euler's constant. We see that DFA allows us to predict satisfactorily the results for times that are an order of magnitude greater than the times used for fitting. Had the data for the first three times (including t = 0) been used to empirically extrapolate the damage, the mean value of the deepest pit would be predicted to be several thousand micrometers after 200 days. The difference is due to the fact that the system under study is strongly passivating ("stifling") and hence all pits die before any attain depths beyond about 800 microns, a feature that is captured by the deterministic model, but not by empirical extrapolation. The practical consequence is that the empirical method would severely over-predict the accumulation of pitting damage on aluminum exposed to tap water.

It is important to note that the evolution of the pit depth with time, for the case of steam turbine disc steel, can be described by Equation with m = 0.38 [6]. However, theoretically, the lowest value of, m, that should be observed for an isolated, open pit corresponds to diffusion control with m equaling 0.5. The hyperbolic forms of parameters h and α can be easily extrapolated using Equation with m < 0.5. This means that observed pit propagation rates with m<0.5 can be explained by influence of repassivation phenomenon. Unfortunately we do not have experimental data for the propagation of localized corrosion damage in turbine steels over a period that is comparable with the service life, so that we do not have the opportunity to evaluate DFA for accumulated pitting damage in real or simulated phase transition zone environments, in the manner that we did above for the pitting of aluminum in tap water. However, Turnbull and Zhou have recently published an extensive data set for the pitting of 3NiCrMoV turbine disc steel in simulated phase transition environments extending over 15758 hours (656.6 days). This set appears to be the most extensive in existence for well-controlled conditions and, while it falls short of service life (40 years or 14600 days), it does provide an opportunity to test the ability of DFA to predict damage over periods that extend beyond normal laboratory times. However, Turnbull and Zhou's paper shows that the form of the damage function is remarkably similar to those predicted by DFA for a strongly repassivating system [3,18]. Furthermore, the damage functions demonstrate that the maximum pit depth is not a strong function of time, which is another characteristic that is consistent with strong repassivation.

Experimental studies are also underway in the Pennsylvania State University and Frumkin Institute (Moscow) and some preliminary results for Type 403 SS turbine blade alloy obtained at PSU for relatively short observation times. Excellent agreement is obtained between the experimental and calculated mean maximum pit depth for the longest time, and the form of the data demonstrates that this, too, is a strongly repassivating system.

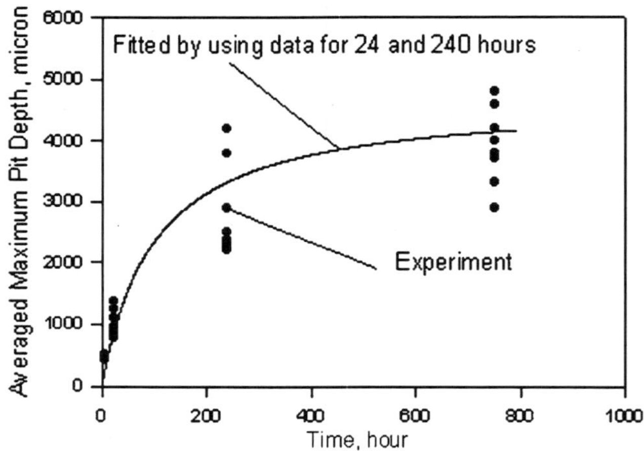

Fig. Variation of the extreme value (EV) in pit depth with exposure time for Type 403 SS in 0.10M NaCl-containing deaerated borate buffer solution at pH = 8.1, T = 25 °C, and at an applied potential of $0.090V_{SCE}$.

It is important to emphasize that the application of DFA for describing and predicting localized corrosion damage has real advantages over the purely statistical approaches. Thus, DFA yields the connection between the CDF (cumulative distribution function) and the physical parameters that determine the accumulation of damage (i.e., with the "mechanism"). This connection allows us to reduce the number of fitting parameters for the CDF, to predict the time dependencies of these parameters, and to predict the influence of the external conditions (corrosion potential, temperature, electrolyte composition, etc.) on the fitting parameters.

In accordance with the theory developed above, calculation of the damage functions requires the determination of three independent functions for each kind of corrosion defect, *k*: The rate of defect nucleation, n_k; the flux density of defect, *jk*; and the rate of transition of one kind of defect into another, R_k, (for example, the transition of a an active pit into a passivated pit or a passivated pit into a crack). Accordingly, the algorithm for estimating the damage function must have at least three modules for handling these functions, but the operation of the algorithm is quite independent of the form of each function. Of course, additional modules are required for determining external parameters, such as the composition of the liquid film, temperature, corrosion potential, and so forth. A general algorithm together with a general, modular computer code for predicting localized corrosion damage in steam turbines has been developed.

Damage Function Module

The DFM is the main module of the code. To start the calculation, input information, such as the initial distribution of damage, i.e. initial values for damage functions, f_{k0}, environmental conditions, etc., are entered. Sometimes, but not in the general case, we can simply assume that all f_{k0} are equal to zero (there is no initial damage in the system). Practically, the function of this module reduces to one of numerically or analytically solving the system of differential Equations for calculating the DF for all kinds of defects in the system. For each time step, this module calls the:

Nucleation Module (NM), which calculates the rate of defect nucleation, nk, at the metal surface as a function of induction time;

Growth Module (GM), which computes the flux of corrosion events, jk, as a function of its depth and, in the general case, at sufficiently small time steps, if the external conditions depend on time;

Transition Module (TM), which computes the rate of transition (transformation) of one defect into another, Rk, (for example pit repassivation or transition pit into crack) as a function of population of defects, depth, and, in the general case, observation time.

In turn, the three modules, NM, GM, and TM, call the Environmental Module (EM) that defines the external conditions for the system, such as the chemical composition, conductivity, and the thickness of the electrolyte film on the steel surface, corrosion potential, temperature, mechanical conditions, and so forth. In the general case, all parameters may depend on time (for example, due to shutdowns and startups and due to transients during operation). After obtaining reliable electrochemical kinetic data for oxygen, hydrogen ion, and water reduction on the steam turbine steel, along with the value of the passive corrosion current density, we will incorporate into the EM the Mixed-Potential sub-Module (MPM), which calculates the corrosion potential of the turbine blades and disks as a function of electrolyte composition, temperature, and electrolyte film thickness. The MPM is based on the Wagner-Traud hypothesis for free corrosion processes and is customized for the case of general corrosion under thin electrolyte films and for the real electrochemical conditions that exist in a steam turbine. Many of the experimental data that are required for the MPM have been obtained and will be published in near future. After determining the damage function for the system, we are in a position to address some important design questions. Thus, the output of the algorithm can be specified in three forms:

- For a specified probability of failure, the algorithm estimates the damage function as a function of exposure time, and computes the observation time at which the depth of the deepest crack exceeds a critical dimension. The calculated observation time is the service life.
- For a specified probability of failure and design life, the algorithm calculates the critical dimension that can be tolerated to ensure acceptable performance.
- For a specified wall thickness and design life, the algorithm estimates the failure probability.

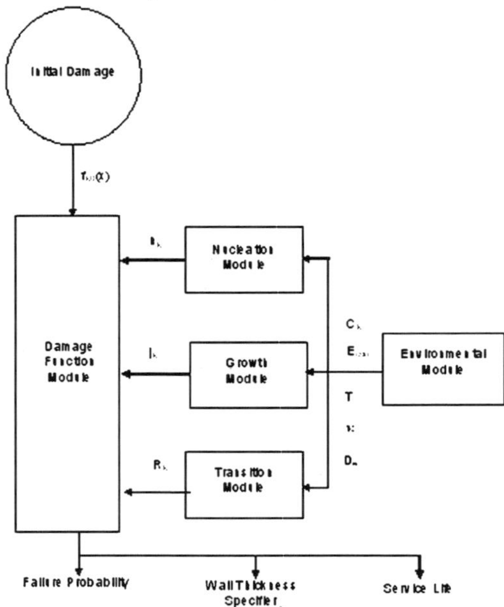

Fig. Structure of algorithm for prediction of damage function.

We have recently modeled the accumulation of damage due to pitting corrosion and corrosion fatigue in disks and blades for a turbine that is subjected to cyclical operation. Thus, in this case, the corrosion evolutionary path comprises multiple cycles of an operational period of 500 hours at 95 oC under cyclic loading followed by 100-hour shutdown period at 25 oC. The transition period between shutdown and operation was considered to be 0.5 hours, during which time the properties of the system change linearly with time. Thus, the total cycle time is 600.5 hours and the total operational time was assumed to be 12,5 years corresponding to 219 cycles. The accumulated damage at any given time (i.e., after the corresponding number of cycles) is obtained by integrating the cavity growth rate over the evolutionary path to that point. The electrochemical parameters assumed in the calculation for the low alloy disc steel and the Type 403 SS blade alloy actually correspond to those for iron and Type 316 SS [the experimental data for electrochemical kinetic parameters for 316 SS are taken from Ref. 30], respectively, because of a lack of data for the actual alloys of interest. Details of the calculations will be published at a later date. We note, however, that the exchange current densities for the redox reactions (evolution of hydrogen and the reduction of oxygen) for Type 316SS and carbon steel are not expected to differ significantly from those for Type 403SS and 3NiCrMoV disk steel, respectively.

Table. Assumed Operating Cycle Parameters for the Development of Corrosion Fatigue Damage in Low Pressure Steam Turbine Blades.

Shutdown	Operation Cycle
t = 100 h	t = 500 h
$\sigma = 0$	σ =sm + 0.5$\Delta\sigma$ sin ($2\pi ft$)
	σ_m = 84 jsum $\Delta\sigma$ = 2 ksi, f = 600 Hz
T = 25 °C	T = 95 °C
$[O_2]$ = 8 ppm	$[O_2]$ = 20 ppb
$[Cl^-]$ = 0.3 M	$[Cl^-]$ = 3 M
pH = 6.96	pH = 6.13

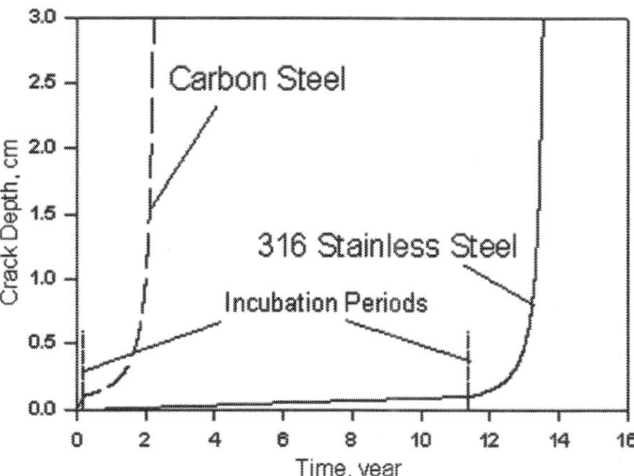

Fig. Plots of accumulated pitting/corrosion fatigue damage in carbon steel and Type 316 SS along a simulated LP steam turbine corrosion evolutionary path comprising multiple cycles of shutdown followed by operation.

The calculated, accumulated damage (the most probable cavity depth as a function of time) for both materials is summarized in Figure 9. In this figure, the incubation period corresponds to pitting corrosion, followed by corrosion fatigue becoming the principal mode of damage accumulation. For the case of the stainless steel, in particular, the failure time is dominated by initiation (incubation) and once a crack nucleates it grows rapidly to a super critical dimension. In the case of iron (disk steel), however, the initiation period is predicted to be short, compared with the total failure time, so that this case corresponds to a growth dominated failure scenario (see below).

A number of important conclusions were drawn from this study:

- At loading frequencies above ca. 100 Hz, fatigue crack growth is due to mechanical fatigue, with negligible impact of the environment.
- At loading frequencies below 0.1 Hz, for low R-ratio loading, advection has little impact on corrosion fatigue crack growth rate. Accordingly, crack propagation can be treated as stress corrosion cracking.
- The service life of a system is very sensitive to the electrochemical (corrosion) properties of the alloys employed and the environment.
- Increasing of the initiation period is the key to increasing of the service life of the system.
- The pit propagation rate is predicted to be much higher during shutdowns than it is during operation for the current mode of operation (aerated conditions during shutdown). Accordingly, it is possible to significantly increase the service life of a system by controlling the environmental conditions (e.g., deaerating) during shutdown.

Using these data, we are ale to estimate the failure probability by specifying a "super critical" crack (i.e., one at which unstable fracture ensues, in this case a 1-cm long crack). The probability of failure is calculated as the likelihood that a crack will become supercritical in the stated observation time. With reference to Figure 10, we see that the failure probability for Type 403 SS blades is reduced by a factor of 104 by simply reducing the oxygen concentration in the condensate on the blades to 1 ppm during shutdown. This is a dramatic effect and is one that is easily enacted in turbine operation.

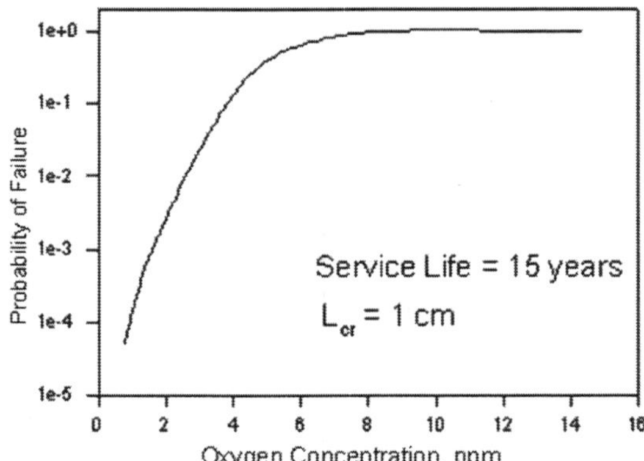

Fig. Probability of failure of a turbine blade as a function of oxygen concentration during the shutdown period.

It is important to emphasize that, while the calculations described above were made for illustrative purposes only (i.e., we still have to introduce parameters for Type 403SS), they do demonstrate the level of sophistication that is now being achieved in corrosion damage prediction.

Table. Assumed Operating Cycle Parameters for the Development of Stress Corrosion Cracking Damage in Low Pressure Steam Turbine Disks.

Shutdown	Operation Cyle
t = 100 h	t = 500 h
$\sigma = 0$	$\sigma = 105$ ksi
T = 25 °C	T = 95 °C
$[O_2] = 8$ ppm	$[O_2] = 20$ ppb
$[Cl^-] = 1.5$ ppm	$[Cl^-] = 1.5$ ppm
pH = 6.96	pH = 6,41

We now turn to consider corrosion-induced failure of LP steam turbine disks; in this case by stress corrosion cracking. Full details of these predictions will be published at a later date, but the corrosion evolutionary path is that specified. The crack growth rate used in these calculations has been estimated by using the standard Equation.

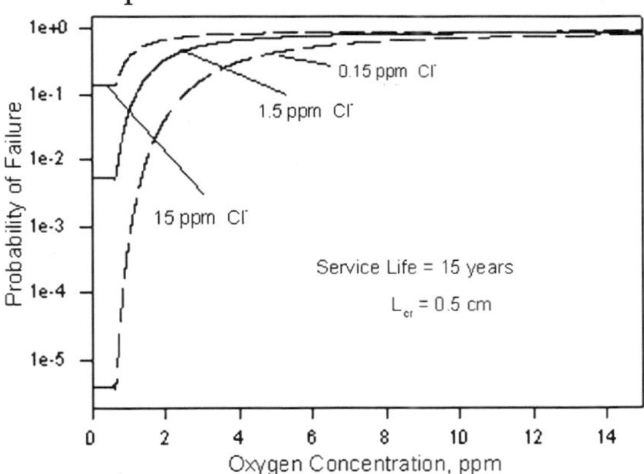

Fig. Calculated failure probability for LP steam turbine disks as a function of oxygen concentration during shutdown for different chloride concentrations in the electrolyte film.

Typical predictions of the failure probability versus oxygen concentration for SCC in LP turbine disk steel for different chloride concentrations in the PTZ condensate for a service life of 15 years. The critical crack length for the onset of unstable fracture is assumed to be 0.5-cm. The calculated failure probability plotted is clearly a sensitive function of both the oxygen concentration and the chloride concentration in the PTZ condensate. Increasing the oxygen concentration under shutdown conditions displaces the corrosion potential in the positive direction and hence increases the pit nucleation and pit growth rates, thereby resulting in an increase in the failure probability. At sufficiently low oxygen concentration, oxygen reduction is no longer the principal cathodic reaction (hydrogen ion and water reduction become dominant), and the failure probability is sensibly predicted to be independent of $[O_2]$ in the condensate. However, the propagation and nucleation rates of pits are also predicted (see above) and found to increase with increasing $[Cl^-]$, so that the failure probability versus oxygen concentration shifts upwards and to the left (i.e., the environment becomes more aggressive). Figure 11 displays the sensitive relationship that is predicted to exist between the oxygen concentration during shutdown and the chloride concentration in the phase transition zone, which is also assumed to persist on the disk surfaces when the turbine is taken out of service. Thus, decreasing the probability of failure over many operating cycles of the turbine is seen to be not only a sensitive function of oxygen concentration during shutdown, but also of the chloride concentration on the steel surface. Clearly, the life of the disk (and blades) can be extended substantially by deaerating the environment and washing the surfaces during and prior to shutdown.

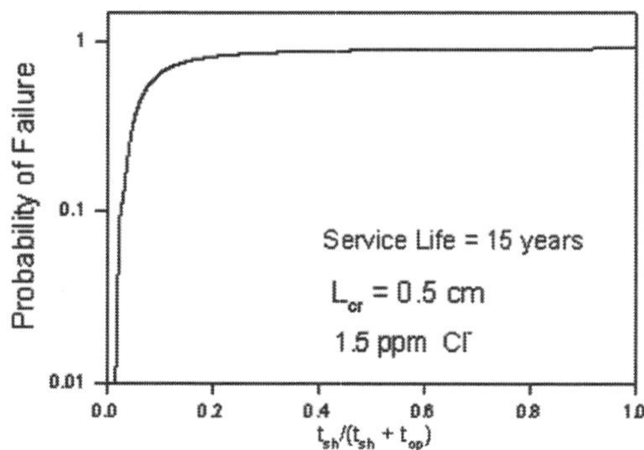

Fig. Probability of failure versus the fraction of the time spent in shutdown under aerated conditions for a surface chloride concentration of 1.5 ppm and a shutdown concentration of oxygen of 8 ppm and σ = 105 ksi.

The modeling work outlined above shows that the tendency toward blade and disk failure depends critically upon the conditions that exist during shut down. Assuming that the shutdown conditions are aerated (8 ppm O_2 in the liquid film on the disk surface) and that the chloride concentration is 1.5 ppm, the probability of failure versus fraction of time spent. This figure shows that the probability of failure increases rapidly with increasing time spent under aerated shutdown conditions and that fractional times of greater than 10 per cent essentially ensure that failure will occur over 190 cycles. Of course, by lowering the oxygen and chloride concentrations during shutdown will pus this curve dramatically to the right, thereby substantially decreasing the probability of failure for increasing time spent under shutdown.

The final issue that we wish to address in this paper is stress. Stress (residual or operating) is an issue, because it determines the depth at which a pit will transition into a crack and it affects the rate of crack growth, particularly in the Stage I region of the crack growth rate versus stress intensity correlation. Extensive laboratory and field observations shows that stress is an important factor and the relationship between the apparent mechanical crack growth rate and the yield stress of the disk steel is given by Equation (18). The calculated relationship between the probability of failure and the stress. One sees that the failure probability is a strong function of stress, but it is also important to note that a threshold stress level exists below which the probability of failure is zero. This threshold value corresponds to the minimum stress that is necessary to ensure that the condition $K_I > K_{ISCC}$ holds, which is one of the two conditions that must be met for the nucleation of a crack at a (passivated) pit.

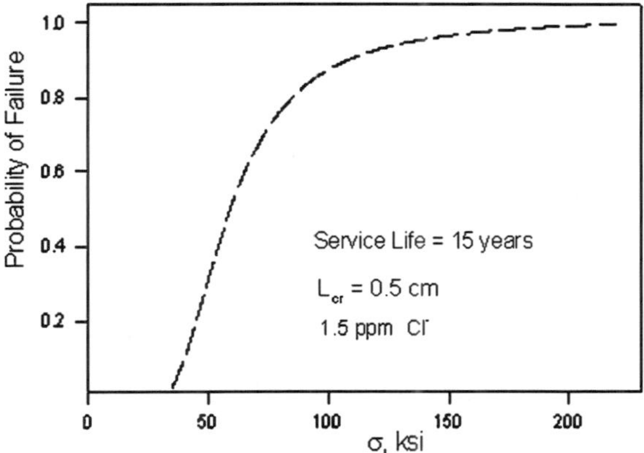

Fig. Probability of failure versus yield strength under aerated conditions for a surface chloride concentration of 1.5 ppm and a shutdown concentration of oxygen of 8 ppm.

Finally, we note that the above is but a small sample of the correlations between the failure probability and the various independent variables that exist in a low pressure steam turbine that can be generated by Damage Function Analysis. Much of the work in the immediate future will be concerned with evaluating these predictions against laboratory and field observation, but it is important to note that the theoretical basis for the deterministic prediction of damage in these systems has now been largely established.

13

Corrosion Thermodynamics

THERMODYNAMIC CORROSION CYCLE:

A further corrosion cycle, often used as introduction to corrosion theory, is a thermodynamic one. Almost all metals and alloys used in service are actually in an unstable thermodynamic state. There is thus a fundamental thermodynamic tendency for them to return to a stable state through corrosion processes. In essence, corrosion reverts the metals/alloys back to the stable state of the ores from which they were derived. Corrosion products tend to be thermodynamically stable species, similar to the original ores.

Thermodynamics is concerned with energy states. The original metallic ores are said to be in a state of low energy. External energy is applied in the conversion of the ores to usable metals and alloys, transforming them to a higher energy state. They tend to revert to a lower (more stable) energy state by reacting with a corrosive environment. While thermodynamics can predict whether a corrosion reaction will take place, it does not provide an indication of the rate of corrosion reactions. The rate of reactions is described by kinetic theory.

One can use thermodynamics, e.g. Pourbaix or E-pH diagrams (Thermo/ironE-pH.htm), to evaluate the theoretical activity of a given metal or alloy in a corrosion situation provided the chemical make-up of the environment is known.

But for practical situations, it is important to realize that the environment is a variable that can change with time and conditions. It is also important to realize that the environment that actually affects a metal corresponds to the micro-environmental conditions this metal really 'sees', i.e. the local environment at the surface of the metal.

It is indeed the reactivity of this local environment that will determine the real corrosion damage. Thus, an experiment that would investigate only the nominal environmental condition without consideration for local effects such as flow, pH cells, deposits, and galvanic effects is useless for lifetime prediction.

FREE ENERGY AND ELECTROCHEMICAL REACTIONS

Electrical work is the product of charge moved Q times the cell potential (E) through which it is moved. If the work done is that of an electrochemical cell in which the potential difference is E, and the charge is that of one mole of reaction in which n moles of electrons are transferred, then the electrical work -w done by the cell must be nE. In this relationship, the Faraday constant F (96485 C mol(e-)-1) is necessary to obtain coulombs from moles of electrons.

In an electrochemical cell operating reversibly, no current flows and:

DG = -nFE

Under standard conditions, the standard free energy of the cell reaction DG0 is directly related to the standard potential difference across the cell, E0:

DG0 = -nFE0

Electrode potentials can be combined algebraically to give cell potential. For a galvanic cell (Aircraft/galvdefi.htm), which operates spontaneously, a positive cell voltage will be obtained if the difference is taken in the usual way, as:

Ecell = Ecathode - Eanode

The free energy change in a galvanic cell, or in a spontaneous cell reaction, is negative and the cell voltage positive. In electrolytic cells, the reaction is driven in the non-spontaneous direction by an external electrical force. The free energy change in an electrolytic cell, or in a non-spontaneous cell reaction, is therefore positive and the cell voltage negative.

Other thermodynamic quantities can be derived from electrochemical measurements. For example, the entropy change (DS) in the cell reaction is given by the temperature dependence of DG:

$$\Delta S = -\left(\frac{\partial \Delta G}{\partial T}\right)_P$$

Hence,

$$\Delta S = nF\left(\frac{\partial F}{\partial T}\right)_P$$

and finally,

$$\Delta H = \Delta G + T\Delta S = nF\left[T\left(\frac{\partial E}{\partial T}\right)_P - E\right]$$

The equilibrium constant (Keq) for the same reaction can be obtained with the following equation:

RT ln Keq = -DG0 = nFE0

NERNST EQUATION

The Nernst equation, named after the German chemist Walther Nernst, can be derived from the equation linking free energy changes to the reaction quotient:

$$\Delta G = \Delta G^0 + RT\ln Q$$

where, for a generalized equation of the form:

aA + bB +..®. mM + nN +..

and the reaction quotient:

$$Q = \frac{a_M^m \cdot a_N^n \cdots}{a_A^a \cdot a_B^b \cdots}$$

where aM, aN..., aA, aB... are the activities of the respective species in the generalized equation. The power terms of these activities correspond to the coefficients in the same equation. Some of the species that take part in electrode reactions are pure solid compounds and pure liquid compounds. In dilute aqueous solutions, water can be treated as a pure liquid. For pure solid compounds or pure liquid compounds, activities are constant and their values are considered to be one.

The activities of gases are usually taken as their partial pressures and the activities of solutes such as ions are usually taken as their molar concentrations: i.e. $a_1 = \gamma_i [i] \approx [i]$ where [i] and γ_i are respectively the molar concentration and the activity coefficient of species i.

In the case of an electrochemical reaction, substitution of the relationships $\Delta G = -nFE$ and $\Delta G^0 = -nFE^0$ into the expression of a reaction free energy and division of both sides by -nF gives the Nernst equation for an electrode reaction:

$$E = E^0 - \frac{RT}{nF}\ln Q$$

Combining constants at 25°C (298.15 K) gives the simpler form of the Nernst equation for an electrode reaction at this standard temperature:

$$E = E^0 - \frac{0.059}{n}\log_{10} Q$$

In this equation, the electrode potential E is the actual potential difference across a cell in which this electrode and a standard hydrogen electrode are present. Alternatively two Nernst equations corresponding to two half-cell reactions can be combined into the Nernst equation for a cell reaction:

$$E_{cell} = \left(E^0_{cathode} - E^0_{anode}\right) - \frac{0.059}{n}\log_{10} Q$$

REFERENCE ELECTRODE POTENTIALS

Name	Nernst Equation	Potential	T coefficient
	(V vs. SHE)	(V vs. SHE)	(mV oC^{-1})
Hydrogen (SHE)	E^0 - 0.059 pH	0.00	..
Silver chloride	E^0 - 0.059 log a chloride	0.2224	-0.6
	0.1 M KCl	0.2881	..
	1.0 M KCl	0.235	..
	saturated (KCl)	0.199	..
	seawater	~0.250	..
Calomel	E^0 - 0.059 log a chloride	0.268	..
	0.1 M KCl	0.3337	-0.06
	1.0 M KCl	0.280	-0.24
(SCE)	saturated (KCl)	0.241	-0.65
Mercurous sulfate	E^0 - 0.0295 log asulfate	0.6151	..
Mercuric	E^0 - 0.059 pH	0.926	..
Copper sulfate	E^0 + 0.0295 log acopper	0.340	..
	saturated	0.318	..

A visual chart was produced to convert the voltages read with various reference electrodes.

STABILITY OF WATER

The following equation describes the equilibrium between hydrogen ions and hydrogen gas in an aqueous environment:

$$2\,H^+ + 2\,e^- = H_2$$

which can be rewritten as following in neutral or alkaline solutions:

$$2\,H_2O + 2\,e^- = H_2 + 2\,OH^-$$

At higher pH than neutral, this second equation is more appropriate. However both equations signify the same reaction for which the thermodynamic behavior can be expressed by a Nernst equation:

$$E_{H^+/H_2} = E^0_{H^+/H_2} + \frac{RT}{nF} \ln \frac{\left[H^+\right]^2}{p_{H_2}}$$

that transforms into the following at 25oC and hydrogen partial pressure (p) of 1 atm:

These equations delineate the stability of water in a reducing environment and are represented in a graphical form by the sloping line (a) on Pourbaix diagrams. Below line (a) the equilibrium reaction indicates that the decomposition of H_2O into hydrogen is favored while it is thermodynamically stable above that line. As potential becomes more positive or noble, water can be decomposed into its other constituent, i.e. oxygen. The equations representing respectively the acidic form and neutral or basic form of this equilibrium are written as:

$$O_2 + 4\ H^+ + 4\ e^- = 2\ H_2O$$

$$O_2 + 2\ H_2O + 4\ e^- = 4\ OH^-$$

And again these equivalent equations can be used to develop a Nernst expression of the potential in standard conditions of temperature and oxygen pressure:

The line labeled (b) in Pourbaix diagrams represents the behavior of E vs. pH for this last equation.

ANTICIPATED CORROSION ENVIRONMENT AND PROCESSES

The research highlights described above indicate that Alloy 22 has remained in the passive state for up to several years under the conditions tested. The metal-dissolution rates in those experiments were in the desired low value range. The localized-corrosion experiments indicated, using a critical potential concept, that crevice corrosion (and by inference, pitting corrosion as well) is very unlikely under the conditions tested. The tests sought to address conditions that were representative of, or more severe than some of the environments expected to develop at the WP surface.To evaluate how well those findings support adequate WP corrosion performance, one should consider whether the information to date covers the necessary breadth of anticipated possible service environments and material conditions and whether any important long-term corrosion mechanisms have been ignored.

Consideration of how adequately relevant conditions are covered has led to ongoing action by DOE to expand the information base, such as the work on the effect of trace elements mentioned earlier. In another example, recent analyses and calculations have shown that highly deliquescent species (e.g., tachyhydrite) may be present in pore water when evaporated to near dryness. Those species, if present in the dust that will deposit on the WP during and after the ventilation period, could promote the formation of liquid-water solutions even at temperatures as high as ~165 °C, which could mean the possibility of aqueous corrosion during much, if not all, of the heat pulse, even in the "hot" repository concept.

As a result, experiments for determining passive dissolution rates and especially susceptibility to localized corrosion now are being planned to cover a higher temperature range than earlier investigated. Such investigations are part of establishing that the WP shell materials satisfy the necessary short-term conditions for adequate performance and should be continued to reduce uncertainty to a level appropriate for decision-making.

RECENT VIEWS ON OTHER POSSIBLE CORROSION PROCESSES

Considering possibly ignored long-term processes deserves special attention because they involve more directly the extrapolation challenges mentioned earlier in this paper. Those challenges cannot be resolved merely by the short-term data available but instead require thorough questioning of possible processes and application (or development) of adequate fundamental understanding. Some of those issues were addressed in a recent Workshop on Long-Term Extrapolation of Passive Behavior, sponsored in July 2001 by the NWTRB. A panel of 14 outstanding corrosion scientists was asked two questions that may be summarized as follows:

In the first question, the premise was that the WP service environment is, as suggested by much of the available evidence, conducive to spontaneous passivation of a recently prepared Alloy 22 surface. The passive regime thus initiated had continued for several hundreds or even thousands of years so that the passive corrosion penetration had reached a substantial depth (e.g., > 10μm). The panelists were asked to propose any mechanism(s) they deemed plausible that would cause the long-term corrosion rate of the Alloy 22 shell to increase, after such prior penetration, so that sustained corrosion rates (maybe no longer uniform) would exceed ~1 ?m/y. The speculative scenarios listed in the "Extrapolation Challenges" section were given as examples for optional consideration.

In the second question, the premise was that short-term evidence indicated that under expected service conditions, the open circuit potential at the package surface would stay significantly more negative (by a few hundred mV or more) than the critical potential deemed necessary for development of stable localized corrosion. The panelists were asked to propose any mechanism(s) they deemed plausible that would cause, over long periods of time, shifts in the open circuit potential and/or the critical potential so that stable localized corrosion could develop. Furthermore, they were asked whether a localized-corrosion process could be proposed that could develop over long times so that initiation and propagation were not amenable to description in terms of a critical potential. For focusing discussion, the scope of the question specifically excluded SAC processes.

For each question, the panelists were asked to assume the repository, WP configuration, and environments indicated earlier in this paper, operating in the "hot" peak temperature mode, and postulating that some electrolyte was present at the WP surface for much of the time. Suggestions of research that could test the validity of any proposed mechanisms were requested as well. The oral deliberations during the Workshop have been transcribed and are available for public view. In addition, most participants have prepared brief papers. The following presents highlights of the panelists' responses.

In responses to the first question, the panelists gave much attention to the possibility of transpassive dissolution as a long-term mechanism for increase in the passive-corrosion rate (resembling speculative scenario (iv)). Most opinions held that the large potential shift to achieve transpassive dissolution of Cr oxide was unlikely but that possible Mo-transpassive behavior, which needs a lower shift in the noble direction, should be examined carefully. Mechanisms for OCP elevation that could trigger transpassive behavior were considered and are addressed via the second question.

Fundamental approaches using variations of the Point Defect Model (PDM) were presented to evaluate the possible accumulation of vacancies followed by spalling at the metal-film interface, the value of the steady-state passive current and its temperature dependence, the development of the OCP, and the extent of Cr-transpassive behavior. Those approaches can make only crude specific predictions at present, but point the way to experiments that could provide the necessary parameter values to refine forecasts. For example, prolonged experiments for measuring the extent of preferential evolution of components in Alloy 22 could provide necessary input for determining whether a vacancy-accumulation mechanism (as in scenario (ii)) could be detrimental in the repository time frame.

Likewise, precise determination of steady-state passive dissolution rates and their temperature dependence under selected regimes could allow for calibration of the PDM-derived models and reduce uncertainty when extrapolating to repository conditions and time frame. Application of the limited information available to date suggested that total outer-shell corrosion loss after 10,000 years service would be less than 2 mm, with modest dependence on peak WP surface temperature and little likelihood of Cr-transpassivity. However, these model projections did not include possible effects of the presence of a correspondingly thick external layer of passive-dissolution debris.

Panelists called attention to how little is presently known about the nature of the passive film on Alloy 22 (for example, whether it is initially crystalline and the relative sizes of its barrier and external layers). Likewise, knowledge of the degree of uniformity that passive dissolution may have over long times is needed because that may influence the roughness of the resulting surface and the onset of mechanisms like vacancy accumulation. Local variations in concentration of critical species like Cr at welds and at other fabrication features could promote undesirable localization of passive behavior as well. The potentially adverse effect of transition from an amorphous to a crystalline layer was reevaluated, with consequent conclusion about the need for experiments for resolving the issue for Alloy 22 in relevant environments. Several panelists indicated that the initial WP surface exposure to the long dry and hot pulse may create a thicker film not necessarily conducive to subsequent formation of an adequately protective layer after cooler and wetter conditions develop on the WP surface.

Those arguments underscored the importance of investigating the behavior under such a surface evolution schedule. Panelists speculated about similar effects resulting from cyclic wet and dry conditions (as in dripping regimes or due to spatial heterogeneity in the repository). Long-term evolution processes during passive dissolution that could lead to localized corrosion were proposed also.

One scenario was that extended periods of passive dissolution could lead to accumulation of impurities, such as S at the passive layer/metal interface (anodic segregation). Rough calculations using assumed parameters suggested that monolayer accumulations of S could take place during the first millenium, clearly indicating the desirability of experiments for evaluating the likelihood of this phenomenon. Other compositional changes beneath the evolving passive layer (e.g., Cr depletion) could act adversely. The possible detrimental effect of long-term passive-dissolution debris (scenario (iii)) also was deemed worthy of experimental test. Some panelists were concerned about detrimental synergisms from the simultaneous enrichment of Cl^- and F^- in the electrolyte on the WP surface, in part because of the possible incorporation of F^- into the passive layer itself.

In response to the second question, several mechanisms for long-term elevation of the OCP were considered, including the effect of radiolytic H_2O_2 generation early in the life of the repository and the need for more-precise quantification and forecasting of this effect. Panelists also indicated that, independent of the nature of the species being reduced, the cathodic reaction could become increasingly catalyzed as the passive layer matures, or by effects of Cl^- and/or F^- interaction with the passive film. Surface area increases by roughening also would increase overall cathodic efficiency under the expected, mostly activation-energy-limited, conditions.

The response of the passive layer, as an extrinsic semiconductor, to low-energy photons (possibly from secondary effects of the higher-energy gamma radiation) was proposed as another candidate mechanism for elevating the OCP. The need for experiments for determining the extent of these possible effects on the long-term evolution of OCP (and of the passive current density) was generally recognized.

Crevice corrosion was considered the most likely localized-corrosion process to occur, because the pitting resistance of Alloy 22 is recognized as very high (although the possible integrated effect of metastable pitting over extremely long times merited attention). Also mentioned were possible aggravating factors, such as gas formation in crevices, which could create extremely narrow gaps and consequently stabilize corrosion under conditions normally neglected.

Downward changes in E_{crit} were considered as possible results from forming passive films after extended dry and high-temperature periods. Various panelists expressed confidence in the applicability of presently available methods for determining E_{crit} (generally understood as the potential low enough to repassivate an initially active deep crevice), given that enough time and sufficient environmental severity is used in the experiment. However, it was proposed that certain processes may not be amenable to use of an E_{crit} value so determined. Among those processes were corrosion promoted by a membrane of former passive-dissolution debris, some forms of deep intergranular corrosion, and corrosion at regions of variation in local composition created by thermomechanical treatments (forming, welding).

THERMODYNAMICS AND ITS APPLICATIONS

Pyrometallurgy, by its very nature, involves high temperatures and the application of energy to materials. For this reason, the study of thermodynamics is one of the most important fundamentals of the subject.

WHAT IS THERMODYNAMICS?

Thermodynamics is a collection of useful mathematical relations between quantities, every one of which is independently measurable. Although thermodynamics tells us nothing whatsoever of the microscopic explanation of macroscopic changes, it is useful because it can be used to quantify many unknowns. Thermodynamics is useful precisely because some quantities are easier to measure than others.

The laws of thermodynamics provide an elegant mathematical expression of some empirically-discovered facts of nature. The principle of energy conservation allows the energy requirements for processes to be calculated. The principle of increasing entropy (and the resulting free-energy minimization) allows predictions to be made of the extent to which those processes may proceed.

Thermodynamics deals with some very abstract quantities, and makes deductions using mathematical relations. In this, it is a little like mathematics itself, which, according to Bertrand Russell, is a domain where you never know a) what you're talking about, nor b) whether what you're saying is true. However, thermodynamics is trusted as a reliable source of information about the real world, precisely because it has delivered the goods in the past. Its ultimate justification is that it works.

Confusion in thermodynamics can easily result if terms are not properly defined. There is no room for the loose use of words in this subject.

Energy or Heat

Many books on thermodynamics contain vague and strange statements, such as 'heat flows', or 'heat is a form of energy', or 'heat is energy in transit', or it is 'energy at a boundary', or it is 'the process', or 'the mechanism by which energy is transferred'. Work is 'being done' or is 'being transformed into heat'. Heat and work, these 'two illegitimate troublemakers', in the words of Barrow1, do not provide a proper base on which to build thermodynamics.

Heat and work seem to float between the system we deal with and its surroundings. They are not properties of the system we are dealing with, or of any other system. The quantities q and w (to be defined later) should not be mixed with actual properties like energy and heat capacity.

In contrast to heat and work, energy is a well-defined property. It has its origin in the ideas of the potential and kinetic energy of simple mechanical systems. Experiments such as Joule's 'mechanical equivalent of heat' let us extend the concept, and definition, to include thermal energy. Then, using the idea of the conservation of energy, changes in the energy of a chemical system of any complexity can be dealt with.

A niche can, however, be found for the action terms 'heating' and 'doing work'. They can be used to indicate that the process by which the energy of the system changes is accompanied by a change in the thermal or the mechanical surroundings.

According to Barrow1, our attachment to 'heat' stems from the caloric theory of the 18th and 19th centuries. That theory held that heat was a manifestation of a material called 'caloric'. This material flowed in and out of objects when the temperature changed. Studies such as those of Mayer, Thompson, and Joule showed that caloric could be created or destroyed and, therefore, that heat was not one of the substances of the material world. The 'caloric period' had come to an end. By continuing to use 'heat', we remain tied to the myths of the past. It is time to give up this awkward remnant and to continue building thermodynamics on the sound foundation of 'energy' alone.

STATE FUNCTIONS (THE BUILDING BLOCKS)

State functions are quantities whose values depend only on their current state, not on how they got to that state. Height above mean sea level is an example of this. It is not necessary to know anything about the path followed to a particular point, as long as one can measure the altitude at the destination. Obviously, the distance covered or the work expended on the journey depend very much on the route chosen and the mode of transport.

For the calculation of state functions, it is possible to use an imaginary route to get to the destination, without changing the final answer. While processes are physical, and occur generally as a series of non-equilibrium stages, paths (used for purposes of convenient calculation) are mathematical abstractions. Fortunately, enthalpy (H), entropy (S), and Gibbs free energy (G) are all state functions.

All good introductions to thermodynamics show the functional dependence of enthalpy, entropy, and Gibbs free energy (of a particular substance) on variables such as temperature and pressure. These thermodynamic functions may be expressed in terms of other variables, such as volume, for example, using straightforward mathematical transformations. However, the state variables that are of primary interest to pyrometallurgists are temperature and pressure.

$$H = H\ (T, P)$$

$$S = S\ (T, P)$$

$$G = G\ (T, P)$$

It is accepted for now that the functions H, S, and G are clearly defined, and that known values for these functions exist or can be measured.

An equation of state (such as $PV = nRT$ for an ideal gas) is used to calculate the interrelation between the measurable properties P, V, and T.

Most processes of interest to pyrometallurgists can be idealized as operating at constant temperature (isothermal) or constant pressure (isobaric). The requirement of constant volume (isochore, or isometric) is less commonplace.

The energy required for any steady-state flow process is essentially the difference in enthalpy between the products and reactants, plus the amount of energy lost to the surroundings. The change in enthalpy over the process is easily calculated if the enthalpies of all the chemical species are calculated relative to the same reference state, namely that of the elements in their standard states at 25°C and 1 atm.

By choosing the elemental reference state, the difference in enthalpy can be calculated without having to take into account any of the chemical reactions which may have taken place. (If there are no reactions, it is alright to use compounds as the basis, but the basis specified above is easy to remember and is always applicable. This is, therefore, strongly recommended.) Because the thermodynamic functions of interest, namely enthalpy, entropy, and Gibbs free energy, are all state functions, their values can be calculated independently of any reaction path.

In general, the partial molal enthalpy of any chemical species is a function of temperature, pressure, and composition. The effect of composition on the enthalpies of individual components is small in most cases, and, in any case, there is very little data available on the variation in enthalpy with composition. The effects of composition on enthalpy are therefore usually ignored, which is equivalent to the treatment of the process streams (at least for this purpose) as ideal solutions (i.e. the enthalpy of mixing is taken to be zero).

Except for gases under high pressure, the dependence of enthalpy on pressure is small. (If deemed important, the effect could be allowed for by the use of equations of state, or reduced property correlations.) In the field of high-temperature chemistry, the enthalpy is often assumed to depend solely on temperature. The total enthalpy of a stream is taken to be equal to the sum of the enthalpies of all the chemical species in the stream. For each chemical species:

$$H = \Delta H^{o}_{f298} + \int_{298}^{T1} C_{P1}\, dT + L_{T1} + \int_{T1}^{T} C_{P2}\, dT$$

$$S = S^o_{298} + \int_{298}^{T1} \frac{C_{P1}}{T} dT + \frac{L_{T1}}{T} + \int_{T1}^{T} \frac{C_{P2}}{T} dT$$

where:

H = enthalpy of chemical species relative to elements in their standard states at 25°C and 1 atm (J/ mol)

D = standard enthalpy of formation of the species at 298K (J/mol)

T1 = phase transition temperature (K)

CP(T) = $a + bT + cT^{-2} + dT^2$ (J/mol/K)
(subscripts 1 and 2 refer to different phases)

T = temperature (K)

LT1 = latent energy of phase transformation (J/mol)

S = absolute entropy of chemical species relative to elements in their standard states at 25°C and 1 atm (J/mol/K)

$S°_{298}$ = standard entropy of the species at 298K (J/mol/K)

Also:

$$G = H - TS$$

where

G = Gibbs free energy of chemical species relative to elements in their standard states at 25°C and 1 atm (J/mol)

Note that equations and may easily be extended to cover the situation where more than one phase transition occurs.

For computational convenience, the enthalpy and entropy can be evaluated by performing a single integration in each case. In order to do this, the first terms of equations for enthalpy and for entropy can be combined as integration constants. In this way, the same equations can be used, with different sets of constants being applicable to each temperature range. The upper temperature of each range is often the temperature of a phase transformation at a pressure of one atmosphere, but this need not necessarily be the case. For example, the data on gases come to an end at a temperature that is determined by the range of the experimental measurements. Also, in cases where it is difficult to adequately represent the C_P term with a four-term expression (for example), the temperature range for a particular phase can be arbitrarily divided, in order to obtain a more accurate fit to the data. The following equations now apply:

$$H = \text{"}\Delta H^o_{f298}\text{"} + \int_{298}^{T} C_P \, dT$$

$$S = \text{"}S^o_{298}\text{"} + \int_{298}^{T} \frac{C_P}{T} dT$$

$$C_P(T) = a + bT + cT^{-2} + dT^2$$

A particular phase (or temperature range), and the constants a, b, c, and d are specific to that range. Note that " ΔH^o_{f298} " and " S^o_{298} " may be considered to be the standard enthalpy of formation and standard entropy of the phase in its metastable state at 298K. To calculate the enthalpy and entropy of a particular compound at a given temperature, it is necessary to obtain the correct values of the constants that pertain to that particular temperature range. By default, the program selects the thermodynamic constants of the most stable phase of each species at the specified temperature.

To explain further, ΔH^o_{f298} refers to the standard isothermal enthalpy change for the formation reaction from the most stable phases of the elements at 25°C and 1 standard atmosphere pressure. S^o_{298} is the absolute (or Third Law) entropy of the phase at 25°C and 1 atmosphere. In cases where there is more than one data set for a given phase, " ΔH^o_{f298} " and " S^o_{298} " refer to the properties that would be reported at 25°C and 1 atmosphere if the C_P function behaved near this standard condition the way it does in the specified temperature range. As we are dealing merely with convenient integration constants, it should not be surprising that the 'absolute entropy' for some unstable phases at 25°C may be negative. The fictitious entropy constants are useful only for purposes of facilitating the computation of the actual entropy at higher temperatures where the phase in question is stable.

As thermodynamic data are not too readily available for minerals as such, it is often necessary to treat these as mixtures of chemical species. This is fairly straightforward for most minerals.

PRINCIPLE OF ENERGY CONSERVATION (FIRST LAW)

The conservation equations for matter in general can sometimes be rather complicated. Fortunately, these can usually be simplified. The First Law of thermodynamics is not a general energy balance, but represents the balance of internal energy for a material with very particular constitutive properties, in particular, the absence of irreversible energy transfer. In essence, the First Law of thermodynamics states that the energy of an isolated system (one that does not exchange matter or energy with its surroundings) remains constant.

For non-nuclear processes, the principle of energy conservation states that the sum of the changes of the extensive properties kinetic energy (E_k), potential energy (E_p), and internal energy (U) is equal to the sum of the modes of energy transfer q (defined as the thermal transfer of energy) and w (defined as the mechanical transfer of energy).

$$\Delta E_k + \Delta E_p + \Delta U = q + w$$

This equation is applicable to all constant-matter systems in general, and to steady flow processes in particular. By assuming mechanical equilibrium for the entering and exiting regions of a hypothetical volume, it is possible to produce a general equation relating the change in enthalpy to the sum of the thermal transfer of energy and the so-called shaft work, ws, and the change in the product of the pressure, P, and volume, V, as shown in equation. One of the fundamental relations of thermodynamics is used in equation to relate the change in enthalpy, H, to the change in internal energy, U.

$$w = w_s + \Delta PV$$

$$\Delta H = \Delta U + \Delta PV$$

Therefore,

$$\Delta E_k + \Delta E_p + \Delta H = q + w_s$$

However, for most chemical systems of interest, the changes in kinetic and potential energy are very small compared to the changes in ΔH. This allows us to simplify equation as follows.

$$\Delta H = q + w_s$$

This equation allows us to calculate the amount of energy transferred to or from any process, simply by calculating the difference in enthalpy before and after. As enthalpy, H, is a state function, its value does not in any way depend on the process itself or on the imaginary path followed during the process. Enthalpy is a function of temperature and pressure only. However, the dependence on pressure is small in most cases, and is usually ignored at reasonable pressures.

A process is said to be endothermic when $\Delta H > 0$, and exothermic when $\Delta H < 0$.

Thermo software

Chemical thermodynamics is a clear, simple, and elegant subject, if one does not get bogged down by the details. Fortunately, readily available computer software is able to provide the tools for performing thermodynamic calculations. Such software, containing suitable data, is able to calculate the standard enthalpy, entropy, and Gibbs free energy of different chemical species at any specified temperature.

Unit conversions, such as from J/mol to kWh/kg, are also able to be performed automatically. The software is able to check the consistency of stoichiometry of reactions, such as $Cr_2O_3 + 3C = 2Cr + 3CO$, and to calculate the standard thermodynamic functions for the reaction, with specified reactant and product temperatures. The thermodynamic functions can be tabulated and depicted graphically.

Energy balances using the Thermo program

Note well that it is correct to talk about an energy balance, but incorrect to talk of a heat balance (as 'heat' does not exist anyway) or an enthalpy balance (as enthalpy is conserved only in very special cases). The First Law of thermodynamics deals with the conservation of energy, and energy alone!

Note also, in the examples that follow, the effects of pressure on enthalpy, and the physical effects of mixing are neglected (justifiably, as they are small relative to the other quantities).

Example 1 – Melting of iron

Consider a small furnace containing 100 kg of iron at room temperature (nominally 25°C) that needs to be melted and heated to a temperature of 1600°C. (It can be assumed that the losses of energy through the walls and roof of the furnace can be ignored for our purposes.) If the furnace is set to supply 160 kW, the time taken for this process can easily be calculated.

Iron undergoes a number of solid-state phase changes on being heated, and then melts around 1536°C. However, this poses no complication for the calculation of the energy balance, as the enthalpy of the iron is a state function (and is therefore independent of the path followed).

The standard enthalpy of formation of Fe as a function of temperature. (Remember that the enthalpy is a function of pressure also, but the dependence is small.) In most compilations of enthalpy tables, pressure is not mentioned, as the small effect can be safely ignored at reasonable pressures.

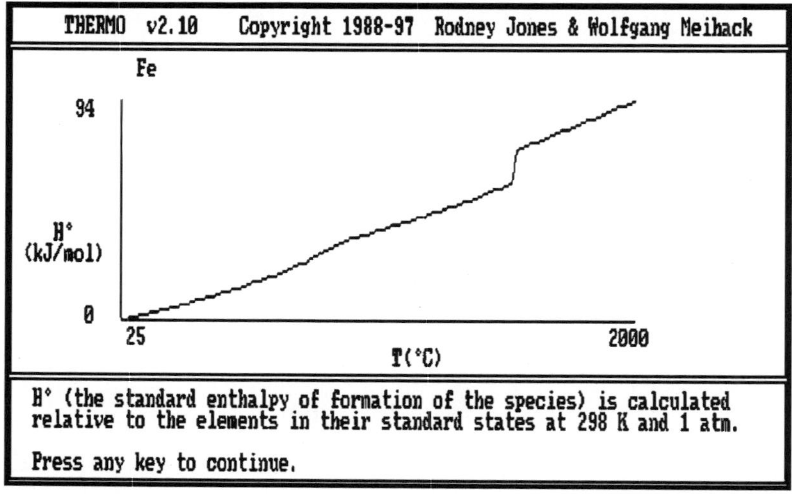

Fig. The standard enthalpy of formation of Fe, as a function of temperature

Relative to the standard state of iron at 25°C and 1 atm having an enthalpy of zero, the enthalpies are readily obtained. As can be seen in Figure, the standard enthalpy of formation of Fe is 0.00 kJ/mol at 25°C, and 75.90 kJ/mol at 1600°C.

The difference of 75.90 J/mol or 0.377 kWh/kg is easily obtained by subtraction. In our case, we would use this information to say that 100 kg of iron requires 100 × 0.377 kWh. At a power rating of 160 kW, this would take (37.7/ 160) hours or 14 minutes.

```
THERMO  v2.10    Copyright 1988-97  Rodney Jones & Wolfgang Meihack

Fe          (mw =   55.85)     Fe: 1
(Phase/data boundaries:   527   727   769   787   911 1392 1536 2862 3327 °C)

T1 (°C)          :          25.0          T2 (°C)          :        1600.0
Phase            :         Solid          Phase            :        Liquid
Cp (J/mol/K)     :         24.96          Cp (J/mol/K)     :         46.02
H° (J/mol)       :          3.74          H° (J/mol)       :      75895.70
S° (J/mol/K)     :         27.29          S° (J/mol/K)     :        101.88
G° (J/mol)       :      -8133.53          G° (J/mol)       :    -114942.98

Delta T  :        1575 K
Delta H°:        75.89 kJ/mol          18.14 kcal/mol          0.377 kWh/kg
Delta S°:        74.59 J/mol/K         17.83 cal/mol/K         0.371 Wh/kg/K
Delta G°:      -106.81 kJ/mol         -25.53 kcal/mol         -0.531 kWh/kg

=========================== Species menu ===========================
Set final temperature
Select         Temperature 1   Temperature 2   Temperature 0   Tabulate
Graph Cp       Graph H°        Graph G°        Menu
Use cursor keys to move and press Enter to choose an option
```

Fig. The heating of Fe from 25° to 1600°C

The thermodynamic functions can also be tabulated, to aid in certain calculations. For instance, if 100 kg of iron at 25°C is placed in a 16 kW furnace for one hour, what temperature will it reach? A tabulation of the energy requirements as a function of temperature. It is a simple task to match the energy supply of 0.16 kWh/kg with a final temperature of 855°C.

THERMO v2.10 Copyright 1988-97 Rodney Jones & Wolfgang Meihack

Fe (mw = 55.85)

T(°C)	Cp(kWh/kg/K)	H°(kWh/kg)	S°(kWh/kg/K)	G°(kWh/kg)	H°(T) - H°(25°C)
25	0.0001242	0.0000	0.000136	-0.040	0.0000
810	0.0002398	0.1498	0.000364	-0.245	0.1498
820	0.0002339	0.1522	0.000367	-0.249	0.1522
830	0.0002287	0.1545	0.000369	-0.252	0.1545
840	0.0002240	0.1568	0.000371	-0.256	0.1567
850	0.0002199	0.1590	0.000373	-0.260	0.1590
860	0.0002163	0.1612	0.000375	-0.263	0.1611
870	0.0002133	0.1633	0.000377	-0.267	0.1633
880	0.0002108	0.1654	0.000378	-0.271	0.1654
890	0.0002088	0.1675	0.000380	-0.275	0.1675
900	0.0002072	0.1696	0.000382	-0.279	0.1696

H°, S°, and G° (the standard enthalpy, entropy, and Gibbs free energy of formation of the species) are calculated relative to the elements in their standard states at 298 K and 1 atm.
Press any key to continue.

Fig. The energy requirement for the heating of Fe, as a function of temperature

Energy requirement for a simple reacting system

Consider a simple reacting system at atmospheric pressure, where 100 moles of oxygen at 200°C is blown onto a 1 kg graphite block initially at 25°C. There is sufficient oxygen that the graphite is completely combusted to CO2. How much energy would be released from the system, if it is assumed that the products reach a temperature of 1200°C?

The first step is to carry out the very simple mass balance in order to specify what and how much material is in each process stream (incoming and outgoing). Values of ΔH_f^o (in kJ/mol) are calculated for each chemical species present at its own temperature.

In:	100 moles O_2 at 200°C	ΔH_f^o = +5.32 kJ/mol
In:	83.3 moles C at 25°C	ΔH_f^o = 0.00 kJ/mol
Out:	83.3 moles CO_2 at 1200°C	ΔH_f^o = -334.51 kJ/mol
Out:	16.7 moles O_2 at 1200°C	ΔH_f^o = +39.11 kJ/mol

It is now a simple matter to calculate the total enthalpy entering and leaving the system. The energy requirement for this process is therefore [(83.3 x –334.51) + (16.7 x 39.11) – (100 x 5.317) - (83.3 x 0.00)] = -27 744 kJ.

Note that we have not needed to use the value of ΔH_r^o at 1200°C (-396.2 kJ/mol) in our calculations, as the energy balance is concerned only with the initial and final states of the system, and not with any reactions that might have taken place along the way.

Figure illustrates how this calculation may be performed directly, using the Thermo computer program.

```
THERMO  v2.10    Copyright 1988-97  Rodney Jones & Wolfgang Meihack

        83.3 C + 100 O2 @ 200 = 83.3 CO2 + 16.7 O2

Default temperatures (°C) : reactants    25.0   products  1200.0

                          J          kWh         J/mol C        kWh/kg C
Delta H°      :-27744042.00    -7.70668      -333061.72          -7.703
Delta S°(/K):       6081.62     0.00169           73.01           0.002
Delta G°      :-59170504.00   -16.43625      -710330.15         -16.429

Edit reaction or enter new reaction
Edit         Reactant temp   Product temp   Tabulate     Graph H°
Graph G°     Lookup          Save           Delete       Menu
 Use cursor keys to move and press Enter to choose an option
```

Fig. Energy balance calculation for a simple reacting system

Complex Balances

A simple procedure suffices for the calculation of energy balances in even the most complex systems. By a sensible choice of a reference state, energy balance calculations can be carried out without reference to the process or reaction paths.

- Calculate a mass balance for the system in terms of molar quantities of each species entering an leaving the system.
- Use the Thermo (or other) program to calculate the value of ΔH_f^o (in J/mol) for each chemical species present at the entering and

leaving temperature. The recommended reference state is one where the elements in their standard state at 25°C and 1 atm have a standard enthalpy of formation of zero.

- Calculate the total enthalpy of each process stream entering and leaving the system. This is done by multiplying the number of moles of each species in the stream by the value obtained for ΔH_f^o for the species at the temperature of the stream. (If deemed important, the effects of pressure and mixing can be included at this point.)
- The energy requirement of the process is equal to the difference in enthalpy between the products and reactants, plus the amount of energy lost to the surroundings.

Roasting of zinc sulphide

The roasting of zinc sulphide is an example of an autogenous process, i.e. one which is able to supply enough energy to sustain the reaction without requiring an additional energy source. In the course of practical operation of the process, it is necessary to know how much of a surplus or deficit of energy there is, under a specified set of conditions, so that the appropriate control actions may be taken.

Suppose the roasting of zinc sulphide can be simplified as follows. ZnS at 25°C is fed to a fluidized-bed reactor, together with pre-heated air at 200°C. We will assume for now that all the ZnS is converted to ZnO, and that all the sulphur is converted to SO_2, as per the reaction:

$$ZnS + 3/2\ O_2 = ZnO + SO_2$$

It is desired to have the products leaving the reactor at 900°C. Air is fed in excess, say 1.5 times the amount of oxygen required by the reaction. Obviously, this introduces N2 into the system in the proportion normally present in air (i.e. 8.46 moles of N_2 to 2.25 moles of O_2). The nitrogen does not take part in the reaction, but affects the energy requirement of the process, as it enters the process at 200°C and leaves at 900°C.

The system needs to be cooled by 0.448 kWh/kg of ZnS. In practice, the reactor loses energy to the surroundings by convection, radiation, and conduction from the walls of the reactor. Additional cooling may be achieved by using a water spray on the walls (or even adding water with the charge to the reactor). Alternatively, more air could be added, or the degree of preheating could be reduced.

Adiabatic flame temperature

Adiabatic conditions are those in which there is no thermal transfer of energy between a system and its surroundings. In the combustion of fuels, the adiabatic flame temperature sets an upper limit on the temperature that may be achieved in a system. The adiabatic flame temperature is the temperature attained in a reacting system that experiences no loss of energy. The adiabatic flame temperature may be calculated by finding the temperature at which the total enthalpy of the products (at the product temperature) is the same as that of the reactants (at their initial temperatures).

In the case of propane being fully combusted by air (initially at ambient temperature), the reaction may be written as shown in Figure. The adiabatic flame temperature is found (by trial-and-error variation of the temperature until the enthalpy difference is approximately zero) to be 2121°C in this case.

PRINCIPLE OF INCREASING ENTROPY (SECOND LAW)

C.P. Snow once suggested the second law of thermodynamics as a test of scientific literacy for the humanist, and said it was 'about the scientific equivalent of: Have you read a work of Shakespeare's?'. Yet, most people in the scientific world also have many misconceptions about the concept of entropy.

It is tempting, but rather a waste of time, to inquire into the 'meanings' of thermodynamic functions such as enthalpy or entropy. Thermodynamics reveals nothing of any microscopic or molecular meaning for its functions.

It is widely assumed that entropy measures the degree of disorder, randomness, or 'mixed-upness' of a system. In fact, the entropy change which occurs when an isolated body moves spontaneously toward equilibrium is, according to thermodynamics, always positive. By the methods of statistical mechanics, the entropy increase in such an isolated body can be simply related to the increase in the number of independent eigenstates to which the isolated body has access. This number can be simply related to the purely geometrical or spatial mixed-upness in only three very special cases, namely mixtures of perfect gases, mixtures of isotopes, or crystals at temperatures near absolute zero, none of which is commonly studied in the ordinary chemical laboratory, let alone in high-temperature furnaces. In all other cases, the entropy change is capable of no simple quasi-geometrical interpretation, even for changes in isolated bodies.

But the situation is even worse than this, for chemists commonly behave not only as if entropy increases in isolated bodies were a measure of disorder, but also as if this were true of entropy changes at constant temperature and pressure, under which conditions, very different from isolation, most chemical reactions are actually carried out. Even if the entropy change were a measure of disorder in an isolated body, the corresponding measure in an isothermal and isobaric experiment would be the Gibbs free-energy function and not the entropy.

Misunderstandings regarding entropy have led to some widely-held misconceptions, such as the so-called 'heat-death' of the universe. McGlashan3 rather eloquently refutes this idea by pointing out some of the misconceptions on which it is built. The argument runs something like this. Any process which actually takes place in the universe increases the entropy of the universe. Increase of entropy implies an increase in disorder. Therefore, the ultimate fate of the universe is chaos. If thermodynamics could be shown (but how?) to be applicable to the universe, and if the universe were known to be a bounded and isolated body, then we might deduce that the universe would eventually reach a state of complete equilibrium. But there is no scientific reason to suppose that the universe is a bounded isolated body. Even if it were, there is no reason to suppose that the experimental science of thermodynamics can be applied to bodies as large as the universe.

The principle of increasing entropy (and the resulting free-energy minimization) allows predictions to be made as to the extent to which those processes may proceed. The concept of entropy is of limited direct use for open systems, and the concept of free energy ($G = H - TS$) was introduced by Willard Gibbs as a criterion indicating the unidirectionality of spontaneous change. Systems will adjust themselves so as to achieve a minimum free energy.

The main application of the Second Law in pyrometallurgy is the use of Gibbs free energy to predict whether a reaction may occur under certain conditions and to what extent it will occur. Reactions proceed spontaneously in such a way as to minimize the overall free energy of the system.

A THERMODYNAMIC STUDY OF THE GASEOUS POTASSIUM CHEMISTRY

The interest of using biomass based fuels in power boilers has increased in recent years. Biomass is considered to be a CO_2 neutral fuel, because the CO_2 emissions released during combustion are absorbed back to the biomass during the growth cycle. However, it must be remembered that the term biomass comprise of many different types of plants with varying lengths in the carbon cycle. Grasses and residues of agricultural crops, like straw, can serve as an example of a very short carbon cycle biomass, whereas the carbon cycle of trees can be several hundred years in length. This must be kept in mind if biomass combustion is considered as a mitigation technology for the human induced global warming. The industrialization period responsible for the human induced global warming is only about 150 years in length, less than the carbon cycle of many forest based biomass fuels. A sudden increase in the use of forest based fuels would still increase the atmospheric CO_2 level for many decades until the young forests would capture the equivalent amount during growth. Therefore, the short carbon cycle biomass fuels should be preferred in energy production. Unfortunately, these fuels are more difficult from a technological point of view.

One of the major problems in biomass combustion is the fouling and corrosion of heat transfer surfaces in the boiler. The high alkali content accompanied with the often low sulfur and silicon contents of the biomass based fuels leads to high amounts of gaseous alkali compounds in the flue gases that condense on the heat transfer surfaces causing fouling and corrosion. Potassium is by far the major alkali in most types of biomass, sodium being present in significant proportions in some biomass fuels, for example in black liquor.

In biomass combustion and co-firing the high fouling and corrosion rates have been explained by several mechanisms. In the most common hypothesis found in the literature, KCl is believed to condense or otherwise deposit on the heat transfer surfaces. Solid potassium chloride is then believed to react with the protective oxide of the steel releasing corrosive Cl_2 gas near the metal-scale interface, possibly forming low melting eutectics with iron or chromium which increase the corrosion rate and promote further sticking of the ash particles.

Another mechanism originating from the studies of recovery boiler corrosion postulates that the melt is formed prior to the reaction of KCl with the tube by formation of low melting point mixtures in the (K,Na) Cl- $(K,Na)_2 CO_3$ - $(K,Na)_2 SO_4$ system and then a molten phase attack towards the protective oxide is initiated.

For high alkali, low chlorine and low silicon fuels, KOH(g) $\Rightarrow$ KOH(l,s) or $K_2 CO_3(g) \Rightarrow K_2CO_3(s)$ condensation have been suggested as the possible initial deposit formation mechanisms. However, these formation routes have not been studied in detail and the majority of the work on fouling and corrosion done so far has focused on the assumption that KCl(g) $\Rightarrow$ KCl(l,s) condensation is the mechanism for the initial deposit growth. In the earlier work by the author and in this paper, the importance of the KOH(g) $\Rightarrow$ KOH(l,s) condensation in deposit formation and its implications to high temperature corrosion in biomass fired boilers is studied in more detail.

CALCULATION METHODS

During combustion, potassium is released in the gas phase mainly as K(g), KOH(g) and KCl(g). Part of the potassium reacts with the fuel ash in the furnace forming K_2O-SiO_2 or Al_2O_3-SiO_2-K_2O silicates. Another route for the gaseous potassium capture is the reaction with the fuel sulfur:

$$2KOH(g) + SO_2(g) + \tfrac{1}{2}O_2(g) \Leftrightarrow K_2SO_4(g) + H_2O(g)$$

$$2KCl(g) + SO_2(g) + \tfrac{1}{2}O_2(g) + H_2O(g) \Leftrightarrow K_2SO_4(g) + 2HCl(g)$$

$$K_2SO_4(g) \Leftrightarrow K_2SO_4(l,s)$$

The saturation vapor pressure of K_2SO_4(l,s) is 0,3 Pa at 1000 °C, so for a reasonable first approximation all the K_2SO_4(g) formed is condensed to solid particles before the convection section of the boiler (although the literature data suggests some kinetic limitations in the K_2SO_4(g) formation, it is assumed here that the reactions have taken place before the convection section of a typical boiler). The remaining reactions of interest for the gaseous potassium in the convection section of a boiler are:

$$KOH(g) + HCl(g) \Leftrightarrow KCl(g) + H_2O(g)$$

$$2KOH(g) + CO_2(g) \Leftrightarrow K_2CO_3(g) + H_2O(g)$$

$$K_2CO_3(g) + HCl(g) \Leftrightarrow 2KCl(g) + CO_2(g) + H_2O(g)$$

$$KCl(g) \Leftrightarrow KCl(l,s)$$

$$KOH(g) \Leftrightarrow KOH(l,s)$$

$$K_2CO_3(g) \Leftrightarrow K_2CO_3(l,s)$$

Multi-component thermodynamic calculations can be used to predict the abundance of the different alkali species. However, it must be emphasized that thermodynamic calculations provide only the final equilibrium concentrations of the species in a perfectly mixed flue gas and any kinetic barriers in the gas phase reactions as well as in the nucleation/condensation steps are ignored.

Gibbs energy minimization program HSC Chemistry Ver. 4.1 from Outokumpu Technology Oy was used in the calculations. The output data was further processed with Microsoft Excel spreadsheets to get the desired graphics.

All the gas phase potassium was initially assumed to be K(g) and all the gas phase chlorine to be Cl(g). The major flue gas components, $N_2(g)$, $H_2O(g)$, $O_2(g)$ and $CO_2(g)$ were fixed in all calculations to be 70 mol-%, 13 mol-%, 4 mol-% and 13 mol-%, except for the case where the effect of $H_2O(g)$ was studied. In this calculation, the $N_2(g)$ partial pressure was still kept at 70 mol-% and the change in $H_2O(g)$ partial pressure effected the $CO_2(g)$ level accordingly. The potassium and chlorine concentrations were varied between 1-100ppm representing typical levels of these components measured in real boilers at the inlet of the convection section. Pressure was fixed in all calculations to 1 atm and the studied temperature range was 400-1000 °C, simulating the temperature range typical in the convection section of a biomass boiler. The compounds included in the Gibbs energy minimization calculations were: $N_2(g)$, $CO_2(g)$, $H_2O(g)$, $O_2(g)$, K(g), Cl(g), $Cl_2(g)$, HCl(g), KCl(g), KOH(g), $K_2CO_3(g)$ and $K_2O(g)$. In one of the calculations, also the condensed phase compounds KOH, KCl, K_2CO_3 and K_2O were added to see their effect on the system behavior.

Results and discussion

The effect of water partial pressure on the potassium distribution when there is 100 ppm of Cl and K (K/Cl molar ratio is one) in the gas. As expected from equation (4) higher water content stabilizes more KOH(g), but the effect is quite small, due to the strong thermodynamic stability of KCl(g). $K_2CO_3(g)$ is not present at all, because of its very low thermodynamic stability and the lack of excess potassium available for the $K_2CO3(g)$ formation after the K(g) or KOH(g) has reacted with the chlorine.

Figure presents the effect of the amount of potassium and chlorine with the K/Cl ratio still being one. The relative proportion of the KOH(g) raises quite remarkably when the amounts of K and Cl are lowered. This shift towards KOH(g) formation can be very significant with fuels having both low alkali and low chlorine content.

The effect of the K/Cl molar ratio in cases where the ratio is lower than unity. Obviously the higher chlorine content increases the relative amount of KCl(g) as expected from equation (4). The excess chlorine is in the form of HCl(g).

Figure shows the most interesting case, where the K/Cl molar ratio is higher than one. This will have a huge influence on the behavior of the system. At high gas temperatures, the increase in the KOH(g) concentration is directly proportional to the added potassium. Also K_2CO_3(g) starts to appear now at low temperatures. It is interesting to notice that K_2CO_3(g) is thermodynamically very unstable in the high temperature range and becomes a relevant potassium species only below 800 °C. This indicates that K_2CO_3(s) vaporizes predominately by dissociation. KOH(g) is actually very stable to very low gas temperatures, if only the gas phase species are accounted for. In reality however, one has to take into account the formation of K_2CO_3(s) on the surfaces.

Therefore, the condensed phase compounds were added to the calculation. Figure shows the results. K_2CO_3(s) starts to decompose at 750 oC and is completely decomposed at 850 °C, therefore K_2CO_3(s) cannot exist on surfaces hotter than 850 °C. On cooler surfaces the deposition of K_2CO_3(s) can follow four routes:

1. Through deposition of K_2CO_3(s) aerosols formed from the homogeneous nucleation of K_2CO_3(g) vapor. The K_2CO_3(s) may be formed from the KOH(g) in the thermal boundary layer adjacent to the surface, according to reaction equations (5) and (9).
2. By heterogeneous condensation of K_2CO_3(g), either directly on the tube surface, or on larger ash particles present in the boundary layer. Again, the K_2CO_3(g) may be formed from the KOH(g) according to reaction equation (5).
3. In the temperature range below 850 oC, but above the dew point of KOH(g), by the adsorption of KOH(g), either directly on the tube surface, or on larger ash particles present in the boundary layer, followed by a heterogeneous reaction of the adspecies with CO_2(g). Because the surface temperature is higher than the dew point of KOH(g), the adsorption is of Langmuir type.
4. At surface temperatures lower than the dew point of KOH(g), the formation of K_2CO_3(s) may be enhanced by the heterogeneous reaction of the condensed KOH(l,s):

$$2KOH(l,s) + CO_2(g) \Leftrightarrow K_2CO_3(s) + H_2O(g)$$

It is interesting to notice that in the conditions, which can be considered the situation that the gas phase has to reach before $K_2CO_3(s)$ formation is possible through the $K_2CO_3(g) \Rightarrow K_2CO_3(s)$ route, KOH(g) is thermodynamically stable all the way to the temperatures prevailing at the surfaces of the superheaters, 400-600 oC. In reality, the flux of KOH(g) at the heat transfer surfaces would be further increased from its equilibrium concentration if any kinetic delay would be evident in reaction (5), because the residence time of KOH(g) molecules in the thermal boundary layer of the heat transfer surfaces is extremely short, in the order of few ms. After the KOH condensation, the heterogeneous carbonization kinetics on the surface may be substantially slower than the homogeneous carbonization in the gas phase and therefore, once condensed, KOH(l,s) can exist on the heat transfer surfaces although there is a high thermodynamic driving force for the transformation to $K_2CO_3(s)$ (and further to KCl(s) and $K_2SO_4(s)$ if there is HCl(g) and $SO_2(g)/SO_3(g)$ available in the flue gases).

The thermodynamic dew points of KOH(g), KCl(g) and $K_2CO_3(g)$ for concentrations typically measured in real boilers are presented in Figure. It can be noted that for KCl(g), the dew point is always below the melting point of KCl, 774 °C, and $K_2CO_3(g)$ is stable only below the melting point of K_2CO_3, 891 °C. Therefore, these compounds will always condense in the solid state on the heat transfer surfaces. On the contrary, the dew point of more than about 1 ppm of KOH(g) in the flue gases is always higher than the melting point of KOH, 406 °C and thus KOH will condense in the molten state if the heat transfer surface temperature is higher than its melting point. This can have significant implications on the corrosion and fouling mechanisms taking place, because molten phase accelerates corrosion reactions and increases the stickiness of the heat transfer surfaces.

APPLICATION OF THE RESULTS TO BIOMASS FUELS

In order to apply the results to biomass based fuels, first an estimation of the gaseous potassium and chlorine levels at the inlet of the convection section is needed. From the elemental analysis of the fuel, the gas phase alkali (˜potassium with high potassium, low sodium fuels) to chlorine molar ratio can be predicted as follows:

$$\frac{K(g)}{Cl(g)} \approx \frac{x \cdot (K + Na) - 2 \cdot S}{Cl}$$

x = conversion of fuel alkali to vaporized alkali (effect of alkali silicate formation and furnace design)

K = potassium content of the fuel, mol/kg

Na = sodium content of the fuel, mol/kg

S = sulfur content of the fuel, mol/kg

Cl = chlorine content of the fuel, mol/kg

Unfortunately, the conversion of fuel alkali to vaporized alkali with different fuels is not known in detail. The percentage vaporized depends on the heating rate of the biomass particle and on the chemical composition. High Si/K ratios and low heating rates tend to reduce the vaporization percentage by formation of alkali silicates. With high heating rates (which are more representative of the combustion conditions than the low heating rate results) alkali release percentages of around 50 per cent or higher for different biomass fuels have been reported.

It must also be emphasized that this type of estimation assumes 100 per cent alkali capture efficiency for fuel sulfur. In reality, reactions of fuel sulfur with fuel calcium compete with the reactions with fuel alkali. Therefore, with high calcium-low silicon fuels, the alkali capture efficiency of fuel sulfur is lower than 100 per cent and therefore the vaporized alkali before the inlet of the convection section may be higher than estimated with this simple methodology. Additionally, measurements of SO_2(g) and HCl(g) in the convection sections in real boilers show non-zero concentrations, values in the range of 1-100 ppm are typical. This means that either the alkali release to the gas phase in these cases has been really low (alkali silicate formation), or the mixing of the flue gases is not perfect, or that there has been a kinetic delay in the formation of K_2SO_4(g,l,s) or KCl(g,l,s) from fuel potassium.

The estimations of the K(g)/Cl(g) molar ratios for a number of different fuels calculated with equation (11), and assuming 0-100 per cent conversion of fuel alkali to vaporized alkali. Generally, if the conversion is higher than 60 per cent, almost all the biomass fuels studied result in the estimated K(g)/Cl(g) ratios above one, which categorizes these fuels in the Figures 4 and 5 cases.

Implications for alloy selection

In principle, the main components of the deposits in the convection superheaters of biomass fired boilers are K_2CO_3, KCl and K_2SO_4. The relative distribution of these differs from boiler to boiler and from fuel to fuel. However, the formation mechanism of the deposits seems to have common features for the different cases.

Namely, initially the formation of a condensation layer, which then serves as a sticky surface promoting the adherence of the impacting fly ash particles. The formation mechanism of the initial condensation layer is the key for understanding the corrosion and also the fouling in these boilers.

The results of the calculations presented suggest that the KOH(g) $\Rightarrow$ KOH(l) condensation may be an important mechanism in biomass fired boilers. First of all, it explains the formation of a melt above 406 °C on the superheater surfaces, which is a prerequisite for the hot corrosion to take place. Secondly, the formed alkali melt will be very basic. The O_2- ion activities of the pure alkali compounds are determined by their dissociation reactions:

Reaction	Equilibrium basicity at 500 °C
$2KOH \Leftrightarrow K_2O + H_2O$ (g)	p[K2O] = 5.45
$K_2CO_3 \Leftrightarrow K_2O + CO_2$ (g)	p[K2O] = 9.25
$2KCl + H_2O(g) \Leftrightarrow K_2O + 2HCl(g)$	p[K2O] = 10.9
$K_2SO_4 \Leftrightarrow K_2O + SO_3(g)$	p[K2O] = 18.32

For the KOH-K_2CO_3-KCl-K_2SO_4 system, the basicity range depends on the relative proportion of the pure substances in the melt, sulfate rich system being the least basic and hydroxide rich system being the most basic. The O_2-activity depends also on the HCl(g), CO_2(g), SO_3(g) and H_2O(g) concentrations above the melt and may also change if the alloy materials are dissolved in the melt. Furthermore, the synergetic effects of the different anions, especially Cl^-, and cations involved in the corrosion reactions may further complicate the corrosion mechanisms taking place in actual service conditions.

However, the KOH(g) ⇒ KOH(l) condensation and the subsequent heterogeneous reactions forming the KOH-K_2SO_4-KCl-K_2CO_3 mixture may explain a basic environment with high O_2- ion activities. This in turn results in the shift in the hot corrosion resistances of the alloying elements compared to the relatively acidic sulfate melts found in coal fired boilers. Figures present the solubilities of different protective oxides in alkali sulfate and chloride melts as a function of melt basicity. It is clear that when the environment on the surface is very basic, alloying with Cr is not beneficial for the improvement of hot corrosion resistance. Instead, NiO, CO_3O_4 or Fe3O4 forming alloys should provide better protection in these service conditions.

Direct experimental evidence to support the presented corrosion mechanism in biomass fired boilers is not available. The problem is experimentally very challenging. First of all, sampling of KOH(g) from the flue gas stream requires care that the KOH(g) does not react to KCl(s) and K_2SO_4(s) particles in the sampling system. Secondly, if the KOH(g) is condensed on the tube surface it has to be detected in-situ, because the KOH condensation layer is continuously reacting and cannot be detected after cool down of the surface. The author is aware of only two studies, which gives some indirect evidence of the hydroxide condensation, the authors own measurements with a galvanic probe, and a very recent electrochemical impedance spectroscopic (EIS) study of the fouling in brown coal fired boiler, which suggests that a similar type of mechanism with NaOH as the condensing substance may take place in some brown coal fired boilers.

Conclusions

Thermodynamic calculations of the potassium chemistry in the flue gases show, that if the K(g)/Cl(g) molar ratio in the gas phase is higher than one, KOH(g) is stable down to typical superheater surface temperatures. Estimations of the K(g)/Cl(g) molar ratios at the inlet of the convection section in biomass fired boilers result in ratios higher than one. Therefore, it is likely that KOH(g) ⇒ KOH(l) condensation followed by a heterogeneous formation of K_2CO_3(s) plays an important role in the fouling and corrosion mechanisms of the heat transfer surfaces in biomass fired boilers. This leads to the possible formation of a basic alkali melt on the heat transfer surfaces and thus Cr_2O_3 forming alloys may be less stable than the NiO, CO_3O_4 or Fe_3O_4 formers in these service conditions.

ENERGY CALCULATIONS

The branch of science that deals with the energy requirements of physical and chemical changes is called thermodynamics. This handout will cover various types of problems encountered in thermodynamics in a freshman chemistry course. The topics will include material found in both semesters (one year) of general chemistry.

There are many ways to look at energy. Let's look at some typical variables used to express energy and what they measure:

ΔE or ΔU	measures the internal energy of a system – i.e. the total energy of a system
q	measures the heat of a system
w	measures the work of a system
ΔH	measures the enthalpy change of a system – i.e. whether the process is exothermic or endothermic
ΔG	measures the Gibb's free energy change of a system – i.e. whether a process is spontaneous or nonspontaneous
ΔS	measures the entropy change of a system or surroundings i.e. whether the randomness of the system or surroundings is increasing or decreasing

We will take each of these thermodynamic quantities in turn and examine them in more detail, as well as look at types of calculations involving these variables. In order to discuss these terms adequately, we need to define what is meant by a state function. A state function is a quantity whose value does not depend on the path used to measure the value. These quantities have upper case letters for symbols, such as, E, U, H, G, or S. Quantities, such as work, w, and heat, q, are not state functions. Their symbols use lower case letters. Also, the term "system" is defined as the part of the Universe being studied. It could be something like a chemical reaction or a phase change. The "surroundings" is everything else (the rest of the Universe).

INTERNAL ENERGY

In most applications, internal energy is measured in terms of work and heat. The following equation relates internal energy, heat and work:

$$\Delta E = q + w$$

This equation is often applied to the first law of thermodynamics. This law states that the energy of the universe remains constant, or energy can be both neither created nor destroyed, only changed from one form of energy to another. The sign of q, heat, or work, w, indicates the direction of the flow of energy. The currently accepted sign convention is that if heat flows out the system to the surroundings, q is negative. If one were carrying out a reaction in a test tube, the test tube would feel warmer. If heat flows into the system from the surroundings, q is positive. If one were carrying out the reaction in a test tube, the test tube would feel colder. If the system does work on the surroundings, w is negative. This means that energy is flowing out of the system. If the surroundings do work on the system, w is positive. Energy is flowing into the system from the surroundings. The way to keep the signs straight is to relate them to what is happening to the system. Heat or energy flowing out of the system is negative; heat or energy flowing into the system is positive. Unfortunately, some areas of science choose to follow the opposite sign convention. It is important to note in a text which sign convention is used. Let's look at a problem involving internal energy, heat and work. Suppose we have a process in which 3.4 kJ of heat flows out of the system while 4.8 kJ of work is done by the system on the surroundings. What is the internal energy?

Applying the equation, $\Delta E = q + w$, one may calculate the internal energy:

$$\Delta E = q + w = -3.4 \text{ kJ} + (-4.8 \text{ kJ}) = -8.2 \text{ kJ}$$

Note the use of appropriate signs for heat and work. Overall, energy has flowed out of the system to the surroundings by -8.2 kJ. There is another way in which work may be calculated, particularly when one is dealing with a gas phase system. The system consists of the gas and its container. Since gases may be expanded or compressed, work may be related by the pressure of the gas and the change in volume of the gas:

$$w = -P\Delta V$$

The change in volume, ΔV is always calculated as the final volume of the gas less the initial volume of the gas:

$$\Delta V = V\text{final} - V\text{initial}$$

To expand a gas, the volume of the gas is increased (ΔV is positive). The gas (part of the system) has to do work on the surroundings, and thus the work must be negative. Similarly, to compress a gas, the volume of the gas is decreased ΔV is negative). The surroundings must do work on the system, and thus the work must be positive.

ENTHALPY

Heat of Formation

Enthalpy may be measured and calculated in many ways. We'll look at "theoretical" ways of calculating enthalpy, first, and then at some experimental methods.

One of the ways enthalpy may be calculated from theoretical data is from a table of the standard heat of formation, $\Delta H_f°$, of a substance. The standard heat of formation is defined as the enthalpy change taking place when a substance is formed from the elements in their standard states. Standard state, in thermodynamics is 1 atmosphere pressure and a temperature of 25 °C (298 K).

Using the tabulated data found in any general chemistry textbook or other references, one may calculate the standard heat of reaction, ?Hrxno, by the following equation:

$$\Delta H°_{rxn} = \sum n(\Delta H°_f \text{ of products}) - \sum n(\Delta H°_f \text{ of reactants})$$

where n represents the moles of each reactant or product as found in the balanced chemical equation. The Greek letter, Σ, means one takes the sum of the variables which follow.

When choosing $\Delta H_f°$ values from a table, make sure you choose the value corresponding to the appropriate state of matter (solid, liquid, aqueous, gas). Also, note that the standard heat of formation for an element in its standard state always has a value of 0 kJ/mol.

Let's look at a problem in which the heat of reaction is calculated from the heats of formation. Suppose you are given the following balanced chemical equation and were asked to find the heat of reaction and determine whether the process is exothermic or endothermic:

$$16\ H_2S(g) + 8\ SO_2(g) \rightarrow 16\ H_2O(l) + 3\ S8(s)$$

From a table of thermodynamic quantities, one can gather the appropriate values for the heats of formation of each the components in the reaction, and set up the equation to calculate the heat of reaction:

$$\Delta H_{rxn}° = [16 \text{ mol}(\Delta H_f° \text{ of } H_2O(l)) + 3 \text{ mol}(\Delta H_f° \text{ of } S8(s))] -$$

$$[16 \text{ mol}(\Delta H_f° \text{ of } H_2S(g)) + 8 \text{ mol}(\Delta H_f° \text{ of of } SO_2(g))]$$

$$\Delta H_{rxn}° = [16 \text{ mol}(-285.8 \text{ kJ/mol}) + 3 \text{ mol}(0 \text{ kJ/mol})] -$$

$$[16 \text{ mol}(-20.2 \text{ kJ/mol}) + 8 \text{ mol}(-296.8 \text{ kJ/mol})]$$

$$\Delta H_{rxn}° = [-4573 \text{ kJ} + 0 \text{ kJ}] - [-323 \text{ kJ} + (-2374 \text{ kJ})]$$

$$\Delta H_{rxn}° = -4573 \text{ kJ} + 323 \text{ kJ} + 2374 \text{ kJ}$$

$$\Delta H_{rxn}° = -1876 \text{ kJ}$$

Since $\Delta H_{rxn}°$ is negative, the reaction is an exothermic process. Heat is released by the system to the surroundings. This particular reaction commonly occurs in the vents of volcanos, and deposits sulfur at the entrance of the vents. We will use this reaction to discuss other thermodynamic quantities, as well, throughout the handout.

Hess' Law

Hess' Law takes advantage of the fact that enthalpy is a state function. Recall that a state function depends only on initial and final states; it does not depend on the path one takes to go from one state to another. With Hess' Law, if an alternative means is found to calculate enthalpy, i.e., a series of reactions whose enthalpies are known, and the overall reaction gives the reaction sought, the sum of the enthalpies of those is the enthalpy of the sought reaction.

Let's look at an application of Hess' Law. Suppose we want to determine the enthalpy of reaction for the reaction:

$$2\ N_2(g) + 5\ O_2(g) \rightarrow 2\ N_2O_5(g)\ \Delta H = ?$$

from the following heats of reaction:

$$\text{Eqn}(1)\ 2\ H_2(g) + O_2(g) \rightarrow 2\ H_2O(l)\ \Delta H = -571.7\ kJ$$

$$\text{Eqn}(2)\ N_2O_5(g) + H_2O(l) \rightarrow 2\ HNO_3(l)\ \Delta H = -92\ kJ$$

$$\text{Eqn}(3)\ N_2(g) + 3\ O_2(g) \rightarrow 2\ HNO_3(l)\ \Delta H = -348.2\ kJ$$

The goal is to manipulate the above equations in such a way that the overall equation sums to the equation sought. First, we have to decide the equation to start with. One usually looks at the compounds in the equation sought, and tries to find each of the compounds in one of the given equations. Nitrogen, N_2, is found only in equation (3), so we could start with this equation. Oxygen, O_2, is found in both equations (1) and (3), so that is not a good compound with which to begin solving the problem. Dinitrogen pentoxide, N_2O_5, is found only in equation (2), so we could also start with equation (3).

Arbitrarily, let's chose nitrogen and equation (3) and see what else we have to do this equation. Equation (3) is written containing only one mole of nitrogen as a reactant, and in the equation we seek, two moles of nitrogen, as a reactant are needed. Since nitrogen is a reactant in both instances, we do not need to reverse the equation, however, we do need to multiply equation (3) by two in order to obtain the required two moles of nitrogen. Whatever is done to manipulate the chemical equation is also applied to the heat of reaction; thus, we also multiply the given heat of reaction by two:

$$2 \times \text{Eqn}(3){:}\ 2\ N_2(g) + 6\ O_2(g) + 2\ H_2O(g) \rightarrow 4\ HNO_3(l)\ \Delta H = 2(-348.7\ kJ)$$
$$= -696.4\ kJ$$

Next, let's introduce dinitrogen pentoxide into the problem, since only equation (2) contained the compound. In the equation sought, N2O5 appears as two moles of product. In equation (2), N_2O_5 appears as one mole of reactant. We must reverse the equation and multiply by 2. This means we will also change the sign of the heat of reaction, and multiply it by two:

$$-2 \times \text{Eqn}(2){:}\ 4\ HNO_3(l) \rightarrow 2\ N_2O_5(g) + 2\ H_2O(l)\ \Delta H = -2(-92\ kJ)$$
$$= +184\ kJ$$

At this point, let's look at the two equations generated. We see that the four moles of nitric acid will cancel. This is "convenient" since this compound is not found in the sought equation.

$$2 \times \text{Eqn(3)}: 2\ N_2(g) + 6\ O_2(g) + 2\ H_2O(g) \rightarrow 4\ HNO_3(l)\ \Delta H = 2(-348.7\ kJ)$$

$$= -696.4\ kJ$$

$$-2 \times \text{Eqn(2)}: 4\ HNO_3(l) \rightarrow 2\ N_2O_5(g) + 2\ H_2O(l)\ \Delta H = -2(-92\ kJ)$$

$$= +184\ kJ$$

We also need to cancel two moles of H_2, two moles of H_2O, and one mole of O_2. This is accomplished using equation (1) in reverse. Remember to reverse the sign for the heat of reaction:

5

$$2 \times \text{Eqn(3)}: 2\ N_2(g) + 6\ O_2(g) + 2\ H_2O(g) \rightarrow 4\ HNO_3(l)\ \Delta H = 2(-348.7\ kJ)$$

$$= -696.4\ kJ$$

$$-2 \times \text{Eqn(2)}: 4\ HNO_3(l) \rightarrow 2\ N_2O_5(g) + 2\ H_2O(l)\ \Delta H = -2(-92\ kJ) = +184\ kJ$$

$$-1 \times \text{Eqn(1)}\ 2\ H_2O(l) \rightarrow 2\ H_2(g) + O_2(g) \rightarrow H = -1(571.7\ kJ) = +571.1\ kJ$$

Summing the chemical equations gives us the sought reaction. Summing the heats of reaction gives us the heat of reaction for the overall process:

$$2\ N_2(g) + 5\ O_2(g) \rightarrow 2\ N_2O_5(g)\ \Delta H = +59\ kJ$$

Calorimetry

The basic principle behind calorimetry centers around heat flow. Heat lost by the system would equal to the heat gained by the surroundings during an exothermic process. Conversely, heat gained by the system from the surroundings would equal to the heat lost by the surroundings in an endothermic process.

A physical property relating the ability of a substance to hold heat is called the heat capacity. Heat capacity is defined as the amount of heat energy required to raise a substance by one degree Celsius. Often the specific heat capacity is used in place of heat capacity.

The specific heat capacity differs from heat capacity in that it relates the amount of heat energy required to raise one gram of a substance by one degree Celsius.

The equation which relates heat, mass, specific heat capacity and temperature is shown below:

$$q = mC\Delta T$$

where q = heat in Joules
m = mass in grams
C = specific heat capacity in Joules/g°C
ΔT = temperature change in °C

In a typical calorimetry experiment, a hot substance (defined as the system) is introduced to a cold substance (defined as the surroundings). The system and surroundings are allowed to reach equilibrium at some final temperature. Heat flows from the hotter substance to the cooler substance. The heat lost by the hot substance (the system) must equal the heat gained by the cooler substance (the surroundings).

Let's look at a typical calorimetry problem. To begin with, in the real world, no calorimeter is perfectly insulating to temperature. The calorimeter itself, as part of the surroundings, will absorb some heat. The simplest and cheapest calorimeter is a styrofoam cup. It is often used in introductory chemistry classes as a calorimeter. Although the temperature insulating properties of styrofoam cup are fairly good, the cup will absorb some heat. Thus, to obtain the most accurate result, the "coffee cup calorimeter" must be calibrated. This is often performed as an experiment in which hot water of known mass and temperature is poured into a coffee cup containing cool water of known mass and temperature. The combined water samples are allowed to equilibrate to a final maximum temperature.

Suppose 60.1 g of water at 97.6 °C is poured into a coffee cup calorimeter containing 50.3 g of water at 24.7 °C. The final temperature of the combined water samples reaches 62.8 °C. What is the calorimeter constant?

To begin this problem, let's consider the basic principle of calorimetry:

heat lost by the hot water = heat gained by the cool water + heat gained by the calorimeter

Next, let's show the mathematical equations corresponding to the premise described:

heat lost by the hot water = heat gained by the cool water + heat gained by the calorimeter

$$- m_{hot}\, C_{water}\, (T_{final} - T_{inti,\ hot}) = m_{cool}\, C_{water} (T_{final} - T_{init,}\ cool) + K_{cal} (T_{final} - T_{init,\ cool})$$

Notice the negative sign in front of the expression for the heat lost by the hot water. Recall that heat lost is negative. Substitute the proper masses and temperatures into the above expression. Note that since the cool water was contained in the cup, the initial temperature of the calorimeter and the cool water are the same. The specific heat capacity of water, Cwater, has a value of 4.184 J/goC.

$$-(60.1\ g)(4.184\ J/g°C)(62.8\ °C - 97.6\ °C) = (50.3\ g)(4.184\ J/°C)(62.8\ °C - 24.7\ °C) + Kcal\ (62.8\ °C - 24.7\ °C)$$

Combining terms gives: 8.75×10^3 J = 8.02×10^3 J + Kcal (38.1 °C)

Solving for K_{cal}:

$$K_{cal} = \frac{8.75 \times 10^3\ J - 8.02 \times 10^3\ J}{38.1\ °C} = 19.2\ J/°C$$

Now that the calorimeter has been calibrated, we can use it to determine the specific heat capacity of a metal. In this experiment, a known quantity of a metal is heated to a known temperature. The heat metal sample is dumped into a coffee cup calorimeter containing a known quantity of water at a known temperature. The temperature is allowed to equilibrate, and a final temperature is measured.

For instance, 28.2 g of an unknown metal is heated to 99.8 °C and placed into a coffee cup calorimeter containing 150.0 g of water at a temperature of 23.5 °C. The temperature of the water equilibrates at a final temperature of 25.0 °C. The calorimeter constant has been determined to be 19.2 J/°C. What is the specific heat capacity of the metal?

Once again set up the problem based on the heat flow:

heat lost by the hot metal = heat gained by the water + heat gained by the calorimeter

$$-(28.2 \text{ g})(C\text{metal})(25.0\ °\text{C} - 99.8\ °\text{C}) = (150.0 \text{ g})(4.184 \text{ J/g°C})(25.0\ °\text{C} - 23.5\ °\text{C}) + 19.2 \text{ J/°C}(25.0\ °\text{C} - 23.5\ °\text{C})$$

Combining terms gives:

$$2.11 \times 10^3 \text{ g°C } (C\text{metal}) = 9.4 \times 10^2 \text{ J} + 29 \text{ J}$$

$$2.11 \times 10^3 \text{ J } (C\text{metal}) = 969 \text{ J}$$

Solving for specific heat capacity:

$$C\text{metal} = \frac{969 \text{ J}}{2.11 \times 10^3 \text{ g°C}} = 0.459 \text{ J/g°C}$$

Metals tend to have low specific heat capacities. Water has a relatively high heat capacity. It is this property of water that helps control global temperatures. Temperature fluctuations during day and night would be much larger if the heat capacity of water was a lower value. This is evident in a desert where water is scarce. Daytime temperatures are very high, yet nighttime temperatures can drop to near freezing.

ENTROPY

Entropy, symbolized with an S, is a measure of the randomness or disorder in a system. The second law of thermodynamics states that for a spontaneous process, the entropy of the Universe will increase. Implicit in this statement is that to create order takes energy. Let's look at an example describing the second law of thermodynamics.

To mimic a spontaneous process, suppose you had a sudden urge to take a piece of paper and tear it up into small pieces, and throw the pieces into the air. What would happen? The pieces of paper will randomly scatter about you. They certainly would not fly back together into the original piece of paper. You have just increased the entropy of the Universe. It did not take much energy to tear of the piece of paper and toss the pieces in the air. Now suppose you want to restore this piece of paper into its original form. How much energy would that take? First you have to expend some energy to go around the room and pick up the pieces of paper. The paper is still not restored to its original state. Next, you have to take the pieces of paper to a pulp mill and have the paper turned back into pulp and repressed into the original piece. These processes certainly took much more energy than the energy required to rip up the paper and toss it in the air.

One way to calculate entropy change is to use tabulated values for the entropy of formation. This calculation will look familiar since it follows the same format as the calculation of the enthalpy change from heats of formation. The equation is described below:

$$\Delta S^\circ_{rxn} = \sum n(S^\circ_f \text{ of products}) - \sum n(S^\circ_f \text{ of reactants})$$

where n represents the moles of each reactant or product, as given in the balanced chemical equation.

Let's use this equation to solve an entropy problem. Suppose you wanted to know the entropy change for the same chemical equation we use to calculate the heat of reaction:

$$16\ H_2S(g) + 8\ SO_2(g) \rightarrow 16\ H_2O(l) + 3\ S8(s)$$

Set up the equation:

$$\Delta S^\circ_{rxn} = [(16\text{ mol})(S^\circ_f \text{ of } H_2O(l)) + (3\text{ mol})(S^\circ_f \text{ of } S8(s))] -$$
$$[(16\text{ mol})(S^\circ_f \text{ of } H_2S(g)) + (8\text{ mol})(S^\circ_f \text{ of } SO_2(g))]$$

From a table of standard entropies of formation, substitute in the quantities for the standard entropies of formation:

$$\Delta S^\circ_{rxn} = [(16\text{ mol})(69.91\text{ J/mol-K}) + (3\text{ mol})(31.80\text{ J/mol-K})] -$$
$$[(16\text{ mol})(205.79\text{ J/mol-K}) + (8\text{ mol})(161.92\text{ J/mol-K})]$$

Collect terms, and make sure that the mathematical signs are accounted for:

$$\Delta S^\circ_{rxn} = 1118\text{ J/K} + 95.40\text{ J/K} - 3292.6\text{ J/K} - 1295.4\text{ J/K}$$

Solving for the entropy of reaction gives:

$$\Delta S^\circ_{rxn} = -3375\text{ J/K}$$

Qualitatively, $\Delta S^\circ < 0$ represents a decrease in entropy. A $\Delta S^\circ > 0$ represents an increase in entropy. Solids have low entropy. The particles forming the solid are packed in an orderly matrix in the crystal. Liquids are intermediate in entropy. Intermolecular forces hold the particles in the liquid together in a slightly organized fashion.

Gases are high in entropy. The gas particles are randomly moving around in their container with very few particles actually arranged in any orderly fashion. If we look at the above reaction, we see that a solid and a liquid are produced from gaseous products. Thus, it is not surprising that the entropy of reaction has a negative value. In the previous example, ΔS°_{rxn} is a measure of the entropy change of a system. The entropy change of the surroundings is related to the heat flow and the temperature, as shown by the equation below:

$$\Delta S_{surr} = -\frac{q}{T} \quad \text{or} \quad -\frac{\Delta H}{T} \quad \text{(at constant pressure)}$$

The negative sign in this equation accounts for the fact that the heat of the surroundings is opposite in sign to the heat of the system. From the previous example, we calculated that the enthalpy of reaction was -1876 kJ at standard temperature, 298 K. The entropy change of this system was calculated to be -3375 J/K (or 3.375 kJ/K). What is the entropy change of the surroundings?

Using the above equation:

$$\Delta S_{surr} = -\frac{(-1876 \text{ kJ})}{298 \text{ K}} = 6.30 \text{ kJ/K or } 6.30 \times 10^3 \text{ J/K}$$

The result implies that the entropy of the surrounding increased. The total entropy change would be the sum of the entropy change of the system and the entropy change of the surroundings:

$$\Delta S_{total} = \Delta S_{sys} + \Delta S_{surr}$$

$$\Delta S_{total} = -3.375 \times 10^3 \text{ J/K} + 6.30 \times 10^3 \text{ J/K} = 2.92 \times 10^3 \text{ J/K}$$

Overall, the entropy has increased. According to the second law of thermodynamics, this implies that the reaction we have been discussing is spontaneous. The best way to determine the spontaneity of a reaction is by looking at the next topic, Gibb's Free Energy.

Gibb's Free Energy

As mentioned, Gibb's free energy, ΔG, determines whether a reaction is spontaneous or nonspontaneous. If $\Delta G < 0$, the reaction is spontaneous; if $\Delta G > 0$, the reaction is nonspontaneous; if $\Delta G = 0$, the reaction is at equilibrium. There are numerous ways to calculate Gibb's free energy. One of the methods will be analogous to previous methods discussed for enthalpy and entropy. The equation is shown below:

$$\Delta G^\circ_{rxn} = \sum n(\Delta G^\circ_f \text{ of products}) - \sum n(\Delta G^\circ_f \text{ of reactants})$$

where n represents the moles of each product or reactant is given by the coefficient in the balanced chemical equation. As with enthalpy, you will find that the standard Gibb's free energy of formation of elements in their standard states is 0 kJ/mol.

Let's determine the change in the Gibb's free energy for the reaction we have discussed throughout the handout:

$$16\ H_2S(g) + 8\ SO_2(g) \rightarrow 16\ H_2O(l) + 3\ S_8(s)$$

Set up the equation:

$$\Delta G^\circ_{rxn} = [(16 \text{ mol})(\Delta G^\circ_f \text{ of } H_2O(l)) + (3 \text{ mol})(\Delta G^\circ_f \text{ of } S_8(s))] -$$

$$[(16 \text{ mol})(\Delta G^\circ_f \text{ of } H_2S(g)) - (8 \text{ mol})(\Delta G^\circ_f \text{ of } SO_2(g))]$$

From tabulated data, substitute for the Gibb's free energies:

$$\Delta G^\circ_{rxn} = [(16 \text{ mol})(-237.13 \text{ kJ/mol}) + (3 \text{ mol})(0 \text{ kJ/mol})] -$$

$$[(16 \text{ mol})(-33.56 \text{ kJ/mol}) + (8 \text{ mol})(-300.19 \text{ kJ/mol})]$$

Combine terms, paying careful attention to the mathematical signs:

$$\Delta G^\circ_{rxn} = -3794.1 \text{ kJ} + 0 \text{ kJ} + 537.0 \text{ kJ} + 2401.5 \text{ kJ}$$

$$\Delta G^\circ_{rxn} = -855.6 \text{ kJ}$$

As predicted, the reaction is spontaneous, since $\Delta G^\circ_{rxn} < 0$.

There is an alternative means for calculating ΔG°_{rxn} in which the Gibb's free energy is related to the enthalpy, entropy and temperature. If these values are known, then Gibb's free energy may be calculated by:

$$\Delta G^\circ = \Delta H^\circ - T\,\Delta S^\circ$$

From previous calculations, we have determined that the enthalpy change was -1876 kJ, the entropy change was -4.065 kJ/K, and the temperature has a value of 298 K (standard temperature). Notice that is important to make sure the units are consistent, so entropy, which is often expressed in units of J/K has been converted to kJ/K. Substituting these quantities into the above equation gives:

$$\Delta G^\circ = -1876 \text{ kJ} - (298 \text{ K})(-3.375 \text{ kJ/K}) = -870.2 \text{ kJ}$$

This answer agrees reasonably well with the result from the calculation from the Gibb's free energies of formation.

Finally, there are two other equations to calculate ΔG° which are often encountered in an introductory level chemistry course. The first of these equations is:

$$\Delta G^\circ = -RT\ln K \text{ or } \Delta G^\circ = -2.303RT\log K$$

where R is the ideal gas law constant, 8.314 J/mol-K, T is temperature in kelvin, and K is the equilibrium constant for the chemical equilibrium taking place.

The second equation is:

$$\Delta G^\circ = -nFE^\circ$$

where n represents the moles of electrons transfered in an electrochemical process, F is Faraday's constant, 96,485 C/mol, and E° is the standard potential for an electrochemical cell. Calculation of the Gibb's free energy from these equations involves substitution of the appropriate quantities into the equation.

Index